Computer–Aided Drug Discovery and Design

Series Editor
Alan Talevi
Facultad de Ciencias Exactas
Universidad Nacional de La Plata
La Plata, Argentina

The series covers all the stages of drug discovery and development that use computational approximations, including bioinformatics, cheminformatics, structure-based approximations, ligand-based approximations and network-approaches. Drug discovery and development stages include target identification and validation, hit identification and hit-to-lead and lead optimization programs.

- The series covers the early drug discovery and development process in a comprehensive manner.
- It explains both the historical background and underlying principles of each methodology, as well as state-of-the-art innovations within each field. For instance, when dealing with structure-based approximations, both classical universal scoring functions and the novel trend in the field (tailored scoring functions) will be considered.
- Accordingly, both experts and students taking their first steps in computer-guided drug discovery will find the series of interest. Students can resort to the series volumes to get familiarized with the basics of the methodologies, and experts can turn to the series for trends in the field.
- No other series covers all the fields and approximations in a deep,updated and comprehensive manner.
- For an integrative perspective, the potential readers might acquire all the books that compose the series; on the contrary, people with a narrower, more specific scope can resort to a particular volume of choice

Marcus T. Scotti • Carolina L. Bellera
Editors

Drug Target Selection and Validation

Springer

Editors
Marcus T. Scotti
Department of Bioactive Natural
Federal University of Paraíba
Joao Pessoa, Paraíba, Brazil

Carolina L. Bellera
Laboratory of Bioactive Compounds
Research and Development
(LIDeB), Department of Biological
Sciences, Faculty of Exact Sciences
University of La Plata (UNLP)
La Plata, Buenos Aires, Argentina

Editorial Contact
Carolyn Spence

ISSN 2730-5457 ISSN 2730-5465 (electronic)
Computer–Aided Drug Discovery and Design
ISBN 978-3-030-95894-7 ISBN 978-3-030-95895-4 (eBook)
https://doi.org/10.1007/978-3-030-95895-4

This Springer imprint is published by the registered company Springer Nature Switzerland AG
The registered company address is: Gewerbestrasse 11, 6330 Cham, Switzerland

M. T. Scotti dedicates this volume to Luciana Scotti.

C.L. Bellera dedicates this volume to Víctor, Amanda, and Mateo

Foreword

This volume inaugurates Springer Nature's Drug Discovery and Design series. Without question, drug discovery is a risky (though potentially very profitable) business characterized by high attrition rates and long development times. Recognizing the key points across the many stages of the drug discovery and development process and addressing them systematically to make informed decisions has become essential to expedite the process, improve investment return, and efficiently provide therapeutic solutions to unmet needs. Whereas this series is expected to include volumes dealing, in depth, with specific aspects of the field, this inaugural book provides a rather general insight into current and future drug discovery paradigms, along with challenges faced in the early stage of the drug discovery and development cycle, particularly at target identification and validation stages.

Chapter "Drug Discovery Paradigms: Target-Based Drug Discovery" describes the main concepts related to target-based drug discovery, including two key steps, target selection and binding site identification, as well as the main features and limitations of commonly applied target-focused computational approaches, namely molecular docking and molecular dynamics. The advantages and disadvantages of phenotypic screening in drug discovery are discussed next, addressing the renewed interest in phenotypic screening in the pharmaceutical community, supported by the flourishing paradigm of network pharmacology, as well as high-content screening and state-of-the-art screening platforms (e.g., organoids, advanced cell culture platforms, and others). Chapter "Target Identification Approaches in Drug Discovery" discusses the more relevant target identification approximations, their classification and working schemes. This chapter is complemented by chapter "Introduction to Target Validation", which focuses on the ideal features of a target and the diverse strategies available to (bio)chemically and biologically validate a molecular drug target, and chapter "Structure-Based Binding Pocket Detection and Druggability Assessment", which reviews bioinformatic tools for binding pocket and druggability prediction.

The second part of the book focuses on emerging paradigms in the field. Chapter "Network-Based Target Identification" deals with the application of network-based approaches for the prediction and appraisal of drug targets, and chapter "The Current State of Precision Medicine and Targeted-Cancer Therapies: Where Are We?" debates the scope, challenges, and limitations of precision medicine in the field. Chapter "Metabolic Control Analysis for Drug Target Selection Against

Human Diseases" provides a thorough review of the principles of Metabolic Control Analysis (linked to the emerging concept of "target vulnerability" and often overlooked when choosing potential new drug targets). Finally, chapter "Progress on Open Chemoinformatic Tools for Drug Discovery" reviews the recent progress in the development of public chemoinformatic resources for different tasks, with special emphasis on drug discovery applications.

Acknowledgments

C.L. Bellera and A. Talevi thank the National University of La Plata (UNLP) and the Argentinean National Council of Scientific and Technical Research (CONICET)

M. T. Scotti thanks the Federal University of Paraíba.

Contents

Contributors

Juan I. Alice Laboratory of Bioactive Compounds Research and Development (LIDeB), Department of Biological Sciences, Faculty of Exact Sciences, University of La Plata (UNLP), La Plata, Buenos Aires, Argentina

Argentinean National Council of Scientific and Technical Research (CONICET)—CCT La Plata, La Plata, Buenos Aires, Argentina

Carolina L. Bellera Laboratory of Bioactive Compounds Research and Development (LIDeB), Department of Biological Sciences, Faculty of Exact Sciences, University of La Plata (UNLP), La Plata, Buenos Aires, Argentina

Argentinean National Council of Scientific and Technical Research (CONICET)—CCT La Plata, La Plata, Buenos Aires, Argentina

Javier Belmont-Díaz Departamento de Bioquímica, Instituto Nacional de Cardiología Ignacio Chávez, Mexico City, Mexico

Gul Bushra Department of Bioinformatics and Biotechnology, Faculty of Life Sciences, Government College University, Faisalabad, Pakistan

Marcelo Comini Laboratory Redox Biology of Trypanosomes, Institut Pasteur de Montevideo, Montevideo, Uruguay

José Alixandre de Sousa Luis Education and Health Center, Federal University of Campina Grande, Cuité, PB, Brazil

Rusely Encalada Departamento de Bioquímica, Instituto Nacional de Cardiología Ignacio Chávez, Mexico City, Mexico

Homero Gómez-Velasco DIFACQUIM Research Group, Department of Pharmacy, School of Chemistry, Universidad Nacional Autónoma de México, Mexico City, Mexico

Rodrigo Gutiérrez-Nieto DIFACQUIM Research Group, Department of Pharmacy, School of Chemistry, Universidad Nacional Autónoma de México, Mexico City, Mexico

Mudassir Hassan Department of Zoology, Faculty of Life Sciences, Government College University, Faisalabad, Pakistan

Chonny Herrera-Acevedo Post-Graduate Program in Natural and Synthetic Bioactive Products, Federal University of Paraíba, João Pessoa, PB, Brazil

Ghulam Hussain Neurochemical Biology and Genetics Laboratory, Department of Physiology, Faculty of Life Sciences, Government College University, Faisalabad, Pakistan

Muhammad Javid Iqbal Department of Zoology, Faculty of Life Sciences, Government College University, Faisalabad, Pakistan

Samreen Gul Khan Department of Chemistry, Faculty of Physical Sciences, Government College University, Faisalabad, Pakistan

Khatereh Khorsandi Department of Photodynamic, Medical Laser Research Center, Yara Institute, ACECR, Tehran, Iran

José L. Medina-Franco DIFACQUIM Research Group, Department of Pharmacy, School of Chemistry, Universidad Nacional Autónoma de México, Mexico City, Mexico

Paul A. M. Michels Centre for Immunity, Infection and Evolution (CIIE) and Centre for Translational and Chemical Biology (CTCB), School of Biological Sciences, The University of Edinburgh, Edinburgh, Scotland, UK

Rafael Moreno-Sánchez Departamento de Bioquímica, Instituto Nacional de Cardiología Ignacio Chávez, Mexico City, Mexico

Cecilia Ortíz Laboratory Redox Biology of Trypanosomes, Institut Pasteur de Montevideo, Montevideo, Uruguay

Laboratory of Biotechnology, Universidad ORT Uruguay, Montevideo, Uruguay

Camilo Perdomo-Madrigal School of Science, Universidad de Ciencias Aplicadas y Ambientales, Bogotá, Colombia

Azhar Rasul Department of Zoology, Faculty of Life Sciences, Government College University, Faisalabad, Pakistan

Ammara Riaz Department of Zoology, Faculty of Life Sciences, Government College University, Faisalabad, Pakistan

Saba Riaz Department of Zoology, Faculty of Life Sciences, Government College University, Faisalabad, Pakistan

Santiago Rodriguez Laboratory of Bioactive Research Compounds and Development (LIDeB), Department of Biological Sciences, Faculty of Exact Sciences, University of La Plata (UNLP), La Plata, Buenos Aires, Argentina

Argentinean National Council of Scientific and Technical Research (CONICET)—CCT La Plata, La Plata, Buenos Aires, Argentina

Santiago Ruatta Laboratory Redox Biology of Trypanosomes, Institut Pasteur de Montevideo, Montevideo, Uruguay

Laboratory of Applied Chemistry, FBCB, UNL, Santa Fe, Argentina

Emma Saavedra Departamento de Bioquímica, Instituto Nacional de Cardiología Ignacio Chávez, Mexico City, Mexico

Ayesha Sadiqa Department of Zoology, Faculty of Life Sciences, Government College University, Faisalabad, Pakistan

Iqra Sarfraz Department of Zoology, Faculty of Life Sciences, Government College University, Faisalabad, Pakistan

Luciana Scotti Post-Graduate Program in Natural and Synthetic Bioactive Products, Federal University of Paraíba, João Pessoa, PB, Brazil

Marcus Tullius Scotti Post-Graduate Program in Natural and Synthetic Bioactive Products, Federal University of Paraíba, João Pessoa, PB, Brazil

Attila A. Seyhan Laboratory of Translational Oncology and Experimental Cancer Therapeutics, The Warren Alpert Medical School, Brown University, Providence, RI, USA

The Joint Program in Cancer Biology, Brown University and Lifespan Health System, Providence, RI, USA

Department of Pathology and Laboratory Medicine, The Warren Alpert Medical School, Brown University, Providence, RI, USA

Cancer Center at Brown University, The Warren Alpert Medical School, Brown University, Providence, RI, USA

Alan Talevi Laboratory of Bioactive Compounds Research and Development (LIDeB), Department of Biological Sciences, Faculty of Exact Sciences, University of La Plata (UNLP), La Plata, Buenos Aires, Argentina

Argentinean National Council of Scientific and Technical Research (CONICET)—CCT La Plata, La Plata, Buenos Aires, Argentina

Yun Tang Shanghai Key Laboratory of New Drug Design, School of Pharmacy, East China University of Science and Technology, Shanghai, China

Citlali Vázquez Departamento de Bioquímica, Instituto Nacional de Cardiología Ignacio Chávez, Mexico City, Mexico

Zengrui Wu Shanghai Key Laboratory of New Drug Design, School of Pharmacy, East China University of Science and Technology, Shanghai, China

Rabia Zara Department of Zoology, Faculty of Life Sciences, Government College University, Faisalabad, Pakistan

Drug Discovery Paradigms: Target-Based Drug Discovery

Chonny Herrera-Acevedo, Camilo Perdomo-Madrigal,
José Alixandre de Sousa Luis, Luciana Scotti,
and Marcus Tullius Scotti

Contents

Abstract

Target-based drug discovery tools have been used with success in the pharmaceutical industry. They have become the fundamental methodologies for discovering new drugs in recent years, with two main advantages over the traditional methodologies: increased speed and greater economic efficiency. Improved computational capacities and new software packages have allowed the diversification and strengthening of these procedures. This chapter describes the main concepts related to target-based drug discovery, including two key steps—target and binding site identification—as well as the main features and limitations of the most common target-based methodologies: de novo drug discovery, molecular docking, and molecular dynamics.

C. Herrera-Acevedo · L. Scotti · M. T. Scotti (✉)
Post-Graduate Program in Natural and Synthetic Bioactive Products, Federal University of Paraíba, João Pessoa, PB, Brazil

C. Perdomo-Madrigal
School of Science, Universidad de Ciencias Aplicadas y Ambientales, Bogotá, Colombia

J. A. de Sousa Luis
Education and Health Center, Federal University of Campina Grande, Cuité, PB, Brazil

M. T. Scotti, C. L. Bellera (eds.), *Drug Target Selection and Validation*, Computer–Aided Drug Discovery and Design,
https://doi.org/10.1007/978-3-030-95895-4_1

Keywords
Target-based · Molecular docking · Molecular dynamics · Drug discovery · Sampling algorithms · Scoring functions · De Novo drug discovery · Binding site

Abbreviations

3D	Tridimensional
CADD	Computer-aided drug design
COCONUT	Collection of Open Natural Products
Cryo-EM	Cryogenic electron microscopy
DL	Deep Learning
DNDD	De novo drug design
DRL	Deep Reinforcement Learning
HIV	Human immunodeficiency virus
Ki	Inhibition constant
LBDD	Ligand-based drug design
MC	Monte Carlo
MD	Molecular dynamics
MM	Molecular mechanics
PDB	Protein databank
QM	Quantum mechanics
ReLeaSE	Reinforcement Learning for Structural Evolution
RL	Reinforcement Learning
SBDD	Structure-based drug design
SBVS	Structure-based virtual screening
vdW	Van der Waals
VS	Virtual screening

1 Introduction

In recent years, computer-aided drug design (CADD) has become the fundamental approach for the discovery, development, and analysis of structures with potential activity against many diseases. Two main types of approaches in CADD have been reported: ligand-based virtual screening (LBSS) and structure-based drug design (SBDD). These two approaches offer versatility and synergy, both in academia and industry (Ferreira et al. 2015). SBDD, also known as target-based drug discovery, uses computational methods and the three-dimensional (3D) structural information of the protein target to investigate the underlying molecular interactions involved in ligand–protein binding and thus interpret experimental results at an atomic level of detail (Lionta et al. 2014).

Advances in structural resolution techniques such as X-ray crystallography, nuclear magnetic resonance (NMR), and cryogenic electron microscopy (cryo-EM) have allowed the development and strengthening of SBDD techniques (Wang 2021). Currently, the number of macromolecular structures registered in the Protein Data Bank (PDB) is close to 180,000 (Berman et al. 2000), significantly increasing the structural information about key macromolecular drug targets (Ferreira et al. 2015). This evolution also drove the development of sophisticated software packages, facilitating in silico calculations of inhibitors into their predicted binding sites as well as the computational analysis of inhibitor binding and information on further enhancements (Van Montfort and Workman 2017).

Various examples of commercially available drugs developed from target-based drug design studies can be identified, including Amprenavir (Fig. 1), which acts against the human immunodeficiency virus (HIV) protease. Molecular dynamics calculations were used to explain the experimental observation that the P1′ amide NH of substrate sequences was not required for binding and productive catalysis. From these results, and supported by in silico modeling, the *N,N*-dialkyl sulfonamide moiety was chosen to bind to the flap water molecule and to act as a scaffold for the P1′ and P2′ groups (Clark 2006).

Other examples of SBDD supporting the development of new drugs include Raltitrexed and Zanamivir (Fig. 1). Raltitrexed acts against thymidylate synthase of HIV (Batool et al. 2019). Zanamivir was the first neuraminidase inhibitor to be marketed for the treatment of influenza. Using the GRID program, it was predicted that in the active site of the target, replacing the hydroxyl group at the 4-position of the ring by an amine will improve the interactions with two neighboring glutamic acid residues, which identified this potent inhibitor of the neuraminidase enzyme ($K_i = 0.2$ nM) (Clark 2006).

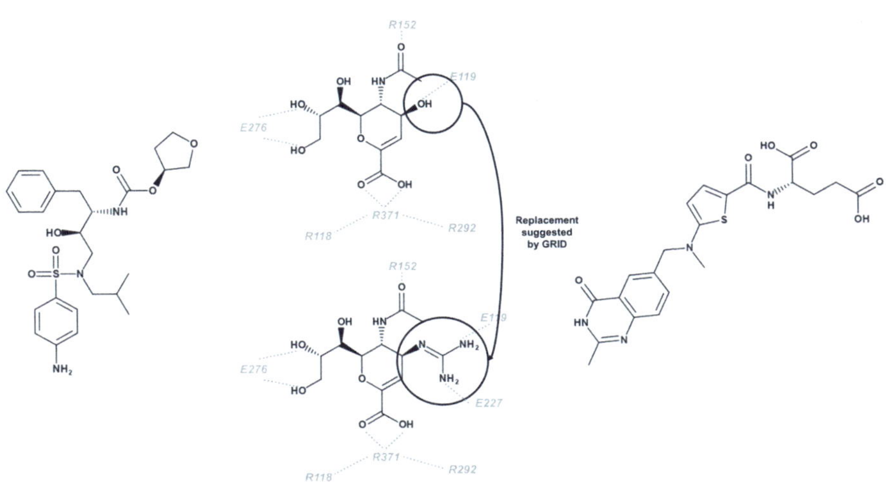

Fig. 1 Examples of drugs commercially available that were developed from target-based drug design. *Left:* Amprenavir. *Center:* Zanamivir – GRID program predicted that the replacement of the hydroxyl group at the 4-position of the ring by an amine, would improve the interaction with two neighboring glutamic acid (E) residues of the target (neuraminidase) (Clark 2006). *Right*: Raltitrexed

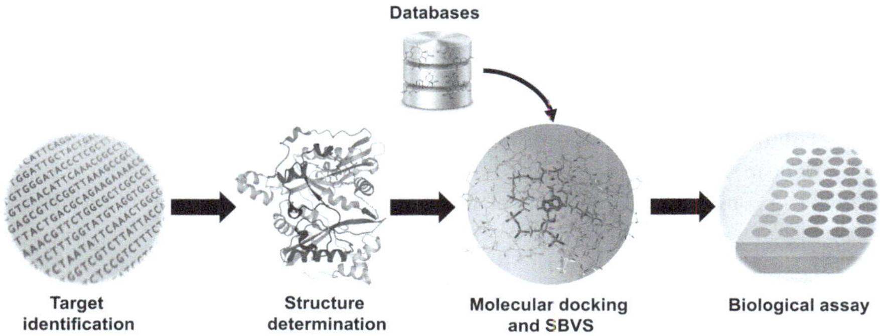

Fig. 2 General workflow of the structure-based drug design (SBDD) (Ramesh et al. 2021)

In summary, the workflow of the SBDD process begins with target identification, which is supported by genetics, molecular biology, and bioinformatics methodologies. Next, protein extraction and purification are carried out. A structural determination of the target is performed, mainly using NMR, X-ray crystallography, and Cry-EM; for those proteins whose crystal structure is not defined, homology models are built using specialized software. Then, the biological assay is performed using different methodologies (Fig. 2). The main three SBDD approaches are molecular docking, molecular dynamics, and de novo drug design. The small molecules to be evaluated in structure-based virtual screening (VS) methodologies are typically selected from databases of drug-like compounds. Finally, the top hits are synthesized, and in-vitro tests are performed to identify the best structures (Batool et al. 2019; Ramesh et al. 2021).

This chapter reviews the main concepts in target-based drug design and describes the two key steps in these studies: identifying the target and the binding site. Also, we summarize the three main SBDD methodologies: molecular docking, molecular dynamics, and de novo drug design, discussing the basic concepts and classifications, as well as the algorithms and functions used for the calculations and the limitations of these methods.

2 Target Identification

When a research group proposes developing a new drug for a given disease, the first step is to understand the physiological and pathological processes of the disease. Once this study is completed, it is possible to visualize possible molecular targets that are components of the human body or a particular pathogen. The main targets are proteins, such as enzymes, metabotropic or ionotropic receptors, antibodies, and nucleic acids, including nuclear and mitochondrial DNA in addition to messenger RNA. Protein targets are the most abundant since they mediate several important metabolic processes (Lappano and Maggiolini 2011; Valeur et al. 2017; Surabhi and Singh 2018).

Drugs that act on enzymatic targets are mostly inhibitors; however, some can act as allosteric activators. In contrast, the drugs that act on membrane receptors can be agonists (when mimicking the effect of the endogenous substance) or antagonists (when blocking the signal promoted by the endogenous substance). Therefore, the mode of interaction determines the type of signal that will be transmitted to effectors located on the cytosolic side of the membrane. These effectors can be enzymes, ion channels, or sites for binding intracellular proteins. Examples of signal transduction pathways are those in which the receptors are coupled to G proteins that in turn activate other enzymes that catalyze the conversion of triphosphate nucleotides into their cyclic variants that will act as second messengers. In other ways, these messengers can activate other enzymes and ion channels. Sometimes the channel itself can be the target, either through an allosteric receptor site located on the same protein or through direct interaction with the ion transport pore, which usually involves blocking the ion channel (Bagal et al. 2013; Zhu et al. 2012).

Drugs that interact with nucleic acids can be obtained in two ways: (i) by planning molecules that have the ability for direct interaction, that is, by forming covalent bonds or cleaving bonds, as well as by intercalation; (ii) indirectly, through proteins that regulate gene replication and expression. These pathways are important for designing drugs to combat diseases caused by disorders in the regulation of the cell cycle, such as cancer (Batool et al. 2019; Surabhi and Singh 2018).

For the development of the SBDD, it is necessary to know the topological arrangement of the molecular targets and, for that, detailed 3D data of the macromolecule, obtained through X-ray crystallography techniques, NMR spectroscopy, and cryo-EM, are used. Such structures, when determined, are deposited in public databases that can be accessed freely. The most popular and widespread among such databases are the PDB, InterPro, ExPASy, and Relibase (Batool et al. 2019; Valeur et al. 2017).

When the specific structure of a given macromolecule is not known, it is also possible to perform SBDD using homology modeling. This method uses in silico approximations, where it is possible to model a 3D structure from a homologous protein with a known structure. The three most well-known methods of predicting structures are comparative modeling, threading, and ab initio modeling, which are reliable and possible to validate. The first step in this process is to determine the registration or alignment of the target sequence that allows insertions and deletions in the experimental structure, which can be accomplished with several dynamic programming algorithms. In the second stage, mutations are made to the amino acid residues of the experimental structure so that they correspond to those of the target protein. This strategy can also be used when more than one experimental structure is available, allowing the creation of a model with a hybrid structure that is closer to the structure of the target protein. The final step is to refine and examine the structure to ensure that it is reasonable, thus validating the model (Oda 2011; Lounnas et al. 2013; Wang et al. 2016a).

In the process of refining the model, one of the steps is the alignment of the structures, where similarities are verified, the conserved regions are observed, and common amino acid residues in the active site are identified. A detailed investigation of these similarities allows the planning of new drugs that are more likely to be effective against the pathology under study. There are tools available for the validation stage, such as the Ramachandran plot, which is useful because it defines the residues found in the most energetically favorable and unfavorable regions and guides the evaluation of the quality of the theoretical or experimental models of the proteins. This graph represents all possible combinations of the dihedral angles Ψ versus φ for each amino acid, except for glycine, which has no side chain. For the model to be considered reliable, at least 90% of the amino acids should fall in the favorable regions (Maia et al. 2020; Potapov et al. 2010).

The next step of SBDD is to identify the binding site on the macromolecular target and the types of interactions necessary for the small molecule to interact effectively and trigger a useful biological response. In this process, information about the free energy of the complex can be obtained, and the types of interactions between the atoms can be observed, leading to the identification of the best pharmacophores for the new ligand. These binding sites are determined experimentally by X-ray crystallography techniques, including co-crystallization. The data obtained from the co-crystallized structures provide robust information about the binding sites and are very useful to understand the interaction between the micro- and macromolecules. In some situations, only the macromolecule structures are deposited in the databases, without the presence of a ligand. There is no information about the binding site in such cases, so servers and online tools must be used to identify the sites. Some examples of servers that can assist are DoGSite Scorer (Volkamer et al. 2012), CASTp (Sahu et al. 2017), NSiteMatch (Sun and Chen 2017), Metapocket (Huang 2009), DEPTH (Tan et al. 2013), LISE (Xie et al. 2013), and MSpocket (Zhu and Pisabarro 2011). After the binding site has been identified, the volume of the binding pocket can be assessed using tools such as TRAPP (Stank et al. 2017) and POVME (Wagner et al. 2017). More specifically, the residues in the macromolecules that favor an optimal interaction to trigger a biological response are identified. Therefore, it is necessary to know the interaction energies, van der Waals (vdW) forces, to obtain an optimal mapping of the connection site. There are several methods for this purpose, one of which is Q-SiteFinder (Laurie and Jackson 2005), which calculates the vdW interaction energies with a methyl probe. This method allows the retention and grouping of those sites with more favorable energies based on their total interaction energies. From this knowledge, simulating the binding of various library compounds or compounds designed de novo in the active site of the protein allows the potential identification of novel drug candidates (Batool et al. 2019).

3 Binding Site Identification

Once the target protein has been identified and its biological effect has been confirmed, the next challenge in any SBDD research is to identify and validate the sites in the structure of the protein that will be more likely to interact with ligands. These binding sites, or cavities, can be classified into two different categories. If the interaction between the ligand and the target protein occurs in the catalytic site of the protein, it is called an orthosteric site; if the interaction takes place in a different region where the interaction produces changes in the protein conformation and the structure of the catalytic site, it is called an allosteric site (Nussinov and Tsai 2012). Regardless of the type, understanding the cavity structure and its interaction energies with ligands is of great importance to any SBDD process. Therefore, several strategies have been developed to identify these binding sites in the target protein. These involve methods and algorithms that make use of information such as similarities of residues between functionally related proteins (evolutionary methods), structural features of the protein (geometrical methods), and interaction energy with probe molecules (energy-based methods) (Harigua-Souiai et al. 2015; Oliveira et al. 2014).

Evolutionary Methods

Based on the idea that homologous proteins (proteins that have a common ancestry) have some conserved residues, it is possible to assume that important regions of those proteins, such as those involved in biological or enzymatic activity, have functionally relevant sequences of residues that are conserved across different organisms to avoid malfunction in their physiological processes (Armon et al. 2001). According to this assumption, information related to the cavities in a protein can be used in an evolutionary method to find the binding sites of any homologous or functionally related protein by comparing the residue sequence of their primary structures. Table 1 lists different algorithms based on this method.

The main drawback of this type of algorithm is the necessity of prior knowledge and the availability of information related to the sequences of existing proteins, which is not always possible considering the huge number and diversity of proteins that exist in different organisms and species. Additionally, a low degree of similarity between the sequences of the target protein and the reported protein will lead to deficient cavity prediction results, interfering with the success of the SBDD study (Harigua-Souiai et al. 2015; Armon et al. 2001).

Geometrical Methods

Cavities are usually conceived as regions of large surface area in the protein and displayed as hole-shaped concave sections in the structure of the protein (Simões et al. 2017). Geometry-based methods use the 3D structure of proteins and locate

Table 1 Examples of binding site identification software classified according to the type of algorithm used for the prediction (Oliveira et al. 2014; Simões et al. 2017)

Evolutionary methods	Geometrical methods	Energy-based methods
ConSurf	LIGSITE	GRID
Rate4Site	CAST	DrugSite
GarLig	SURFNET	QsiteFinder
	PocketPicker	MCSS
	VOIDOO	
	KVFinder	

binding sites by considering their size, shape, and chemical properties (e.g., ionizability, intermolecular interaction tendency, and covalent bonding sites) (Laskowski 1995). In these methods, the protein structure and surface are analyzed using probe objects, typically spheres, that fit in the protein cavities, depending on their size (Oliveira et al. 2014; Laskowski 1995). Other geometric techniques involve tessellation and 3D gridding of the protein structure in order to identify voids of a significant size on the protein surface. Thus, geometry-based methods can be classified as grid system scanning, probe sphere filling, and alpha-shape modeling (Yu et al. 2010).

Geometrical methods are the most frequently used for cavity detection in different software and algorithms (Oliveira et al. 2014). Their greatest advantage is that these methods depend only on the availability of the 3D structure of the protein, and no additional prior information is needed. Because these algorithms use crystal structures from databases such as the PDB, the protein structure is treated as static, disregarding its intrinsic flexibility and plasticity (Schmidtke et al. 2011). The most significant disadvantages for this category of methods come from: (i) the freedom and randomness considering the tessellation and gridding parameters that are needed to probe the protein surface, and (ii) the inaccuracies that may arise from protein orientation effects and conformational changes that may occur (Simões et al. 2017).

Energy-Based Methods

As stated previously, in order to study the interaction energy of the molecules and the target protein, it is essential to predict their bonding probability and, consequently, to predict if the ligand will be an eligible potential drug candidate. Hence, another family of methods has been proposed to identify binding sites in target proteins, which is related to evaluating the interaction energy between different regions of the protein and some small molecules.

Energy-based methods identify cavities using only energetic criteria. In these algorithms, the vdW interaction energy, the hydrogen bond energy, or the Lennard–Jones potentials of a probe (a small molecule or fragment, such as water or a methyl group) and the target protein are evaluated (Laurie and Jackson 2005; Yu et al. 2010; Schmidtke et al. 2011). These methods are usually more computationally demanding, which is a considerable disadvantage because of the interest in reduced calculation times in SBDD research (Schmidtke et al. 2011).

4 Target-Based Methodologies

Molecular Docking

Molecular docking is the most common target-based drug discovery methodology. More than 52,000 papers related to target-based drug discovery have been published recently, and more than 50% of them have been published in the past 5 years (2017–2021) (Web of Science, Copyright Clarivate 202). Molecular docking techniques aim to predict the binding mode of a ligand that best matches a macromolecular partner (i.e., proteins). These methodologies aim to accurately predict the structure of a ligand within the constraints of a receptor binding site and correctly estimate the strength of binding (Yuriev and Ramsland 2013). In protein–ligand docking, the process consists in generating several possible conformations and orientations, also called poses, of the ligand within the protein binding site (Salmaso and Moro 2018).

Identifying the most likely binding conformations requires two steps: first, the exploration of a large conformational space representing the various potential binding modes; second, the accurate prediction of the interaction energy associated with each of the predicted binding conformations. All docking programs perform these two steps through a cyclical process, in which the ligand conformation is evaluated by specific scoring functions until the results converge to a solution having minimum energy (Ferreira et al. 2015).

There are three types of molecular docking techniques, categorized by the flexibility of the molecules involved in the molecular docking calculations:

- *Rigid docking*: Both the ligand and the protein are considered rigid entities, and only the three translational and three rotational degrees of freedom are considered during sampling (Salmaso and Moro 2018).
- *Semi-flexible docking*: One of the molecules, the ligand, is flexible, while the protein is rigid. Thus, the conformational degrees of freedom of the ligand are sampled, in addition to the six translational and rotational degrees of freedom (Salmaso and Moro 2018).
- *Flexible docking*: The protein is not a passive rigid entity during binding, and both the ligand and the protein are considered flexible counterparts (Salmaso and Moro 2018).

A molecular docking program has two essential components: sampling the conformations of the ligand in the binding site of the protein and ranking these conformations via a scoring function (Meng et al. 2011; Tripathi and Bankaitis 2017). Sampling describes the generation of putative ligand-binding orientations and conformations near a binding site. It can be further divided into two aspects, ligand sampling, and protein flexibility. Scoring explains the prediction of the binding tightness for individual ligand orientations and conformations using a physical or empirical energy function (Huang and Zou 2010). The computational cost required in the docking calculations increases according to the number of degrees of freedom. For this reason, both sampling and scoring should be optimized to give a good balance between accuracy and speed (Salmaso and Moro 2018).

Sampling Algorithms

Sampling algorithms evaluate the capability of each docking program to predict the ligand-binding poses (Wang et al. 2016b). Sufficient sampling of ligand and protein states in docking is essential, as well as accurate evaluation of the binding energies

Table 2 Summary of the most common sampling algorithms for molecular docking

Algorithm	Description	Examples	Ref.
Shape matching	Considers the geometrical overlap between two molecules, being able to predict docking conformations at a fast speed; however, at lower accuracy rate	ZDOCK SYSDOCK	Dias et al. (2008)
Systematic search			
Exhaustive search	Explore the values of each degree of freedom in a combinatorial manner, rotating all dihedral angles of the ligand according to a predetermined range of values and a set of initial restraints	Glide eHiTS	Guedes et al. (2014)
Fragmentation	The ligand is separated in smaller fragments, followed by the selection and docking of a base fragment into the receptor binding site. The ligand is then reconstructed incrementally by covalently linking the other fragments to the base group	FlexX	Guedes et al. (2014)
Conformation ensemble	Rigidly docks a set of previously generated ligand conformations into the binding site	DOCK 4.0 FLOG	Guedes et al. (2014)
Stochastic search			
Monte-Carlo	The method involves applying random Cartesian moves to the system and accepting or rejecting the move based on a Boltzmann probability	LigandFit, rDock	Taylor et al. (2002)
Genetic algorithm	Evaluating the evolution of a population of possible solutions via genetic operators to a final population, optimizing a predefined fitness function. Degrees of freedom are encoded into genes or binary strings and the collection of genes, or chromosome, being assigned a fitness based on a scoring function	rDock AutoDock GOLD	Taylor et al. (2002)
Tabu search methods	An iterative procedure designed to obtain solutions to optimization problems. The technique is defined as a Meta-Heuristic methodology that can move from one solution to another being able to save in memory the already visited solutions	PSI-DOCK	Dias et al. (2008)
Swarm optimization methods	In each iteration, a particle moves based on the knowledge of other particles and its own experience to speculate about the promising region to explore. One global best solution is kept updated by all particles and each individual particle also keeps a record of its own best solution	PLANTS PSO-VINA	Ng et al. (2015)

of potential protein–ligand complexes. A key issue is whether the docking program samples the possible states sufficiently and how increased sampling relates to improved scoring and outcomes. This includes sampling the internal degrees of freedom within the ligand, as well as sampling the poses between the ligand and the protein receptor (Coleman et al. 2013). The sampling algorithms are classified into three main categories (Table 2) (Wang et al. 2016b):

1. Shape matching: These approaches consider the geometrical overlap between the two molecules. They can predict docking conformations rapidly; however, their accuracy is rather low (Dias et al. 2008).

2. Systematic search: These approaches explore all ligand degrees of freedom during the search (Guedes et al. 2014).
3. Stochastic search algorithms: These approaches randomly change all the degrees of freedom of the ligand (translational, rotational, and conformational) at each step, generating very diverse solutions. However, multiple independent runs of the algorithm are required to maximize the probability of finding the global energy minimum (Guedes et al. 2014).

Scoring Algorithms

In molecular docking calculations, the scoring functions are responsible for distinguishing the correct poses (binders) from the incorrect poses (inactive compounds) in a reasonable computation time. The scoring functions estimate the binding affinity between the protein and the ligand by adopting various assumptions and simplifications (Meng et al. 2011).

Two theoretical aspects of these functions dominate their operational performance. The first is the degree to which a scoring function has a global extremum within the ligand pose landscape at the proper location. The second is the degree to which the magnitude of the function at the extremum is accurate (Jain 2006).

In the docking process, the scoring algorithms have three aims:

1. Pose prediction: The scoring function should be able to distinguish the experimental binding modes from all other modes explored through the searching algorithm (Taylor et al. 2002).
2. Virtual screening: The capacity to classify binders from non-binders (Guedes et al. 2014).
3. Binding affinity estimation: Prediction of the affinity constants and correctly rank several compounds according to their potency (Guedes et al. 2014).

The scoring functions are commonly classified into three general groups: force fields (FF), knowledge-based, and empirical. Recently, a fourth group has appeared and become relevant, machine learning-based functions (Table 3) (Batool et al. 2019).

- *Force field scoring functions*: These employ energy functions from classical molecular mechanics (MM), which is the binding free energy of protein–ligand complexes defined by the sum of the vdW forces and the electrostatic

Table 3 Examples of scoring functions in molecular docking calculations (Huang et al. 2010; Wang et al. 2003; Li and Yang 2017; Gabel et al. 2014)

Force field-based	Empirical	Knowledge-based	Machine learning-based
AutoDock	ICM-Score	ICM-PMF	RF-Score
DOCK/FF	SYBYL/F-Score	SYBYL/PMF-Score3	RF-IChem
SYBYL/G-Score	SYBYL/ChemScore	DrugScore	SVM-IChem
SYBYL/D-Score	LUDI		
	X-Score		

interactions. The solvation is determined as a distance-dependent dielectric function, and the nonpolar contributions are assumed to be proportional to the solvent-accessible surface area (Ferrara et al. 2004).

- *Knowledge-based scoring functions*: These are based on the inverse Boltzmann statistic principle, which assumes that the frequency of different pairs of atoms at different distances is related to the interaction of the two atoms and converts their frequency into the distance-dependent potential of the mean force. Knowledge-based scoring functions have great advantages of reduced computing cost and predictive accuracy relative to other types of scoring functions (Li et al. 2019).
- *Empirical scoring functions:* These functions compute the fitness of a protein–ligand binding by summing up the contributions of a number of individual terms, each representing an important energetic factor in protein–ligand binding. Multivariate linear regression (MLR) or partial least-squares (PLS) analysis are used in these empirical functions to determine the different related factors that affect the final binding score (Liu and Wang 2015).
- *Machine learning-based functions:* dynamic techniques for constructing and optimizing models to predict a binding pose and affinity (Batool et al. 2019). These functions have emerged in the past few years as potential rescoring tools for structure-based VS. The machine learning-based approaches can implicitly learn the function form from the training data and use adjustable parameters to improve performance further, thus offering higher performance flexibility and greater convenience (Shen et al. 2021). The most common machine learning algorithms used as scoring functions are the support vector machine, random forest, neural network, and deep learning algorithms (Li et al. 2019).
- *Consensus scoring.* None of the four types of scoring functions mentioned above have general applicability or are perfectly accurate. Consensus scoring is used to improve the probability of finding correct solutions by combining the scores from multiple scoring functions, using the advantages of the scoring functions while reducing the limitations. Suitable selection of the individual scores is fundamental to the design of the consensus analysis (Huang et al. 2010).

Molecular Docking Limitations

Although molecular docking calculations have been a fundamental tool in target-based drug discovery and many successful cases have been reported, several limitations remain that mainly affect the accuracy of the calculations and their computational cost (Shen et al. 2020). The ability to computationally predict the thermodynamics of these molecular recognition processes has been relatively poor until recently (Foloppe and Hubbard 2006) because of a lack of confidence in the scoring functions used to provide accurate binding energies (Ferreira et al. 2015).

The majority of docking studies have been performed using rigid conformations of the protein due to the high computational cost added by increasing the flexibility of the macromolecules; therefore, although the development of computers with greater calculation power has improved performance in flexible systems, integration of conformational changes with the scoring functions remains a problem to be solved (Guedes et al. 2018).

Additionally, in molecular docking calculations, the solvation phenomenon is not explicitly considered, impeding the estimation of the desolvation energies. Water molecules have an important role in the ligand–target binding process because they are necessary to correct the free energy of binding associated with the ligand displacement of water molecules. Although some docking software such as Autodock4 and GOLD have implemented some approaches, solvation is an important challenge to resolve in the currently available docking scoring functions (Ferreira et al. 2015; Guedes et al. 2018).

The algorithms based on quantum mechanics (QM) and mixed methodologies such as the semiempirical and QM/MM methods have emerged as alternatives for improving the accuracy of binding energy calculations. The development of graphics processing units (GPUs) has supported an increase in the use of these types of techniques (Guedes et al. 2018). In general terms, QM-based scoring functions are better able to predict ligand affinities than MM-based functions. However, this is not observed in all cases: the agreement between experimentally measured activities and calculated binding energies is highly dependent on the chemical series under study (Crespo et al. 2017).

Finally, the limitations related to the X-ray crystallographic data, which is the basis of the molecular docking calculations, are very important. Uncertainties in the atomic model can have significant consequences when this model is used as the basis for manual design, docking, scoring, and VS efforts (Davis et al. 2003). The redocking procedure, where a known ligand is docked to the "induced-fit" form of the target (Morris et al. 2009), is often the only method of validating the docking parameters used; it is important to develop additional validation procedures, especially those that include biological information, in order to decrease potential errors related to the X-ray crystallographic data and improve the robustness of the molecular docking calculations.

Molecular Dynamics Simulations

One of the main limitations of the molecular docking calculations is related to the flexibility of the targets. A high percentage of protein–ligand systems evaluated by docking require rigid conditions to test a large number of molecules quickly with low computational cost by VS. To evaluate the physical movements of these systems, molecular dynamics (MD) simulations are used. MD is an in silico technique which aims to derive statements about the structural, dynamical, and thermodynamical properties of the molecular systems (Salo-Ahen et al. 2021).

Leimkuhler and Matthews define the MD method as developing quantitative predictions of molecular size and shape, flexibilities, interactions with other molecules, behavior under pressure, and the relative frequency of one state or conformation compared to another (Leimkuhler and Matthews 2016). Historically, MD was developed in the early 1950s and has evolved constantly since then. Rahman and Verlet (Rahman 1964; Verlet 1967) refined the technique by implementing the method for all states of matter. Part of the evolution and relevance of these techniques is related to the foundation of powerful programs founded by Martin Karplus, Michael Levitt, and Arieh Warshel that were used to understand and predict chemical processes. These three researchers were recognized with the Nobel Prize in Chemistry in 2013 for the development of multiscale models for complex chemical systems (The Nobel Prize 2013). Currently, more than 20,000 papers related to MD are published annually, with close to 6% annual growth in the number of works from 2016 to 2020 (Web of Science, Copyright Clarivate 202).

Algorithms in Molecular Dynamics Simulations

For an MM system, MD consists of iterations of the instant forces present and the consequent movements of that system. The MM system is described as a set of particles that move in response to their interactions according to Newton's equations of motion, and the MD simulation computes the movements of atoms with time by integrating these equations, as given below (Salmaso and Moro 2018; Adcock and McCammon 2006):

$$\frac{d^2 r_i(t)}{dt^2} = \frac{F_i(t)}{m_i}$$

where $Fi(t)$ is the force exerted on atom i at time t, $r_i(t)$ is the vector position of the atom i at time t, and m_i is the mass of the atom (Salmaso and Moro 2018).

The essential function of the MD algorithms is to derive Newton's equations in a time differential (dt) for each atom of the system (Salmaso and Moro 2018).

- Position: $r_i(t) = (x_i(t), y_i(t), z_i(t))$
- Velocity: $v_i(t) = \frac{dr_i(t)}{dt}$
- Acceleration: $a_i(t) = \frac{d^2 r_i(t)}{d^2 t} = \frac{f_i(t)}{m_i}$
- Force: $f_i(t) = -\frac{dV(r(t))}{dr_i(t)}$
- Potential Energy: $V(r(t))$

where, x_i, y_i, and z_i are the coordinates of the i atom in the time (t).

The algorithms used in MD are classified into five types: integrators, short-range interaction, long-range interaction, parallel computing, and ab initio (Table 4).

Table 4 Main classification of the most common MD algorithms

Short-range interaction (Wang et al. 2007)	Large-range interaction (Sagui and Darden 1999)	Integrators (Adcock and McCammon 2006)	Paralleling computing (Koradi et al. 2000)	Ab initio (Alfe 1999)
Verlet cell-linked list (VCL)	Ewald summation PME P3M MSM	SHAKE Beeman's Verlet-Stoermer Leapfrog	Point-centered domain decomposition	Car-Parrinello

Two types of algorithms depend on the type of nonbonded interactions evaluated, short-range and large-range interactions. The short-range algorithms are based on the Lennard–Jones potential that represents the van der Waals interactions. Meanwhile, the large-range algorithms are based on the treatment of the long-range electrostatic potentials (Sagui and Darden 1999). In the ab initio algorithms, the interactions between ions and electrons are treated fully by QM, and the ions are moved using the classical Newton equations of motion. Carr–Parrinello is one of the most common ab initio algorithms, which was developed to calculate the ab initio forces on the ions and keep the electrons close to the Born–Oppenheimer surface while the atoms move (Alfe 1999). In parallel computing, MD simulations are performed on parallel computers: the molecular system is divided into clusters assigned to individual processors (Koradi et al. 2000). Finally, integrators consist of different algorithms to integrate the equations of motion. Many of these are different methods, in which the integration is partitioned into small steps, each separated by a defined period because the continuous potentials describing atomic interaction preclude an analytical solution. Among the most common MD integrator algorithms identified are Verlet, Leapfrog, and Beeman (Adcock and McCammon 2006).

Force Field in Molecular Dynamics Simulations

As observed previously in molecular docking calculations, MM techniques use a force field to describe the dependence of the energy on the atomic coordinates of the system (González 2011). In MD, the accuracy of the force field is critical to the validity and stability of the simulations of proteins and, in fact, all macromolecules (Adcock and McCammon 2006). The main FFs used in MD include MM2, MM3, MM4, CHARMM, AMBER, GROMOS, OPLS, and COMPASS. Many versions of these force fields exist to perform MD simulations, such as GROMOS96, GROMOS45A3, and GROMOS53A5 (González 2011).

The formal expression of a force field is divided into two groups of terms, bonded and nonbonded. The bonded interactions account for the stretching of bonds, the bending of valence angles, and the rotation of dihedral angles. The nonbonded interactions capture electrostatics, dispersion, and the Pauli exclusion forces (Guvench and MacKerell 2008).

$$E_{\text{Total}} = E_{\text{Bonded}} + E_{\text{nonbonded}} + E_{\text{others}}$$

$$E_{\text{bonded}} = \sum_{\text{Bonds}} K_b \left(b - b_o\right)^2 + \sum_{\text{angles}} K_\theta \left(\theta - \theta_o\right)^2 + \sum_{\text{dihedrals}} K\chi\left[1 + \cos\left(n\chi - \sigma\right)\right]$$

$$E_{\text{nonbonded}} = \sum_{\text{nonbonded pairs}\, ij} \left(_{ij}\left[\left(\frac{R_{\text{min},ij}}{r_{ij}}\right)^{12} - 2 * \left(\frac{R_{\text{min},ij}}{r_{ij}}\right)^{6} \right] + \frac{q_i q_j}{r_{ij}} \right)$$

The first part of the energy equation is related to the bonded interactions and has three terms. The first describes the stretching of bonds, where b is the interatomic distance (bond length) and K_b and b_0 are the parameters describing the stiffness and the equilibrium length of the bond, respectively. In agreement with the assumption in MM treating the bonded interactions as springs, this term also has a quadratic form known as Hooke's Law. For the calculation of the second term, the involvement of three atoms is necessary to describe the bending of angles, where θ is the angle formed by the two bond vectors; K_θ and θ_0 are the parameters describing the stiffness and equilibrium geometry of the angle, respectively. These parameters are similar to the terms for bond stretching, K_b and b_0. The last term is related to the energy associated with the rotation of dihedral angles defined by four atoms, where χ is the dihedral value, $K\chi$ is the energetic parameter that determines the barrier heights, n is the periodicity or multiplicity, and σ is the phase (Salmaso and Moro 2018; Guvench and MacKerell 2008).

Nonbonded interactions have two terms. The first is known as the Lennard–Jones equation. The second term relates the models of attractive dispersion and repulsive Pauli exclusion interactions and is commonly referred to as the van der Waals term (Salmaso and Moro 2018; Guvench and MacKerell 2008). Calculating these nonbonded interactions in biomolecular simulations is a key issue and one of the main challenges in the area (Monticelli and Tieleman 2013).

Since the early 1980s, when the most common protein FFs, Amber, CHARMM, and OPLS, were developed, protein FFs have continuously evolved and improved (Ponder and Case 2003). Currently, protein FFs are highly advanced, even with respect to other macromolecules. Guvench and Mackerell performed a comparison among the most common protein FFs for MD and found that all the studied force fields (Amber, CHARMM, GROMOS, and OPLS-AA) treat proteins at an often satisfactory level of accuracy (Guvench and MacKerell 2008). Developments in these methods have produced multi-microsecond simulations of two proteins, ubiquitin and Protein G, using a number of different FFs. Four FFs (CHARMM22*, CHARMM27, Amber ff99SB-ILDN, and Amber ff99SB*-ILDN) showed a good agreement between NMR data and MD simulations (Martín-García et al. 2015).

Limitations of Molecular Dynamics Simulations

Gonzalez chose an interesting definition of the utility of the MD simulations: the aim is not to reproduce an experimental result but to understand the microscopic origin of the physical properties observed or to predict qualitatively the behavior expected at conditions that cannot be accessed experimentally (González 2011). In protein FFs, the constant evolution has allowed a high level of accuracy in the predictions. However, the same level of accuracy is not observed for other macromolecules because only a few specialized force fields are regularly used today for sugars, nucleic acids, and lipids (Monticelli and Tieleman 2013). For example, Ricci performed a comparison of FF and terminal nucleotide definitions because, despite the importance of DNA as a target for several proteins and drugs, molecular dynamics simulations with nucleic acids still face many challenges, such as the reliability of the chosen force fields (Ricci et al. 2010).

The computational cost is one of the main limitations of MD simulations. The conformational sampling of biological systems is in many cases limited by the capabilities of the computational hardware (Platania and Bucolo 2021). Even though other techniques have been developed to overcome the limitations of MD in classical molecular dynamics, such as enhanced sampling MD simulations, a low computational cost requires some approximations, which decrease the accuracy of the predictions. Ab initio methods produce more realistic simulations of complex molecular systems and processes but are computationally complex. To overcome this difficulty, electronic interactions are approximated with an effective pseudopotential, and the orbitals are expressed in terms of a suitable functional basis. Hybrid QM/MM methods follow a similar approach (Paquet and Viktor 2018).

Moreover, the size of the biological system is a key factor in MD simulations. Some biomolecular processes, including ligand binding and conformational change, often take place on timescales longer than those accessible with a classical all-atom MD simulation. For some systems with approximately 50,000 atoms, one GPU takes a few days to simulate a single microsecond (Hollingsworth and Dror 2018). Through Markov State Models (MSMs), these limitations have been overcome. MSMs are based on an ensemble view of the dynamics, from which statistical properties, such as the probability of a state being occupied and the probability of jumping from one state to another, are computed (Salmaso and Moro 2018).

De Novo Drug Design

The development of new molecular structures and entities with therapeutic uses is an enormous challenge in producing new drugs for commercial use. Consequently, different strategies have been proposed to find and design new candidates to fulfill this demand. Considering the gigantic amount of information available to explore the extant chemical space associated with different diseases, it is possible to approach the drug design task, where this information is valuable and convenient, by proposing new molecular entities from small molecular fragments assembled to maximize the interaction of the ligand with the active site, by means of computational growth algorithms, through a method called de novo drug design (DNDD) (Mouchlis et al. 2021).

De novo design can be defined as a method where a molecule is designed to satisfy the constraints and characteristics needed to achieve the required biological or therapeutic activity (Schneider and Baringhaus 2013). There are two useful ways to start a de novo methodology. First, in the structure-based approach, the construction of the molecules can be started using the knowledge acquired from the structure of the active site: the 3D structure and the possible interaction sites are known. The second approach is the ligand-based approach, in which the construction of molecules is started from a known active ligand, but the active site information is unavailable and hard to obtain (Mouchlis et al. 2021). Once the methodology has been chosen, the next step is to build and evaluate the molecules that satisfy the established restrictions (Mouchlis et al. 2021; Schneider and Baringhaus 2013). Some popular software for these approaches is shown in Table 5 and classified according to the methodology used.

Table 5 Examples of de novo software classified according to the type of methodology used to construct and score molecules (Mouchlis et al. 2021; Schneider and Fechner 2005)

Fragment-based software		
	Ligand-based	Structure-based
LUDI		X
SPROUT		X
Chemical genesis	X	X
PRO_LIGAND	X	X
TOPAS	X	
ADAPT		X
Atom-based software		
Diamond Lattice		X
LEGEND		X
MCDNLG		X
DLD		X
RASSE		X

Sampling, Scoring, and Optimization

In order to assemble a list of candidate molecules, there are two possible methodologies. First, in the atom-based methodology, the transformation of the molecules is achieved by modifying one atom into another to explore the molecular possibilities. This is a very intuitive method, which results in a wider ensemble of novel structures. However, it suffers from the possibility of creating new molecules with unfavorable structures, unstable hetero–hetero atomic bonds, or difficult synthetic accessibility (Mouchlis et al. 2021; Kawai et al. 2014; Liu et al. 2021). Second, in the fragment-based approach, the design of the molecules is achieved by the mutation of pre-defined molecular fragments, which causes a vast reduction of the possibilities within the chemical space, is less time-consuming, and the chemical feasibility of the proposed molecules is higher (see Table 5) (Kawai et al. 2014; Liu et al. 2021).

During the construction of the molecules, they are scored according to a set of suitable molecular descriptors to determine their druggability using different algorithms. If the study is done using a structure-based methodology, the algorithm calculates the score based on the interaction of the fragments with the active site by docking procedures, and those fragments are then used as seeds to build the rest of the molecule. In contrast, in ligand-based approaches, molecules or fragments are compared to a reference ligand by considering its similarity to their descriptors.

Optimization processes are typically based on evolutionary computation, which is inspired by biological evolution, involving mutation, crossover, and selection of the candidates more likely to "survive" based on their performance in the defined biological function. Evolutionary computation is divided into four categories: genetic algorithms, genetic programming, evolution strategies, and evolutionary programming (Mouchlis et al. 2021; Abraham et al. 2006). In general, all of these algorithms begin with a population of candidates, within which the most promising molecules are selected, finally proposing a new generation of candidates to replace the initial one, and then the algorithm restarts. The main advantages of these algorithms are their simplicity, adaptability, and efficacy in exploring the chemical space (Abraham et al. 2006).

Machine Learning in De Novo Drug Design

Regarding the vast amount of information available about existing compounds and their biological activity in databases such as ChEMBL, COCONUT (Sorokina et al. 2021), and ZINC, several methods have been developed to analyze and predict the chemical behavior of these bioactive compounds, causing a considerable reduction in time and allowing a more efficient exploration of the chemical space in de novo studies (Meyers et al. 2021). In particular, due to the advances achieved in machine learning, the possibility of allowing computers to learn to select a promising molecule by starting from chemical and biological information, as well as giving them the ability to find structural fingerprints that can lead to the identification of the most important characteristics of molecules to enhance their activity, has had a remarkable effect on de novo methodologies.

Deep Learning (DL) (Krishnan et al. 2021; Schissel et al. 2020; Chen et al. 2018), Reinforcement Learning (RL) (Putin et al. 2018; Born et al. 2021), Deep Reinforcement Learning (DRL) (Ståhl et al. 2019) and Reinforcement Learning for Structural Evolution (ReLeaSE) (Popova et al. 2018) are some of the machine learning methods developed to assist with de novo design of molecules in the past decade (Mouchlis et al. 2021; Popova et al. 2018). For all these methods, it is necessary to start from molecular information, usually in the SMILES format, and the biological activity measurements of the selected target. The main differences between these methods are the type of architecture developed for data analysis and evaluation and the evaluated molecular properties.

Limitations in De Novo Drug Design

Despite all of the advantages of de novo methodologies, especially the possibility of extensively exploring the chemical space for lead structure identification, some disadvantages and limitations exist. The most relevant limitation is related to the accuracy of the scoring functions. De novo software does not consider the conformational modifications that may take place in the structure of the target protein (particularly in the active site) or the entropic effects caused by the solvent–ligand interaction, which limits the calculation of the ligand–pocket interaction energy and thus reduces the accuracy of the molecular design (Hartenfeller and Schneider 2010).

Additionally, de novo designed molecules might have good predicted selectivities and interaction energies with targets; however, these methodologies do not allow the prediction of other important physicochemical properties of the designed structures (e.g., solubility, permeability to cells, and affinity to transport proteins) involved in pharmacokinetic studies and relevant to choosing the most promising therapeutic candidates (Mouchlis et al. 2021; Schneider and Fechner 2005). These limitations need to be overcome in order to increase the success of de novo methodologies in the design of effective drug molecules.

5 Conclusions and Future Directions

In this chapter, we reviewed the main aspects of target-based drug design to explain the development and importance of these types of techniques in drug discovery. Various structures have been successfully discovered through this class of methods, but it is still necessary to overcome some existing limitations of these methodologies.

The constantly growing and updated databases are the main source of structures for methodologies such as VS and online software for molecular docking calculations present a promising future for target-based drug design. However, for the continued growth of these databases to be useful, the correct use of these techniques is necessary, which begins with understanding the main concepts of the methodologies.

For the molecular docking calculations, the development of scoring functions is critical for flexible receptor docking to improve the energy prediction and the correct spatial position of the ligands. In the same way, it is fundamental to develop new scoring algorithms to improve the accuracy, such as those based on QM or mixed methodologies, including the semiempirical and QM/MM methods and the corresponding validation procedures of the calculations.

Moreover, the importance of MD simulations as a key technique in the drug development process is likely to grow substantially with increasing computer power and advances in the development of FFs and enhanced MD methodologies (Salo-Ahen et al. 2021). In addition, improving and developing the FF for macromolecules such as sugars, nucleic acids, and lipids is vital to expand the applications of the MD simulations in biological systems.

In all cases, the main challenges in target-based drug design methodologies are related to improving the accuracy of the predictions while reducing the computational cost of the calculations.

Acknowledgments We thank the CNPq and Capes for financial Support, Grant Numbers 309648/2019-0 and 431254/2018-4.

References

Abraham A, Nedjah N, de Macedo ML. Evolutionary computation: from genetic algorithms to genetic programming. In: Genetic systems programming. Springer; 2006. p. 1–20.

Adcock SA, McCammon JA. Molecular dynamics: survey of methods for simulating the activity of proteins. Chem Rev. 2006;106:1589–615.

Alfe D. Ab initio molecular dynamics, a simple algorithm for charge extrapolation. Comput Phys Commun. 1999;118:31–3.

Armon A, Graur D, Ben-Tal N. ConSurf: an algorithmic tool for the identification of functional regions in proteins by surface mapping of phylogenetic information. J Mol Biol. 2001;307:447–63.

Bagal SK, Brown AD, Cox PJ, Omoto K, Owen RM, Pryde DC, Sidders B, Stevens SSE, EB, Storer RI. Ion channels as therapeutic targets: a drug discovery perspective. J Med Chem. 2013;56:593–624.

Batool M, Ahmad B, Choi S. A structure-based drug discovery paradigm. Int J Mol Sci. 2019;20:2783.
Berman HM, Westbrook J, Feng Z, Gilliland G, Bhat TN, Weissig H, Shindyalov IN, Bourne PE. The protein data bank. Nucleic Acids Res. 2000;28:235–42.
Born J, Manica M, Oskooei A, Cadow J, Markert G, Martínez MR. PaccMannRL: De novo generation of hit-like anticancer molecules from transcriptomic data via reinforcement learning. Iscience. 2021;24:102269.
Chen H, Engkvist O, Wang Y, Olivecrona M, Blaschke T. The rise of deep learning in drug discovery. Drug Discov Today. 2018;23:1241–50.
Clark DE. What has computer-aided molecular design ever done for drug discovery? Exp Opin Drug Discov. 2006;1:103–10.
Coleman RG, Carchia M, Sterling T, Irwin JJ, Shoichet BK. Ligand pose and orientational sampling in molecular docking. PLoS One. 2013;8:e75992.
Crespo A, Rodriguez-Granillo A, Lim VT. Quantum-mechanics methodologies in drug discovery: applications of docking and scoring in lead optimization. Curr Top Med Chem. 2017;17:2663–80.
Davis AM, Teague SJ, Kleywegt GJ. Application and limitations of X-ray crystallographic data in structure-based ligand and drug design. Angew Chem Int Ed. 2003;42:2718–36.
Dias R, de Azevedo J, Walter F. Molecular docking algorithms. Curr Drug Targets. 2008;9:1040–7.
Ferrara P, Gohlke H, Price DJ, Klebe G, Brooks CL. Assessing scoring functions for protein–ligand interactions. J Med Chem. 2004;47:3032–47.
Ferreira LG, Dos Santos RN, Oliva G, Andricopulo AD. Molecular docking and structure-based drug design strategies. Molecules. 2015;20:13384–421.
Foloppe N, Hubbard R. Towards predictive ligand design with free-energy based computational methods? Curr Med Chem. 2006;13:3583–608.
Gabel J, Desaphy J, Rognan D. Beware of machine learning-based scoring functions on the danger of developing black boxes. J Chem Inform Model. 2014;54:2807–15.
González MA. Force fields and molecular dynamics simulations. École thématique de la Société Française de la Neutronique. 2011;12:169–200.
Guedes IA, de Magalhães CS, Dardenne LE. Receptor–ligand molecular docking. Biophys Rev. 2014;6:75–87.
Guedes IA, Pereira FSS, Dardenne LE. Empirical scoring functions for structure-based virtual screening: applications, critical aspects, and challenges. Front Pharmacol. 2018;9:1089.
Guvench O, MacKerell AD. Comparison of protein force fields for molecular dynamics simulations. Mol Model Proteins. 2008:63–88.
Harigua-Souiai E, Cortes-Ciriano I, Desdouits N, Malliavin TE, Guizani I, Nilges M, Blondel A, Bouvier G. Identification of binding sites and favorable ligand binding moieties by virtual screening and self-organizing map analysis. BMC Bioinformatics. 2015;16:1–15.
Hartenfeller M, Schneider G. De novo drug design. Chemoinform Comput Chem Biol. 2010:299–323.
Hollingsworth SA, Dror RO. Molecular dynamics simulation for all. Neuron. 2018;99:1129–43.
Huang B. MetaPocket: a meta approach to improve protein ligand binding site prediction. OMICS A J Integr Biol. 2009;13:325–30.
Huang S-Y, Zou X. Advances and challenges in protein-ligand docking. Int J Mol Sci. 2010;11:3016–34.
Huang S-Y, Grinter SZ, Zou X. Scoring functions and their evaluation methods for protein–ligand docking: recent advances and future directions. Phys Chem Chem Phys. 2010;12:12899–908.
Jain AN. Scoring functions for protein-ligand docking. Curr Protein Peptide Sci. 2006;7:407–20.
Kawai K, Nagata N, Takahashi Y. De novo design of drug-like molecules by a fragment-based molecular evolutionary approach. J Chem Inform Model. 2014;54:49–56.
Koradi R, Billeter M, Güntert P. Point-centered domain decomposition for parallel molecular dynamics simulation. Comput Phys Commun. 2000;124:139–47.
Krishnan SR, Bung N, Bulusu G, Roy A. Accelerating de novo drug design against novel proteins using deep learning. J Chem Inform Model. 2021;61:621–30.

Lappano R, Maggiolini M. G protein-coupled receptors: novel targets for drug discovery in cancer. Nat Rev Drug Discov. 2011;10:47–60.
Laskowski RA. SURFNET: a program for visualizing molecular surfaces, cavities, and intermolecular interactions. J Mol Graph. 1995;13:323–30.
Laurie ATR, Jackson RM. Q-SiteFinder: an energy-based method for the prediction of protein–ligand binding sites. Bioinformatics. 2005;21:1908–16.
Leimkuhler B, Matthews C. Molecular dynamics. Springer; 2016.
Li Y, Yang J. Structural and sequence similarity makes a significant impact on machine-learning-based scoring functions for protein–ligand interactions. J Chem Inform Model. 2017;57:1007–12.
Li J, Fu A, Zhang L. An overview of scoring functions used for protein–ligand interactions in molecular docking. Interdiscip Sci: Comput Life Sci. 2019;11:320–8.
Lionta E, Spyrou G, Vassilatis K, D, Cournia Z. Structure-based virtual screening for drug discovery: principles, applications and recent advances. Curr Top Med Chem. 2014;14:1923–38.
Liu J, Wang R. Classification of current scoring functions. J Chem Inform Model. 2015;55:475–82.
Liu X, Ijzerman AP, van Westen GJP. Computational approaches for de novo drug design: past, present, and future. In: Artificial neural networks. Springer; 2021. p. 139–65.
Lounnas V, Ritschel T, Kelder J, McGuire R, Bywater RP, Foloppe N. Current progress in structure-based rational drug design marks a new mindset in drug discovery. Comput Struct Biotechnol J. 2013;5:e201302011.
Maia MS, Nunes TAL, Sousa JMS, Rodrigues GCS, Monteiro AFM, Tavares JF, Rodrigues KAF, Mendonça-Junior FJB, Scotti L, Scotti MT. Virtual screening and the in vitro assessment of the antileishmanial activity of lignans. Molecules. 2020;25:2281.
Martín-García F, Papaleo E, Gomez-Puertas P, Boomsma W, Lindorff-Larsen K. Comparing molecular dynamics force fields in the essential subspace. PLoS One. 2015;10:e0121114.
Meng X-Y, Zhang H-X, Mezei M, Cui M. Molecular docking: a powerful approach for structure-based drug discovery. Curr Comput-Aid Drug Des. 2011;7:146–57.
Meyers J, Fabian B, Brown N. De novo molecular design and generative models. Drug Discov Today. 2021;26(11):2707–15.
Monticelli L, Tieleman DP. Force fields for classical molecular dynamics. Biomol Simul. 2013:197–213.
Morris GM, Huey R, Lindstrom W, Sanner MF, Belew RK, Goodsell DS, Olson AJ. AutoDock4 and AutoDockTools4: automated docking with selective receptor flexibility. J Comput Chem. 2009;30:2785–91.
Mouchlis VD, Afantitis A, Serra A, Fratello M, Papadiamantis AG, Aidinis V, Lynch I, Greco D, Melagraki G. Advances in de novo drug design: from conventional to machine learning methods. Int J Mol Sci. 2021;22:1676.
Ng MCK, Fong S, Siu SWI. PSOVina: The hybrid particle swarm optimization algorithm for protein–ligand docking. J Bioinforma Comput Biol. 2015;13:1541007.
Nussinov R, Tsai C-J. The different ways through which specificity works in orthosteric and allosteric drugs. Curr Pharm Des. 2012;18:1311–6.
Oda A. Development and validation of programs for ligand-binding-pocket search. Yakugaku zasshi: J Pharm Soc Japan. 2011;131:1429–35.
Oliveira SHP, Ferraz FAN, Honorato RV, Xavier-Neto J, Sobreira TJP, de Oliveira PSL. KVFinder: steered identification of protein cavities as a PyMOL plugin. BMC Bioinformatics. 2014;15:1–8.
Paquet E, Viktor HL. Computational methods for Ab initio molecular dynamics. Adv Chem. 2018;2018:9839641.
Platania CBM, Bucolo C. Molecular dynamics simulation techniques as tools in drug discovery and pharmacology: a focus on allosteric drugs. In: Allostery. Springer; 2021. p. 245–54.
Ponder JW, Case DA. Force fields for protein simulations. Adv Protein Chem. 2003;66:27–85.
Popova M, Isayev O, Tropsha A. Deep reinforcement learning for de novo drug design. Science Adv. 2018;4:eaap7885.
Potapov V, Cohen M, Inbar Y, Schreiber G. Protein structure modelling and evaluation based on a 4-distance description of side-chain interactions. BMC Bioinformatics. 2010;11:1–17.

Putin E, Asadulaev A, Ivanenkov Y, Aladinskiy V, Sanchez-Lengeling B, Aspuru-Guzik A, Zhavoronkov A. Reinforced adversarial neural computer for de novo molecular design. J Chem Inform Model. 2018;58:1194–204.

Rahman A. Correlations in the motion of atoms in liquid argon. Phys Rev. 1964;136:A405.

Ramesh S, Vallinayagam S, Rajendran K, Rajendran S, Rathinam V, Ramesh S. Computer-aided drug designing–modality of diagnostic system. In: Biomedical signal processing for healthcare applications. CRC Press; 2021. p. 195–218.

Ricci CG, de Andrade ASC, Mottin M, Netz PA. Molecular dynamics of DNA: comparison of force fields and terminal nucleotide definitions. J Phys Chem B. 2010;114:9882–93.

Sagui C, Darden TA. Molecular dynamics simulations of biomolecules: long-range electrostatic effects. Annu Rev Biophys Biomol Struct. 1999;28:155–79.

Sahu A, Patra PK, Yadav MK, Varma M. Identification and characterization of ErbB4 kinase inhibitors for effective breast cancer therapy. J Receptors Signal Transduct. 2017;37:470–80.

Salmaso V, Moro S. Bridging molecular docking to molecular dynamics in exploring ligand-protein recognition process: an overview. Front Pharmacol. 2018;9:923.

Salo-Ahen OMH, Alanko I, Bhadane R, Bonvin AMJJ, Honorato RV, Hossain S, Juffer AH, Kabedev A, Lahtela-Kakkonen M, Larsen AS. Molecular dynamics simulations in drug discovery and pharmaceutical development. PRO. 2021;9:71.

Schissel CK, Mohapatra S, Wolfe JM, Fadzen CM, Bellovoda K, Wu CL, Wood JA, Malmberg AB, Loas A, Gómez-Bombarelli R. Interpretable deep learning for de novo design of cell-penetrating abiotic polymers. bioRxiv; 2020.

Schmidtke P, Bidon-Chanal A, Luque FJ, Barril X. MDpocket: open-source cavity detection and characterization on molecular dynamics trajectories. Bioinformatics. 2011;27:3276–85.

Schneider G, Baringhaus KH. De novo design: from models to molecules. De Novo Mol Des. 2013:1–55.

Schneider G, Fechner U. Computer-based de novo design of drug-like molecules. Nat Rev Drug Discov. 2005;4:649–63.

Shen C, Ding J, Wang Z, Cao D, Ding X, Hou T. From machine learning to deep learning: advances in scoring functions for protein–ligand docking. Wiley Interdiscip Rev: Comput Mol Sci. 2020;10:e1429.

Shen C, Hu Y, Wang Z, Zhang X, Pang J, Wang G, Zhong H, Xu L, Cao D, Hou T. Beware of the generic machine learning-based scoring functions in structure-based virtual screening. Brief Bioinform. 2021;22:bbaa070.

Simões T, Lopes D, Dias S, Fernandes F, Pereira J, Jorge J, Bajaj C, Gomes A. Geometric detection algorithms for cavities on protein surfaces in molecular graphics: a survey, vol. 36. Wiley Online Library; 2017. p. 643–83.

Sorokina M, Merseburger P, Rajan K, Yirik MA, Steinbeck C. Coconut online: collection of Open Natural Products database. J Cheminform. 2021;13:1–13.

Ståhl N, Falkman G, Karlsson A, Mathiason G, Bostrom J. Deep reinforcement learning for multiparameter optimization in de novo drug design. J Chem Inform Model. 2019;59:3166–76.

Stank A, Horn KDB, M, Sizikova E, Neil R, Panecka, J, Richter S, Wade RC. TRAPP webserver: predicting protein binding site flexibility and detecting transient binding pockets. Nucleic Acids Res. 2017;45:W325–30.

Sun J, Chen K. NSiteMatch: prediction of binding sites of nucleotides by identifying the structure similarity of local surface patches. Comput Math Methods Med 2017; 2017.

Surabhi S, Singh BK. Computer aided drug design: an overview. J Drug Deliv Ther. 2018;8:504–9.

Tan KP, Nguyen TB, Patel S, Varadarajan R, Madhusudhan MS. Depth: a web server to compute depth, cavity sizes, detect potential small-molecule ligand-binding cavities and predict the pKa of ionizable residues in proteins. Nucleic Acids Res. 2013;41:W314–21.

Taylor RD, Jewsbury PJ, Essex JW. A review of protein-small molecule docking methods. J Comput Aided Mol Des. 2002;16:151–66.

The Nobel Prize. The Nobel Prize in Chemistry 2013. https://www.nobelprize.org/prizes/chemistry/2013/advanced-information/ (June 22th, 2021).

Tripathi A, Bankaitis VA. Molecular docking: from lock and key to combination lock. J Mol Med Clin Appl. 2017;2(1) https://doi.org/10.16966/2575-0305.106.

Valeur E, Guéret SM, Adihou H, Gopalakrishnan R, Lemurell M, Waldmann H, Grossmann TN, Plowright AT. New modalities for challenging targets in drug discovery. Angew Chem Int Ed. 2017;56:10294–323.

Van Montfort RLM, Workman P. Structure-based drug design: aiming for a perfect fit. Essays Biochem. 2017;61:431–7.

Verlet L. Computer "experiments" on classical fluids. I. Thermodynamical properties of Lennard-Jones molecules. Phys Rev. 1967;159:98.

Volkamer A, Kuhn D, Rippmann F, Rarey M. DoGSiteScorer: a web server for automatic binding site prediction, analysis and druggability assessment. Bioinformatics. 2012;28:2074–5.

Wagner JR, Sørensen J, Hensley N, Wong C, Zhu C, Perison T, Amaro RE. POVME 3.0: software for mapping binding pocket flexibility. J Chem Theory Comput. 2017;13:4584–92.

Wang W. Applications of biophysical methods in small-molecule modulators targeting protein function. (Doctoral dissertation. University of Groningen; 2021.

Wang R, Lu Y, Wang S. Comparative evaluation of 11 scoring functions for molecular docking. J Med Chem. 2003;46:2287–303.

Wang DB, Hsiao FB, Chuang CH, Lee YC. Algorithm optimization in molecular dynamics simulation. Comput Phys Commun. 2007;177:551–9.

Wang T, Wu M-B, Zhang R-H, Chen Z-J, Hua C, Lin J-P, Yang L-R. Advances in computational structure-based drug design and application in drug discovery. Curr Top Med Chem. 2016a;16:901–16.

Wang Z, Sun H, Yao X, Li D, Xu L, Li Y, Tian S, Hou T. Comprehensive evaluation of ten docking programs on a diverse set of protein–ligand complexes: the prediction accuracy of sampling power and scoring power. Phys Chem Chem Phys. 2016b;18:12964–75.

Xie Z-R, Liu C-K, Hsiao F-C, Yao A, Hwang M-J. LISE: a server using ligand-interacting and site-enriched protein triangles for prediction of ligand-binding sites. Nucleic Acids Res. 2013;41:W292–6.

Yu J, Zhou Y, Tanaka I, Yao M. Roll: a new algorithm for the detection of protein pockets and cavities with a rolling probe sphere. Bioinformatics. 2010;26:46–52.

Yuriev E, Ramsland PA. Latest developments in molecular docking: 2010–2011 in review. J Mol Recogn. 2013;26:215–39.

Zhu H, Pisabarro MT. MSPocket: an orientation-independent algorithm for the detection of ligand binding pockets. Bioinformatics. 2011;27:351–8.

Zhu F, Shi Z, Qin C, Tao L, Liu X, Xu F, Zhang L, Song Y, Liu X, Zhang J. Therapeutic target database update 2012: a resource for facilitating target-oriented drug discovery. Nucleic Acids Res. 2012;40:D1128–36.

Drug Discovery Paradigms: Phenotypic-Based Drug Discovery

Alan Talevi and Carolina L. Bellera

Contents

Abstract

A drug discovery and development project typically starts with the identification of novel active scaffolds, i.e., core chemical structures with a desired biological effect. Beyond serendipitous discoveries and findings based on ethnopharmacology/traditional medicine, drug discovery in the modern age has been guided by two fundamental screening philosophies (implemented whether through in silico, in vitro or less often, in vivo approximations). Occasionally, novel chemotypes can be designed de novo by searching for complementary features to a binding site in a predefined drug target. Historically, systematic screening for new active compounds comprised

A. Talevi (✉) · C. L. Bellera
Laboratory of Bioactive Compounds Research and Development (LIDeB), Department of Biological Sciences, Faculty of Exact Sciences, University of La Plata (UNLP), La Plata, Buenos Aires, Argentina

Argentinean National Council of Scientific and Technical Research (CONICET)—CCT La Plata, La Plata, Buenos Aires, Argentina
e-mail: atalevi@biol.unlp.edu.ar

M. T. Scotti, C. L. Bellera (eds.), *Drug Target Selection and Validation*, Computer-Aided Drug Discovery and Design,
https://doi.org/10.1007/978-3-030-95895-4_2

phenotypic screening assays (e.g., against a collection of microorganisms, animal models of disease, or cellular models). Later, the interest of the pharmaceutical companies experienced a substantial shift toward target-focused approximations in which exquisitely selective compounds were sought, usually through high-throughput screening. There, the test compounds were typically confronted with some biological entity, usually a protein, to identify those which could modulate such biomolecule. Nevertheless, as target-focused approximation failed to deliver the expectations, especially when pursuing therapies for complex disorders, renewed interest in phenotypic screening was observed in the pharmaceutical community, supported by a network pharmacology paradigm, high-content screening, small animal models, and organoids and other advanced cell culture platforms.

Phenotypic screening is advantageous in several respects. Remarkably, it can detect drugs with novel, unsuspected modes of action and/or complex pharmacology (e.g., multi-target drugs), and it can also provide hits with an adequate balance of an array of pharmaceutically relevant features, including efficacy, safety, and bioavailability, which in turn could lead to better translatability.

Keywords

Phenotypic screening · High-content screening (HCS) · High-content analysis · Target-focused approximations · High-throughput screening · Target deconvolution · Drug discovery · Hit identification

1 Introduction

Drug discovery and development is a lengthy, expensive, and challenging process, but similar to any other process, it requires a starting point. From a chemical perspective, such starting point is an active compound exhibiting a novel *active scaffold*, this is, a core molecular structure with the pursued biological activity (Hu et al. 2016). Active scaffolds can be detected by empirical serendipitous observations or, more frequently, from organized, systematic exploration of chemical libraries. Alternatively, active chemotypes can be designed de novo by building compounds with molecular features complementary to those in the binding site of a predefined drug target (as occurs, for instance, in fragment-based approximations (Hartenfeller and Schneider 2011)).

Systematic screening of chemical libraries is mainly undertaken under two different paradigms: *phenotypic-based drug discovery* (PBDD) and *target-focused drug discovery*.

Phenotypic drug discovery does not rely on knowledge of the identity of specific drug targets or hypotheses on their role in disease (Moffat et al. 2017). Historically speaking, PBDD preceded target-focused approximations. It consists in physiology-based models to select candidate compounds based on the phenotypic response that they elicit on integrated organismic systems (cellular models—including co-cultures, tissue cultures, ex vivo systems, whole animals) (Margineanu 2016). The first precedent of systematic massive phenotypic-based screens comes from Selman Waksman's team, back in the 1940s (National Historic Chemical Landmarks 2005). Waksman and his coworkers tested the inhibition of targeted pathogenic bacteria by isolated soil microbes, which led to the discovery of several antibiotic compounds, among them streptomycin and neomycin. The phenotypic approximation to drug discovery later led to the discovery of many other therapeutic agents that reached the pharmaceutical market, including drugs to treat disorders of complex etiology, such as many approved antiepileptic drugs found using animal models of seizure (Margineanu 2014; Löscher 2017) or neuropsychiatric treatments discovered through in vivo behavioral studies or fortuitous preclinical or clinical observations (Margineanu 2016). During the past decades of the twentieth century, because of the progress in molecular biology (recombinant DNA technology allows producing and isolating large quantities of putative molecular targets), protein crystallography, automated assay technologies, and computational chemistry, the physiologic-based approximation lost ground to the targeted-centered approximation, in which the starting point is a pre-defined molecular target with a validated or hypothesized relevant role in disease. The target-centered approximation, often synthesized under the "one disease, one gene, one drug" proposition, is a rather reductionist approach that pursues highly selective "clean" drugs in line with the Ehrlich's magic bullet notion. Under this paradigm, it is assumed that exquisitely selective drugs will result in safer medications because undesired off-target events would be avoided (in contrast with drugs emerging from phenotypic drug discovery, which often display a complex pharmacology and have been for some time pejoratively regarded as "dirty"). The targeted-focused paradigm is indeed attractive, as it provides a framework to what has been called "rational" (hypothesis-driven) drug discovery, i.e., the target-guided search and optimization of active scaffolds by enhancing the complementarity to the target protein (Reddy and Parrill 1999; Mavromoustakos et al. 2011). However, in many ways the target-centered paradigm has failed to deliver the expectations, at least in its initial conceptualization. First, many (if not most) of the drugs developed under this approximation have been shown to have serious toxic side effects (Mandal et al. 2009; Talevi 2016). Moreover, the productivity in the pharmaceutical industry did not improve (in relation to the increasing investment) under the domination of the target-centered paradigm. Defining whether the phenotypic or target-focused approximations have contributed the most to drug discovery during the past decades has been a source of controversy. Swinney and Anthony scrutinized the strategies used to discover the drugs approved in the 1999–2008 decade and concluded that the contribution of phenotypic screening to the discovery of first-in-class small-molecule drugs surpassed that of target-based approximations (Swinney and Anthony 2011). Such conclusion was however later challenged by Eder et al., who expanded the analysis to those drugs approved between 1999 and 2013 and included chemocentric-based discoveries (encompassing the identification of an active ingredient from a microbial or plant extract with known

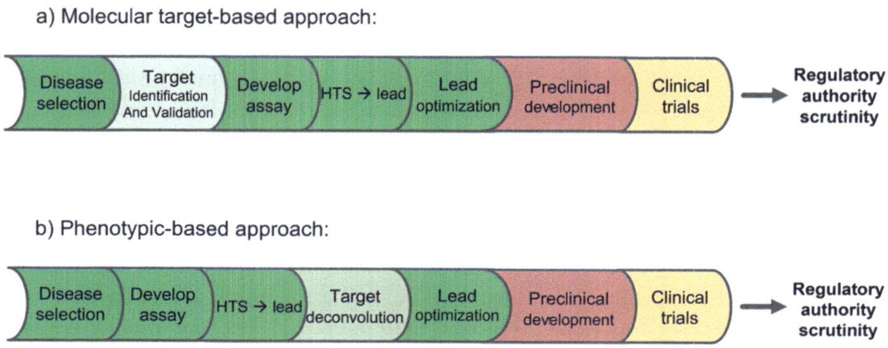

Fig. 1 Comparison of the general workflows for target-based drug discovery (**a**), where the drug target is defined before the screening, and the phenotypic-based approach (**b**), a target-agnostic approximation, where the target of the hits emerging from the screen are investigated a posteriori. In the light green stages, the differences in the two approaches can be observed

pharmacological activity, or the derivatization of an active natural substance or synthetic chemical, based on serendipitous findings) as an additional strategy (Eder et al. 2014). According to their surveillance, 70% of the drugs in that period (including biologicals) were discovered though target-based approximations. The workflows for target- and phenotypic-based drug discovery are schematically presented in Fig. 1.

In any case, a renewed interest toward phenotypic screening has recently arisen (Moffat et al. 2017; Lee and Berg 2013; Zheng et al. 2013; Warchal et al. 2016; Szabo et al. 2017; Childers et al. 2020) owing to the confluence of a multiplicity of factors that will be reviewed separately, including the emergence of small animal models and powerful in vitro techniques with reduced cost and improved throughput and translatability, the development of high content screening (HCS) platforms, the advent of omics and fast and cost-efficient gene-editing technologies, and the implementation of network pharmacology principles in the drug discovery arena.

2 Advantages of Phenotypic Drug Discovery

Appropriate target validation is the key to a successful target-based drug development project, which in principle implies establishing a clear link between a given molecular target and the pathophysiology of a target disease (Lee and Berg 2013), assuming that modulation of a single target could modify the progress or symptomatology of the disease, which might not always be the case. Good drug target prospects often gather other desirable properties, such as druggability, assayability, or vulnerability (Talevi et al. 2019). As target-based approaches depart from a previously established and validated hypothesis, it is unlikely (though not impossible) that an unexpected pharmacological profile arises from this approximation. Pharmacological novelty is pursued at the target identification stage, not later.

Contrariwise, phenotypic drug discovery is, *in principle*, target-agnostic, which makes it possible to find compounds that elicit the desired phenotypic response through unexpected modes of action and, often, *through multiple modes of action* (Margineanu 2016; Bianchi et al. 2009), an aspect that could prove advantageous in the light of systems pharmacology. This strength of phenotypic discovery is possibly its greatest weakness: whereas full understanding of the molecular mechanism of a drug is not essential to gain approval (Swinney 2013), mechanistic knowledge is indeed useful. For instance, in might serve to anticipate whether a phenotypic response at the preclinical level will translate to humans, to improve the pharmacological profile of a hit or lead compound through rational optimization programs, to predict on-target toxicity and differentiate from off-target toxicity, and to select clinical doses based on exposure measures at the site of action. Therefore, target deconvolution (i.e., the retrospective identification of the molecular target(s) that underlie the observed phenotypic responses) is a key step of the drug development process (Terstappen et al. 2007; Jung and Kwon 2015). Diverse methods are available to execute this task, which will be overviewed in a separate section. It is interesting to note that, after his analysis of the 1999–2008 approvals (Swinney and Anthony 2011), Swinney realized a follow-up study and compared the format and mechanistic information used to establish the phenotypic assays that led to first-in-class small-molecule drugs approved by the U.S. Food and Drug Administration (FDA) in the period 1999–2008, and he compared it with those drugs approved in 2012. He observed a relative increase in the number of phenotypic assays that included a pathway/biochemical marker (which were devised to identify drugs with a particular mode of action, in comparison with other models where only functional markers, e.g., cell death, were considered, what we could call unbiased phenotypic screens). This shows that, although usually target-based and physiology-based approaches are presented as complementary, and mutually exclusive paradigms, today they are quite integrated: candidates identified through target-based approximations will be eventually assayed in phenotypic models ("from target to phenotype"), and candidates emerging from phenotypic screening will be later subjected to laborious target deconvolution ("from phenotype to target"). The fundamental difference is *where the focus is set initially*, but it is far from being a trivial difference, as different starting points can certainly lead to different fates.

Drug discovery and development is a challenging multi-objective problem, where numerous pharmaceutically relevant properties (e.g., potency, selectivity, stability, solubility, permeability, etc.) should be addressed simultaneously (Nicolaou and Brown 2013; Talevi 2018; Lambrinidis and Tsantili-Kakoulidou 2021), a problem that is further complicated by the fact that some of these objectives are often in conflict (e.g., increasing potency can be detrimental for compound solubility and, in turn, bioavailability). Balance is a key concept in the drug development field, and it is possibly better or more immediately addressed by phenotypic screens. A second advantage of phenotypic screening is therefore that it provides an integrative, holistic view of many of these relevant aspects for pharmaceutical performance. Cell, tissue and/or whole organism models capture the simultaneous influence of various parameters, including absorption, distribution, metabolism, and elimination parameters, and also environmental factors and even, in some cases, cross talk between components of a complex system (Szabo et al. 2017). For instance, a prodrug would provide a negative result if assayed against an isolated target but could have a detectable effect in more complex physiology-based systems because it requires prior biotransformation to an intrinsically active metabolite. On the other hand, as the response in complex systems depends on many simultaneous factors, establishing a structure-activity relationship could be challenging or unfeasible if the molecular target of a series of compounds is not elucidated.

3 Target Deconvolution

As said before, once a bioactive compound has been identified phenotypically, target deconvolution will retrospectively elucidate the molecular target(s) of such candidate. Identifying direct targets is a time-consuming and challenging step, although it has recently been expedited owing to the development of genomics and proteomics, and advances in genetic engineering and bioinformatics. While an exhaustive description of drug target deconvolution techniques is out of the scope of this chapter, we would like to provide a glimpse of the complexity of the task. Deconvolution approximations can be broadly classified into *direct* and *indirect approaches* (Jung and Kwon 2015; Kubota et al. 2019).

Direct approaches aim to demonstrate the physical interaction between proteins and the test compounds and are mostly based on proteomics. They may be subclassified into the expression cloning-based approaches, protein microarray, and chemical biology approaches. In expression cloning-based approaches, each candidate molecular target is, in principle, individually overexpressed in microorganisms (e.g., yeast) or mammalian systems by introducing into the cells an expression vector coding the protein; later, a compound binding assay is performed. For instance, Bidlingmaier and Liu described the construction of libraries displaying human protein fragments on the surface of yeast cells, which were used to study small molecule/protein interactions (Bidlingmaier and Liu 2007). Alternatively, pooled library approaches such as phage display or ribosomal display are available (Omidfar and Daneshpour 2015; Wada et al. 2014). Phage display technology is based on the presentation of heterogeneous protein or peptide libraries on the surface of phages by fusion of a library of genetic sequences and bacteriophage coat protein genes, so that the different proteins are expressed on the surface of the phage. The phage library is later incubated with the immobilized small molecules. Phages expressing the target proteins bind to the immobilized small molecule; the others are washed off. The interacting phages are subsequently eluted, amplified by transfection into host bacterial cells and, after several rounds, enriched Similarly, in ribosomal display the mRNAs that encode for human proteins are constructed and translated in vitro to prepare pools of protein-ribosome-mRNA complexes used for ribosome display. In protein microarray, meanwhile, each protein is purified after overexpression and spotted in a microarray format for compound binding assays (Berrade et al. 2011). A shortcoming of these approximations is that many hydrophobic small molecules bind non-specifically to non-target proteins, hence producing false positives (Kubota et al. 2019). This disadvantage is solved by modern chemical biology approximations based on chemical proteomics. In the past, biochemical purification based on radioactive compound-binding assays was used for target deconvolution. Protein identification was performed by Edman sequencing, which required a large amount of highly purified protein. Nevertheless, advances in proteomics and in particular mass spectrometry-based proteomics, have facilitated higher throughput while simultaneously requiring much smaller amounts of sample. Affinity purification is possibly the most well-known chemical proteomics target deconvolution method (Saxena et al. 2009). The compound is modified to present a reactive group that serves to immobilize it to beads. A protein mixture (e.g., a protein extract from a cell) is then presented to the compound-immobilized beads; the non-interacting proteins are washed out, and only then the interacting proteins are eluted with adequate buffer solutions. At the end, the interacting proteins are identified by MS-based proteomics. Another relatively popular chemical proteomics target deconvolution method is photoaffinity labeling (Smith and Collins 2015), where a derivative of the drug is also obtained, in this case by adding two functional groups: a photo-crosslinked group that forms a covalent bond under UV radiation, and a purification tag. Protein mixtures or even live cells are incubated with the derivatized compound, and the compound–protein complex is fixed by applying UV radiation, which induces the formation of a covalent bond. The so-bonded complex is later purified using the affinity tag and analyzed though MS-based proteomics. A general drawback of deconvolution methods that require derivatizing and/or immobilizing the test compound is that, upon derivatization and/or immobilization, some compounds will have their affinity to the target changed, in some cases even losing their ability to bind the protein (thus potentially leading to false negatives). To reduce the chances of such inconvenience, structure-activity relationships should be laboriously elucidated to assure that the sites used for affinity tags or immobilization are not essential to the interaction with the target (Jung and Kwon 2015). Alternatively, this issue can be avoided by using label-free methods (Saxena 2016). Non-labeled methods rely on the fact that the physicochemical stability of proteins is often increased upon ligand binding (ligand-bound states are thermodynamically favorable owing to hydrogen bonding, electrostatic, and hydrophobic interactions, among others). For example, cellular thermal shift assay (CETSA®) is based on the fact that the ligand–protein complex commonly shows enhanced thermal stability (Friman 2020). Live cells or cell lysates are incubated with or without the test compound and heated. Denatured proteins are then removed by centrifugation, whereas soluble proteins are analyzed by MS-based proteomics to identify those whose thermal stability has changed due to engagement with a ligand molecule. Similarly, drug affinity responsive target stability (DARTS) relies on the protection against proteolysis conferred to the target by interaction with a small molecule (Pai et al. 2015). Consequently, differences in proteolysis patterns between protease-digested compound-treated and control protein samples can be analyzed by LC/MS/MS. Stability of proteins from rates of oxidation (SPROX) is another affinity-based label-free methodology (Strickland et al. 2013). In SPROX, stability changes induced by small-molecule binding are detected by the rate of methionine oxidation: ligand-induced differences in the target unfolding impact the rate of selective oxidation of methionine by hydrogen peroxide. One of the major limitations of this method is that it requires the detection of methionine residues, for which similar methods that exploit the use of other residues (e.g, lysine or tryptophan) have been developed. Examples of labeled and label-free methods are presented in Fig. 2.

Contrariwise, *indirect target deconvolution approaches* provide circumstantial and often ambiguous evidence on the possible drug target(s) by means of functional associations. For instance, similar mRNA expression profiles elicited on a cell by a test compound and a known reference compound might provide evidence of a similar mode of action, as implemented in large-scale drug perturbation databases such as the Connectivity Map or LINCS (Lamb et al. 2006; Keenan et al. 2018). The fundamental premise is that, owing to compensatory responses, the expression of elements of the target pathway of a drug will be modified upon exposure to it. Some of these adaptive effects might be, however, rather unspecific (e.g., upregulation of efflux transporters to promote drug clearance from the cell), while others would be related to the specific interactions of a drug with a targeted pathway (Sbaraglini and Talevi 2017). Monitoring mRNA expression to study the direct effects of a drug has its own and significant limitations (Kubota et al. 2019; Sbaraglini and Talevi 2017;

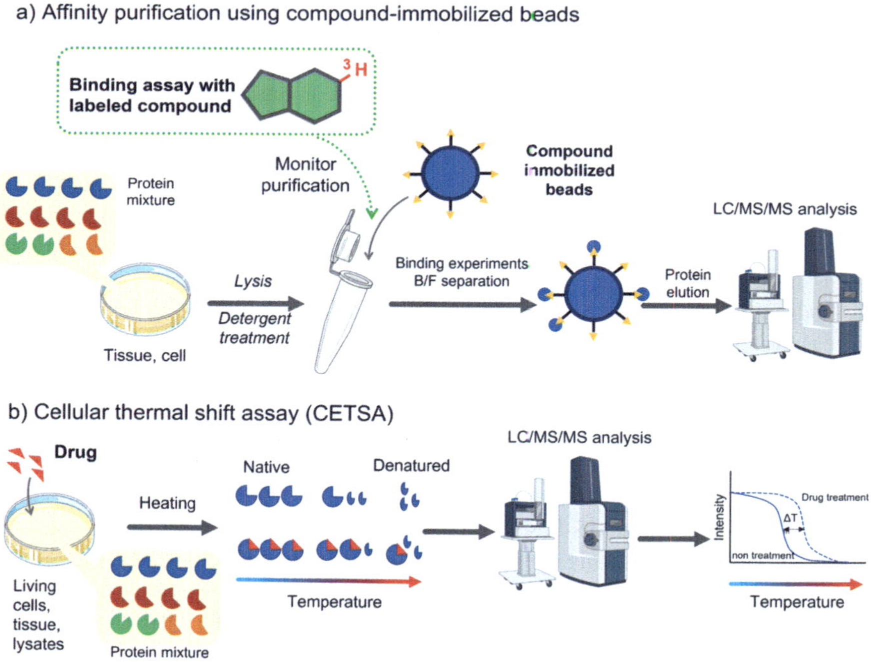

Fig. 2 Examples of labeled (**a**) and label-free (**b**) direct target deconvolution approximations. In label-free methods, the chemical/biochemical stability of the drug–protein complex is exploited to identify the molecular target of a drug

Chan et al. 2010): genes can be regulated at multiple levels beyond the transcriptional level (e.g., translational and post-translational); there is no necessarily simple correlation between mRNA and protein levels; drugs targeting two different proteins of the same pathway might have similar expression profiles and transcription. Other genetic screening methodologies are based on investigating how a gene knockdown can mimic the phenotypic response to a drug, or how gene knockout, knockdown, or upregulation can modify the response to a drug (either increasing or decreasing sensitivity to the drug). For example, gene dose reduction approaches commonly result in increased sensitivity. Yeasts have been used as valuable model systems for developing genome-wide assays to elucidate the modes of action of small molecules, owing to their rapid generation time, facile genetics, and cost-efficient cultivation, which results in well-characterized genome and proteome. In haploinsufficiency profiling, a heterozygous deletion strain is sensitized to a drug that targets the product of the heterozygous locus, which is reflected in decreased fitness and/or growth rate (Smith et al. 2010). All possible heterozygous deletion strains are screened in parallel to identify the drug target. Interestingly, the assay can simultaneously identify a hit compound and its candidate targets without previous knowledge of either. Alternative gene dose reduction methods include those that decrease the abundance of the potential target by mRNA perturbation (Chan et al. 2010) screening, (e.g., genome-wide RNAi) or, more recently, CRISPR-Cas9 screening (Neggers et al. 2018). A limitation of indirect methods is that the readouts used to detect the drug-target interaction are often far upstream or downstream of the real target and that, as previously insinuated, interaction of the drug with several components of the same biological pathway could produce a similar readout, which in some cases makes the results rather ambiguous.

4 A New Paradigm: Network Pharmacology

The behavior of biological systems results from the complex interactions between its numerous constituents, e.g., DNA, RNA, proteins, and small molecules. The large number of components, the high degree of interconnectivity and the complex control of biological networks have now become evident. Biological networks have been described, based on their topologies, as scale-free/small-world networks (Albert 2005; Telesford et al. 2011), although the most proper characterization of the topology of biological networks remains a matter of debate (Khanin and Wit 2006; Broido and Clauset 2019). Generally speaking, biological networks possess a high clustering coefficient (with presence of cliques or near cliques) and are characterized by short paths from any node to another, many nodes with few connections (e.g., bridge and peripheral nodes) and a few nodes with many connections (hubs), and hubs enriched with essential/lethal nodes (Khanin and Wit 2006). As the proportion of peripheral nodes in a small-world network is much larger than the proportion of hubs, the probability of randomly deleting a key node is quite low and deletion of a random node, per se, rarely results in a significant loss of function: it may be said that, owing to their network topology, biological networks are resilient to random perturbations. Redundant functions and compensatory mechanisms contribute to the robustness of the phenotype (Hopkins 2008). Disease often results from the breakdown of robust physiological systems due to multiple simultaneous genetic and/or environmental insults, leading to the establishment of robust disease conditions (Talevi 2015). For instance, synthetic lethality and synthetic disease are more likely to be observed in dual than single knockout organisms, or when a single genetic alteration is coupled to stress due to changes in environmental conditions or exposure to a chemical insult (Hopkins 2008; Nijman 2011). The corollary is that, whereas single-target agents emerging from target-focused approximations might be suitable to treat Mendelian disorders, complex disorders arising from a multiplicity of factors/perturbations might be better addressed by polytherapy (i.e., drug combinations) or by multi-target drugs, which often emerge from phenotypic drug discovery. Moreover, drugs with multiple modes of action are less susceptible to the emergence of drug resistance issues linked to target modification/regulation (Groenendijk and Bernards 2014; Makhoba et al. 2020).

Alternatively, tailored multi-target drugs (Tailored 2016) might provide the best of two worlds, combining the advantages of network pharmacology with the "rationality" of target-focused discoveries (by designing a drug of complex pharmacology by balancing the affinities of a chemical compound for at least two pre-defined drug targets).

5 High-Content Screening

One of the early advantages of target-driven approximations over phenotypic screening was that the biochemical assays required to perform a target-oriented screen were more easily integrated into automated, high-throughput platforms than phenotypic assays. However, this has progressively changed over the past 20 years, under the advent and development of high-content screening technologies.

HCS, also known as high-content analysis, is usually defined as a combination of medium- to high-throughput screening platforms with advanced cellular imaging acquisition technologies (automated microscopy), which allows system-level analysis of biological systems (individual cells in culture, tissue samples, organoids, or whole multicellular organisms) by utilizing high-throughput image phenotyping; however, in a broader sense, it may also encompass proteomics, label-free, and/or flow cytometry (Fraietta and Gasparri 2016). A key aspect of HCS is multiplexing: it produces a multiparametric, quantitative readout of changes in cellular or subcellular features (e.g., morphology, differentiation, motility, cell death, etc.) upon treatment with candidate drugs or genetic interventions (e.g., RNA interference). The later might be used to proposed new disease-modifying drug targets that could be starting points for target-focused drug discovery. In other words, HCS can be exploited for both phenotypic screening and hypothesis generation for target-oriented projects. The assayed biological sample must be processed through appropriate staining with fluorescent dyes, antibodies, reporter genes, induced expression of fluorescent proteins, etc., to be acquired by automated microscopy or other approaches.

HCS is, of course, not excluded from specific challenges and disadvantages that add to those general to any phenotypic screen (Kepiro et al. 2018). They require sophisticated imaging equipment and specialized technicians and software algorithms, which represents significant costs of operation and maintenance. The depth and rate of data acquisition also demands external storage support for terabytes of digital information.

6 Advanced 3D Cell Culture Platforms

Cell culture experiments constitute a fundamental aspect of phenotypic drug discovery. So far, they have been primarily performed in suspension or using two-dimensional platforms, such as Petri dishes, tissue culture flasks, and multi-well plates, where cells are plated onto flat rigid surfaces or in liquid media. In their natural environment, however, nearly all cells reside within an extracellular matrix, i.e., a three-dimensional network that provides structural and biochemical support. In 2D cell cultures, the loss of 3D cues and cell–cell and cell–extracellular matrix interactions, along with the comparatively reduced diffusion barriers to soluble components (e.g., oxygen) determines a considerably different behavior from in vivo behavior, including faster proliferation, modified gene expression profile and, occasionally, *differential response to drug treatment* (for instance, unreliable efficacy and cytotoxicity data that do not translate to in vivo systems, and reduced incidence of drug resistance phenomena) (Koban et al. 2018; Perut et al. 2018; Mirbagheri et al. 2019; Białkowska et al. 2020). The emergence of advanced three-dimensional cell culture platforms, including 3D-scaffold based models, spheroids, organoids, and assambloids, provides models that better mimic the natural cell microenvironment, e.g., accounting for cell heterogenicity and better reproducing intra-cellular and extracellular biochemical signals.

Spheroids are dense cell aggregates that self-assemble in an environment that prevents attachment to a planar surface (Białkowska et al. 2020). They are increasingly used to screen for novel antineoplastic therapies, and they may be based on co-cultures or bioprinting technologies to better reflect the tumor microenvironment.

Organoids, on the other hand, are stem cell-derived, three-dimensional cultured structures that are artificially generated. They include several cell types that develop from stem cells (or organ progenitors) and self-organize owing to cell sorting and spatially restricted lineage commitment, mimicking organ development and morphological characteristics (Takahashi 2019). They are a valuable tool not only to reveal promising drug candidates but also to assist precision medicine, revealing which patients may benefit from treatment with certain drugs. For instance, in 2016, rectal organoids were grown from a rectal biopsy from two patients with cystic fibrosis, and they were used to predict the potential benefit of ivacaftor administration; both patients showed a good outcome after receiving the treatment (Dekkers et al. 2016). Organoids resembling distinct areas can be combined into more complex systems called assembloids (Marton and Paşca 2020; Vogt 2021).

The incorporation of next generation cell culture platforms in the context of high-throughput screening campaigns remains challenging, as 3D cultures have so far been trickier to automate than 2D cultures (Langhans 2018; Booij et al. 2019). To mention just a few obstacles, the cell types used in a 3D cell culture can require extensive optimization to be grown in a miniaturized format, and these cultures can suffer from lower reproducibility. Hydrogels, either of natural or synthetic origin, are commonly used as scaffolds to mimic the extracellular matrix, resulting in viscous liquids that require rapid liquid handling to avoid premature polymerization and blocked pipette tips. For the detection of absorption or fluorescent signals, the 3D culture matrix frequently interferes with measurement; poor light penetration with lengthy imaging acquisition times also makes image acquisition more complicated. 3D cell culture assays require more time than 2D cell cultures owing to the time required for the cells to assemble into multicellular structures and the already mention slower growth of cell lines in 3D. In any case, next generation cell culture platforms promise to revolutionize phenotypic drug discovery by better mimicking physiological architectures and thus providing improved translatability.

7 Small Animal Models

Whereas some level of cell–cell crosstalk and some insight into tissue level can be captured by advanced cell culture technologies, especially by organoids and assembloids, whole animal models undoubtedly provide the most convincing and integrative evidence of efficacy and safety before a drug candidate reaches clinical trials. Although not free of limitations, animal models of disease and animal tests remain the best way to mimic complex disorders in the laboratory and assess the effect of a drug or drug candidate on a complex response (e.g., a behavioral response). For instance, animal models have played a significant role as primary screens or as part of back-translational psychopharmacology strategies to discover many drugs of clinical use in the fields of neurology and psychiatry (Löscher 2017; Moore 2010; Cryan and Sweeney 2011).

A diversity of small animals has recently been integrated in the field of phenotypic drug discovery, besides the well-known *Mus musculus* and *Rattus norvegicus*. The animal models of disease based on such small animals can be advantageous in several ways. On the one hand, they usually have much improved throughput in comparison with rodents, also requiring smaller compound samples for the initial screen. Also importantly, the use of lower animals for primary screening can alleviate, to some extent, bioethical concerns. Prominently, we can mention the growing use of invertebrates such as *Drosophila melanogaster* (common fruit fly) or *Caenorhabditis elegans* (roundworm) (Su 2019; Ségalat 2007) and lower vertebrates such as *Danio rerio* (zebrafish) (Strange 2016). Among some general advantages of these species, we may mention short life cycles, low cost, the amenability to high-throughput screens, and the ease of disease model development owing to forward and reverse genetic tractability, fully sequenced and well annotated genomes, and availability of extensive genetic tools for their genetic manipulation. Other valuable lower organisms that have been used for phenotypic drug discovery include *Saccharomyces cerevisiae, Dictyostelium discoideum, Artemia salina, Oryzias latipes,* and *Xenopus leavis*.

8 Conclusions

Owing to a perception of relatively small investment return in the pharmaceutical sector that has been occasionally associated with the limitations of the excessively reductionist *one gene, one drug* paradigm, the interest in phenotypic drug discovery has been resurrected in the past decades, in the light of a more holistic, systems biology perspective, and also supported by the advances of a diversity of related technologies, including the introduction of high-content screening, low cost–high throughput animal models amenable to genetic engineering, and improved target deconvolution approximation.

Whereas we have provided a glimpse of the advantages and shortcomings of physiologically based drug discovery, we would like to emphasize that while target-oriented and target-agnostic approximations have sometimes been contrasted as mutually exclusive philosophies, they are highly complementary approaches that can be integrated to extract the best of two worlds. For instance, tailored multi-target agents couple the complex pharmacology often observed in hits that emerge from phenotypic screening (sometimes ascribed as the key to their efficacy) with the rationality of hypothesis-driven, target-focused approximations. Network pharmacology, which provides theoretical support to phenotypic drug discovery, can also be used to select an appropriate drug target or drug target combination to develop efficacious and safe drugs (by including network topology analysis among the target selection criteria). Disease-relevant assays can now be modeled using the wide spectrum of animal species used in drug discovery in association with technologies such as transcription activator-like effector nuclease or CRISPR/Cas9. Therefore, it is possible the next generation of therapeutic agents will emerge from a broad multidisciplinary approach where advances in animal models of disease are informed by detailed pathophysiological considerations, and complex phenotypic responses and molecular-level interactions are assessed in parallel.

References

Albert R. Scale-free networks in cell biology. J Cell Sci. 2005;118:4947–57.

Berrade L, Garcia AE, Camarero JA. Protein microarrays: novel developments and applications. Pharm Res. 2011;28:1480–99.

Białkowska K, Komorowski P, Bryszewska M, Miłowska K. Spheroids as a type of three-dimensional cell cultures-examples of methods of preparation and the most important application. Int J Mol Sci. 2020;21:6225.

Bianchi MT, Pathmanathan J, Cash SS. From ion channels to complex networks: magic bullet versus magic shotgun approaches to anticonvulsant pharmacotherapy. Med Hypotheses. 2009;72:297–305.
Bidlingmaier S, Liu B. Interrogating yeast surface-displayed human proteome to identify small molecule-binding proteins. Mol Cell Proteomics. 2007;6:2012–20.
Booij TH, Price LS, Danen EHJ. 3D cell-based assays for drug screens: challenges in imaging, image analysis, and high-content analysis. SLAS Discov. 2019;24:615–27.
Broido AD, Clauset A. Scale-free networks are rare. Nat Commun. 2019;10:1017.
Chan JN, Nislow C, Emili A. Recent advances and method development for drug target identification. Trends Pharmacol Sci. 2010;31:82–8.
Childers WE, Elokely KM, Abou-Gharbia M. The resurrection of phenotypic drug discovery. ACS Med Chem Lett. 2020;11:1820–8.
Cryan JF, Sweeney FF. The age of anxiety: role of animal models of anxiolytic action in drug discovery. Br J Pharmacol. 2011;164:1129–61.
Dekkers JF, Berkers G, Kruisselbrink E, Vonk A, de Jonge HR, Janssens HM, et al. Characterizing responses to CFTR-modulating drugs using rectal organoids derived from subjects with cystic fibrosis. Sci Transl Med. 2016;8:344ra84.
Eder J, Sedrani R, Wiesmann C. The discovery of first-in-class drugs: origins and evolution. Nat Rev Drug Discov. 2014;13(8):577–87.
Fraietta I, Gasparri F. The development of high-content screening (HCS) technology and its importance to drug discovery. Expert Opin Drug Discov. 2016;11:501–14.
Friman T. Mass spectrometry-based cellular thermal shift assay (CETSA®) for target deconvolution in phenotypic drug discovery. Bioorg Med Chem. 2020;28:115174.
Groenendijk FH, Bernards R. Drug resistance to targeted therapies: déjà vu all over again. Mol Oncol. 2014;8:1067–83.
Hartenfeller M, Schneider G. De novo drug design. Methods Mol Biol. 2011;672:299–323.
Hopkins AL. Network pharmacology: the next paradigm in drug discovery. Nat Chem Biol. 2008;4:682–90.
Hu Y, Stumpfe D, Bajorath J. Computational exploration of molecular scaffolds in medicinal chemistry. J Med Chem. 2016;59:4062–76.
Jung HJ, Kwon HJ. Target deconvolution of bioactive small molecules: the heart of chemical biology and drug discovery. Arch Pharm Res. 2015;38:1627–41.
Keenan AB, Jenkins SL, Jagodnik KM, Koplev S, He E, Torre D, et al. The library of integrated network-based cellular signatures NIH program: system-level cataloging of human cells response to perturbations. Cell Syst. 2018;6:13–24.
Kepiro M, Varkuti BH, Davis RL. High content, phenotypic assays and screens for compounds modulating cellular processes in primary neurons. Methods Enzymol. 2018;610:219–50.
Khanin R, Wit E. How scale-free are biological networks. J Comput Biol. 2006;13:810–8.
Koban R, Neumann M, Daugs A, Bloch O, Nitsche A, Langhammer S, et al. A novel three-dimensional cell culture method enhances antiviral drug screening in primary human cells. Antivir Res. 2018;150:20–9.
Kubota K, Funabashi M, Ogura Y. Target deconvolution from phenotype-based drug discovery by using chemical proteomics approaches. Biochim Biophys Acta Proteins Proteom. 2019;1867:22–7.
Lamb J, Crawford ED, Peck D, Modell JW, Blat IC, Wrobel MJ, et al. The connectivity map: using gene-expression signatures to connect small molecules, genes, and disease. Science. 2006;313:1929–35.
Lambrinidis G, Tsantili-Kakoulidou A. Multi-objective optimization methods in novel drug design. Expert Opin Drug Discov. 2021;16:647–58.
Langhans SA. Three-dimensional in vitro cell culture models in drug discovery and drug repositioning. Front Pharmacol. 2018;9:6.
Lee JA, Berg EL. Neoclassic drug discovery: the case for lead generation using phenotypic and functional approaches. J Biomol Screen. 2013;18:1143–55.

Löscher W. Animal models of seizures and epilepsy: past, present, and future role for the discovery of antiseizure drugs. Neurochem Res. 2017;42:1873–88.

Makhoba XH, Viegas C Jr, Mosa RA, Viegas FPD, Pooe OJ. Potential impact of the multi-target drug approach in the treatment of some complex diseases. Drug Des Devel Ther. 2020;14:3235–49.

Mandal S, Moudgil M, Mandal SK. Rational drug design. Eur J Pharmacol. 2009;625:90–100.

Margineanu DG. Systems biology, complexity, and the impact on antiepileptic drug discovery. Epilepsy Behav. 2014;38:131–42.

Margineanu DG. Neuropharmacology beyond reductionism—a likely prospect. Biosystems. 2016;141:1–9.

Marton RM, Paşca SP. Organoid and assembloid technologies for investigating cellular crosstalk in human brain development and disease. Trends Cell Biol. 2020;30:133–43.

Mavromoustakos T, Durdagi S, Koukoulitsa C, Simcic M, Papadopoulos MG, Hodoscek M, et al. Strategies in the rational drug design. Curr Med Chem. 2011;18:2517–30.

Mirbagheri M, Adibnia V, Hughes BR, Waldman SD, Banquyc X, Hwa DK. Advanced cell culture platforms: a growing quest for emulating natural tissues. Mater Horiz. 2019;6:45.

Moffat JG, Vincent F, Lee JA, Eder J, Prunotto M. Opportunities and challenges in phenotypic drug discovery: an industry perspective. Nat Rev Drug Discov. 2017;16:531–43.

Moore H. The role of rodent models in the discovery of new treatments for schizophrenia: updating our strategy. Schizophr Bull. 2010;36:1066–72.

National Historic Chemical Landmarks program of the American Chemical Society Selman Waksman and Antibiotics, commemorative booklet. 2005. Available at https://www.acs.org/content/acs/en/education/whatischemistry/landmarks/selmanwaksman.html. Last assessed June 2021.

Neggers JE, Kwanten B, Dierckx T, Noguchi H, Voet A, Bral L, et al. Target identification of small molecules using large-scale CRISPR-Cas mutagenesis scanning of essential genes. Nat Commun. 2018;9:502.

Nicolaou CA, Brown N. Multi-objective optimization methods in drug design. Drug Discov Today Technol. 2013;10:e427–35.

Nijman SM. Synthetic lethality: general principles, utility and detection using genetic screens in human cells. FEBS Lett. 2011;585:1–6.

Omidfar K, Daneshpour M. Advances in phage display technology for drug discovery. Expert Opin Drug Discov. 2015;10:651–69.

Pai MY, Lomenick B, Hwang H, Schiestl R, McBride W, Loo JA, et al. Drug affinity responsive target stability (DARTS) for small-molecule target identification. Methods Mol Biol. 2015;1263:287–98.

Perut F, Sbrana FV, Avnet S, De Milito A, Baldini N. Spheroid-based 3D cell cultures identify salinomycin as a promising drug for the treatment of chondrosarcoma. J Orthop Res. 2018;36:2305–12.

Reddy MR, Parrill AL. Overview of rational drug design. In: ACS symposium series, vol. 719. Washington, DC: American Chemical Society; 1999. p. 1–11.

Saxena C. Identification of protein binding partners of small molecules using label-free methods. Expert Opin Drug Discov. 2016;11:1017–25.

Saxena C, Higgs RE, Zhen E, Hale JE. Small-molecule affinity chromatography coupled mass spectrometry for drug target deconvolution. Expert Opin Drug Discov. 2009;4:701–14.

Sbaraglini ML, Talevi A. Hybrid compounds as anti-infective agents. Curr Top Med Chem. 2017;17:1080–95.

Ségalat L. Invertebrate animal models of diseases as screening tools in drug discovery. ACS Chem Biol. 2007;2:231–6.

Smith E, Collins I. Photoaffinity labeling in target- and binding-site identification. Future Med Chem. 2015;7:159–83.

Smith AM, Ammar R, Nislow C, Giaever G. A survey of yeast genomic assays for drug and target discovery. Pharmacol Ther. 2010;127:156–64.

Strange K. Drug discovery in fish, flies, and worms. ILAR J. 2016;57:133–43.

Strickland EC, Geer MA, Tran DT, Adhikari J, West GM, DeArmond PD, et al. Thermodynamic analysis of protein-ligand binding interactions in complex biological mixtures using the stability of proteins from rates of oxidation. Nat Protoc. 2013;8:148–61.
Su TT. Drug screening in drosophila; why, when, and when not? Wiley Interdiscip Rev Dev Biol. 2019;8:e346.
Swinney DC. The contribution of mechanistic understanding to phenotypic screening for first-in-class medicines. J Biomol Screen. 2013;18:1186–92.
Swinney DC, Anthony J. How were new medicines discovered? Nat Rev Drug Discov. 2011;10:507–19.
Szabo M, Svensson Akusjärvi S, Saxena A, Liu J, Chandrasekar G, Kitambi SS. Cell and small animal models for phenotypic drug discovery. Drug Des Devel Ther. 2017;11:1957–67.
Tailored TA. Multi-target agents. Applications and design considerations. Curr Pharm Des. 2016;22:3164–70.
Takahashi T. Organoids for drug discovery and personalized medicine. Annu Rev Pharmacol Toxicol. 2019;59:447–62.
Talevi A. Multi-target pharmacology: possibilities and limitations of the "skeleton key approach" from a medicinal chemist perspective. Front Pharmacol. 2015;6:205.
Talevi A. Computational approaches for innovative antiepileptic drug discovery. Expert Opin Drug Discov. 2016;11:1001–16.
Talevi A. Computer-aided drug design: an overview. Methods Mol Biol. 2018;1762:1–19.
Talevi A, Carrillo C, Comini M. The thiol-polyamine metabolism of *Trypanosoma cruzi*: molecular targets and drug repurposing strategies. Curr Med Chem. 2019;26:6614–35.
Telesford QK, Joyce KE, Hayasaka S, Burdette JH, Laurienti PJ. The ubiquity of small-world networks. Brain Connect. 2011;1:367–75.
Terstappen GC, Schlüpen C, Raggiaschi R, Gaviraghi G. Target deconvolution strategies in drug discovery. Nat Rev Drug Discov. 2007;6:891–903.
Vogt N. Assembloids. Nat Methods. 2021;18:27.
Wada A, Hara S, Osada H. Ribosome display and photo-cross-linking techniques for in vitro identification of target proteins of bioactive small molecules. Anal Chem. 2014;86:6768–73.
Warchal SJ, Unciti-Broceta A, Carragher NO. Next-generation phenotypic screening. Future. Med Chem. 2016;8:1331–47.
Zheng W, Thorne N, McKew JC. Phenotypic screens as a renewed approach for drug discovery. Drug Discov Today. 2013;18:1067–73.

Target Identification Approaches in Drug Discovery

Azhar Rasul, Ammara Riaz, Iqra Sarfraz, Samreen Gul Khan, Ghulam Hussain, Rabia Zara, Ayesha Sadiqa, Gul Bushra, Saba Riaz, Muhammad Javid Iqbal, Mudassir Hassan, and Khatereh Khorsandi

Contents

A. Rasul (✉) · I. Sarfraz · R. Zara · A. Sadiqa · S. Riaz · M. J. Iqbal · M. Hassan
Department of Zoology, Faculty of Life Sciences, Government College University, Faisalabad, Pakistan
e-mail: azharrasul@gcuf.edu.pk

S. G. Khan
Department of Chemistry, Faculty of Physical Sciences, Government College University, Faisalabad, Pakistan

G. Hussain
Neurochemical Biology and Genetics Laboratory, Department of Physiology, Faculty of Life Sciences, Government College University, Faisalabad, Pakistan

G. Bushra
Department of Bioinformatics and Biotechnology, Faculty of Life Sciences, Government College University, Faisalabad, Pakistan

K. Khorsandi
Department of Photodynamic, Medical Laser Research Center, Yara Institute, ACECR, Tehran, Iran

A. Riaz
Department of Zoology, Faculty of Life Sciences, Government College University, Faisalabad, Pakistan

Department of Life Sciences, Khwaja Fareed University of Engineering and Information Technology (KFUEIT), Rahim Yar Khan, Pakistan

M. T. Scotti, C. L. Bellera (eds.), *Drug Target Selection and Validation*, Computer-Aided Drug Discovery and Design,
https://doi.org/10.1007/978-3-030-95895-4_3

Abstract

Target identification is the most critical step in chemical genetics and drug discovery. A variety of novel bioactive chemical compounds have been discovered by phenotypic-based screening methods. However, identifying molecular targets for these bioactive compounds is a very laborious process. With the development and advancements in new methodologies and techniques for identification and biological analysis, various target identification methods have been established such as affinity-based methods, genetic-based techniques, computational approaches, and chemical proteomics. In this chapter, we will classify the target identification approaches, various molecular target identification methods, their work schemes, and the applications of target identification techniques in discovery of targets of small molecules, while keeping in view that each and every method is associated with some advantages as well as disadvantages. The efficient execution of every scientific process involves significant infrastructure set up and expertise in multiple disciplines (chemistry, proteomics, genomics, and bioinformatics). The collaborative work scheme comprising integrated knowledge of all of these disciplines will ensure the maximum potential is utilized in academic and industrial grounds.

Keywords
Target identification · Protein targets · Affinity-based methods · Phenotypic-based methods

1 Introduction

Drug discovery refers to the field of medicinal sciences that is based upon a cascade of integrated events carried out to search for the "cure" for "incurables". Drug discovery comprises different complex procedures that are focused on the chemical and pharmaceutical optimization and advancement of novel compounds for the cure of human diseases (Ator et al. 2006; Hughes et al. 2011). The discovery of a drug starts with the identification of the biological origin of the disease and potential targets for the drug. Paul Ehrlich is thought to be the founder of modern drug discovery and chemotherapy. He conceived the notion of selective drugs that could be used to fight against infectious diseases. He also paved the way toward the discovery of synthetic compounds or drugs for the cure of bacterial diseases (Bosch and Rosich 2008).

Target identification is a first and foremost step in drug discovery. It is the process of identifying the direct molecular targets of small molecules, such as nucleic acid and proteins. Target identification can be approached by computational methods, direct biochemical methods, and genetic interactions. A good target should meet several criteria to assure clinical efficacy and safety when modulated through small molecules.

It has been challenging to shift the formerly predominant paradigm of "one target, one drug" to the current "one drug, multiple targets" (Schenone et al. 2013). Both include, however, the validation of one or more targets on the basis of their ability to regulate biological functions through binding with small molecules. Once the targets have been validated, hit identification might be pursued; hits are usually used as starting points to obtain derivatives with the desired potency (Hughes et al. 2011; Macalino et al. 2015).

Among the 259 agents approved by the FDA from 1999–2008, 75 were first-in-class drugs with new molecular mechanisms of action. Contribution of phenotypic screening in the development of the first-in-class small molecule agents surpassed that of target-based approaches. Phenotypic based screening led to the approval of 28 drugs, while the contribution of target-based screening yielded 17 drugs (Swinney and Anthony 2011).

The current therapies are based upon approximately 500 molecular targets such as G-protein coupled receptors (45%), enzymes (28%), hormones (11%), ion channels (5%), and nuclear receptors (2%) (Drews 2000). For cancer treatment, 154 cancer drugs approved by the FDA have been classified into three main categories: five drugs which act through non-protein or unknown targets, 64 drugs are cytotoxic, and 85 drugs can be assigned to protein targets (Santos et al. 2017).

The identification of disease-associated targets is then essential for the screening of small molecules under the target-based paradigm. Association of drug targets with a disease usually relies on genome wide information from a broad array of data sources (Koscielny et al. 2017; Floris et al. 2018). The identification and verification of small molecules' target(s) are, on the other hand, key to understanding their cellular mechanism of hits identified through target-agnostic approaches (phenotypic screening), (Jung and Kwon 2015), which is known as target deconvolution.

2 Approaches for Drug Target Identification

There are several approaches that have been proposed for drug target identification of bioactive molecules. These approaches might be classified into four groups;

1. Affinity-based approaches that rely on direct interaction of proteins with small molecules (Tulloch et al. 2018).
2. Phenotype-based approaches refer to the methods and techniques that are meant to compare the biological profiles of small molecules with already known reference drugs (Schenone et al. 2013).
3. The genetic-based approaches are used to identify the genes that are responsible for drug resistance produced by small molecules, as well as small molecule-sensitive clones (Zheng et al. 2004).
4. Computational approaches predict the drug targets on the basis of chemical complementarity between the proteins and molecules of interest (Katsila et al. 2016; Sliwoski et al. 2014).

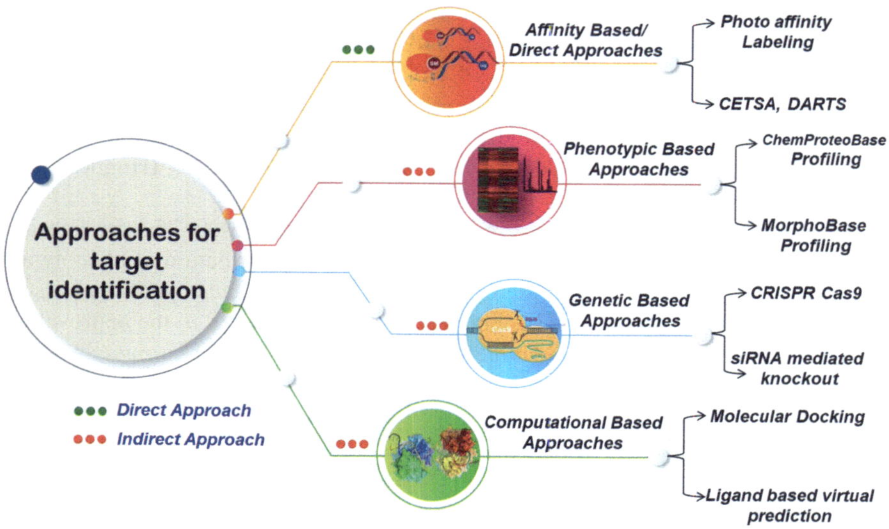

Fig. 1 Approaches for target identification

Out of all of these techniques, affinity-based approaches directly identify the small molecule targets, while other approaches identify the targets indirectly (Fig. 1).

3 Direct or Affinity-Based Approaches

Identification of targets for small molecules is a crucial step in the discovery of drugs. Direct or affinity-based approaches remain the most commonly used approaches for molecular target identification. During recent years, various techniques based on affinity purification, protein stability, and ligand-based protein labeling have been developed to increase the sensitivity of target identification. Here, we will discuss three approaches of target identification by direct or affinity-based approaches, including photo affinity labeling (PAL), drug affinity responsive target stability (DARTS), and thermal shift assays.

Target Identification by Photo Affinity Labeling

PAL is a recent technique in drug discovery and medicinal chemistry used to study the protein–ligand interactions for identification of ligand targets. PAL is employed to identify the targets of hit compounds/drugs obtained from phenotypic screening in early drug discovery (Smith and Collins 2015).

The concept of PAL was introduced by Westheimer in the 1960s by incorporating an aliphatic diazo group in the enzyme chymotrypsin using the process of

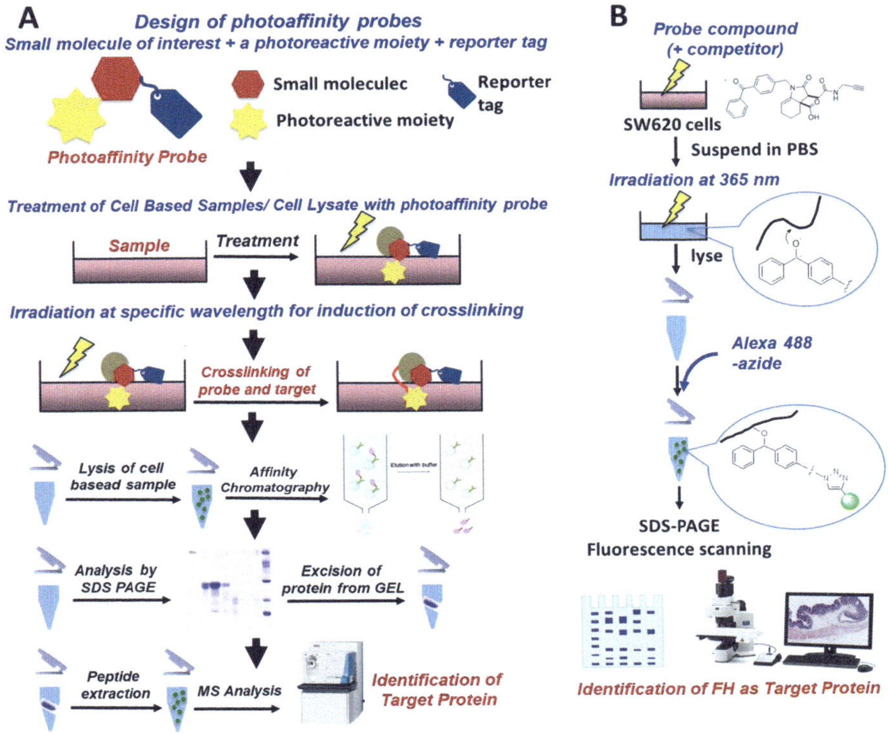

Fig. 2 Schematic representation of target identification by PAL (**a**) Workflow for the identification of molecular targets of drugs by PAL (**b**) Identification of FH as molecular target of pyrrolidinone

acylation. In this technique, a chemical probe is used that can form covalent bonds with its target when activated in the presence of light. This can be done by incorporating a photoreactive moiety within the probe compound. When irradiated with a light of specific wavelength, the photoactive group makes a reactive intermediate that reacts with a nearby molecule (target protein) (Fig. 2a). The main photoactive groups that can be used for PAL include phenyldiazirines, benzophenones, and phenylazides and can make intermediates such as carbine, diradical, and nitrene upon irradiation by specific wavelengths. PAL can be used to identify a wide range of small molecules target proteins in many diseases (Smith and Collins 2015).

There are several successful examples of PAL in the identification of targets of drugs. Here, we present some recent ones.

One successful application of effective PAL was the identification of the molecular target of pyrrolidinone and its structural analogues. Pyrrolidinone was shown to possess highly selective nutrient-dependent cytotoxicity against a number of cancer cell lines in a phenotypic screening of 6000 small molecules, which sparked the interest for the identification of their molecular target(s). SAR studies were

Table 1 Target identification of small molecules by PAL

Ligand	Source	Target	Probe	Validation technique	References
Tamoxifen	Synthetic	Carboxylesterase (ES10), liver fatty acid binding protein (L-FABP)	4-(2-morpholinoe-thoxy)benzophe-none	Immunoblotting	Mesange et al. (2002)
Pladienolide	Natural product	SF3b	3H-labeled, fluorescence-tagged and photoaffinity/biotin (PB)-tagged probes	Immunoblotting	Kotake et al. (2007)
Pyrrolidinone 1	Synthetic	Fumarate hydratase (FH)	–	In vitro fumarate hydratase activity assays	Takeuchi et al. (2015)
3,5-diaryl-1,2,4--oxadiazole	Synthetic	Tail-interacting protein 47(TIP47)	Azido- and tritium-labeled photo affinity agent	RNA interference assays with siRNA duplexes	Jessen et al. (2005)
Suberoylanilide hydroxamic acid	Synthetic	Ribosomal Protein S3	$[^3H]498$	–	Webb et al. (1999)

conducted to enhance compound potency, prior to target identification and validation. Photoaffinity labeling was performed and the cancer cells were treated with the resulting photoaffinity probe. The resulting proteins were separated on SDS-PAGE and visualized. A 50 kDa protein was identified as a possible target protein of the hit compounds. Excision of the identified 50 kDa protein band from gel, followed by digestion and analysis, established the target protein as fumarate hydratase (FH). Thus, PAL identified FH as the pharmacological target of the hits (Fig. 2b). Further biochemical studies and in vitro enzymatic assays confirmed dose-dependent, competitive inhibition of FH by hit compounds that, in turn, validated the results of target identification by PAL.

Other molecular targets of ligands/drugs identified through PAL are shown in Table 1.

Target Identification by DARTS

DARTS is a widely applicable technique used to identify small molecule-protein interactions without the need of prior modification in the drug (Pai et al. 2015). This particular technique is performed by treating the cell lysate with the drug, after which the digestion with proteases is performed. This process is followed by the separation of samples with the help of SDS PAGE, staining, and MS analysis (Fig. 3) in order to identify the protein of interest (Moller et al. 1974).

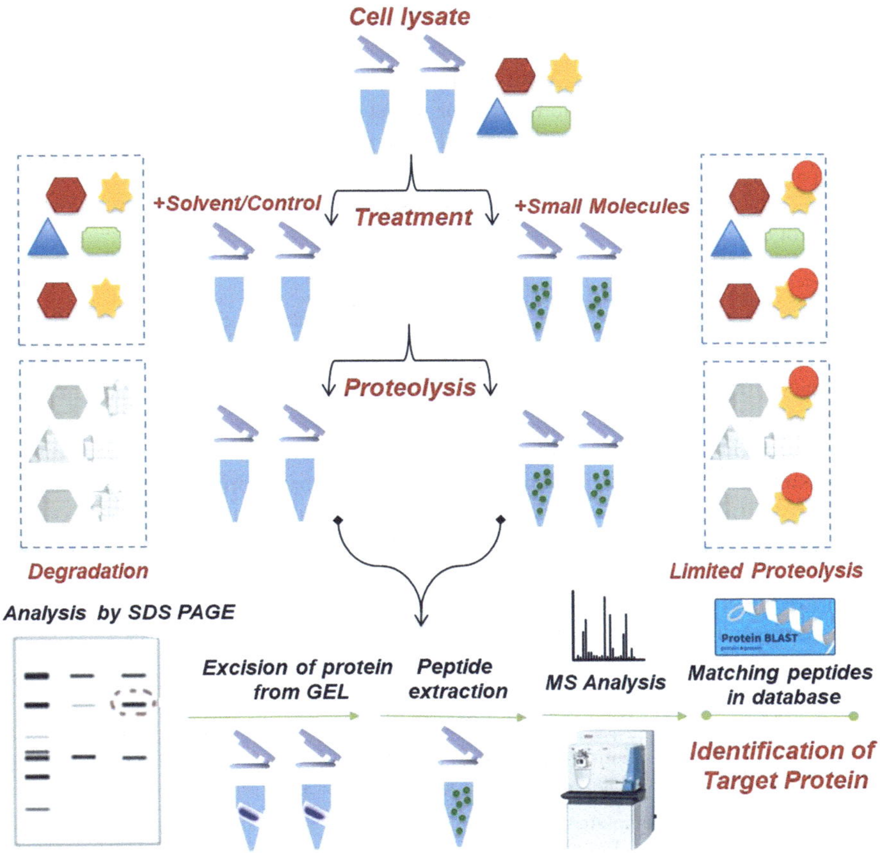

Fig. 3 Schematic presentation of target identification by DARTS

DARTS presents a great efficacy in identifying new proteins targeted by compounds. Similar to affinity chromatography, it is also an affinity-based model starting from complex protein samples and selectively enriching the target proteins while performing the depletion of non-target proteins that offer less resistance against proteases. This technique has an advantage over affinity chromatography in terms of analyzing lower affinity interactions and does not require washing (Lomenick et al. 2011).

Voacangine is a natural compound isolated from *Voacanga Africana*, which has been known to possess antiangiogenic effects. Voacangine have the potential to effectively reduce VEGF-induced chemoinvasion activity on HUVECs. Kim et al. investigated the mechanism of action of voacangine by using label-free DARTS, which led to the successful identification of VEGFR2 as a target protein. HUVECs cells were lysed, and cell lysates were incubated at room temperature with voacangine at the appropriate doses, DARTS was performed by proteolytic digestion of cell lysates, and immunoblotting with primary antibodies was used to validate the obtained results (Kim et al. 2020).

There are many other successful examples of DARTS for identification of bioactive protein targets such as the molecular target of resveratrol, a compound in red grapes, has been identified as eIF4A (Lomenick et al. 2009), identification of Cathepsin D and Thioredoxin-like protein 1 as target proteins of Dichloroacetate (Gong et al. 2013), or identification of actin as the main target for 5-*epi*-sinuleptolide (Morretta et al. 2017).

Target Identification by Cellular Thermal SHIFT ASSay (CETSA)

Thermal shift assay refers to biochemical methods used to investigate the stabilization of target proteins upon binding of ligands and the increased probability of crystal formation in biological samples (Elgert et al. 2020). This technique is also meant to measure the alterations in denaturation temperature of proteins under variable drug concentrations, ionic strength, buffer pH, sequence mutation, and redox potential. Commonly employed methods of assessing thermal shifts include thermofluor or differential scanning fluorimetry (DSF) (Dart et al. 2018). Thermal shift assay performed in cellular format is known as CETSA (Jafari et al. 2014). The diagrammatic representation of CESTA is provided in Fig. 4.

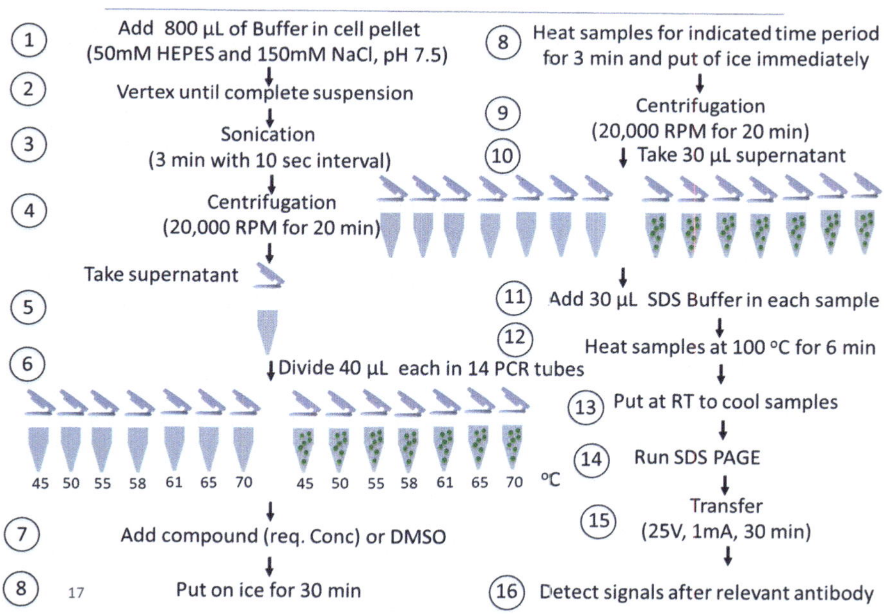

Fig. 4 Brief illustration of protocol for target identification by CETSA

CETSA has been successfully used to identify protein targets of bioactive drugs/compounds. Nagasawa et al. discovered NPD10084 as an anti-proliferative compound during screening of compounds against colorectal cancer cells. The 2DE-CETSA technique was used for profiling of the thermally shifted proteins after treatment with NPD10084. HCT116 cell lysates were either left unheated or heated with variable temperatures after being treated with DMSO or NPD10084. After being tagged, each protein sample was subjected to 2D DIGE analysis. The thermally stabilized spots matched the PKM protein, and MS analysis confirmed the obtained results (Nagasawa et al. 2020).

4 Phenotype-Based Methods

The limited knowledge about complex interactions among genomic and chemical space has narrowed down the process of identification of novel drugs and their targets (Dobson 2004). Precise identification of drug target interactions followed by drug target validation is considered as the initial step in the path of drug discovery. To date, many potential drug target interactions have been reported, but many more remain undiscovered (Yu et al. 2012). Phenotype-based assays are increasingly gaining importance among other drug discovery processes. The action mechanisms of drugs can be determined by making the comparison between the treated and untreated cells. The modes of comparison may include small and focused alterations in the phenotype of cells, while in other cases the potential drug targets may lead to prominent alterations in phenotypes, e.g., cell death and apoptosis. Physical alterations in the cells including the changes in the morphology of cells during cell cycle are very quick and need careful checks and balances at the molecular level. In recent years, advancements in "omics" fields such as proteomics and genomics have initiated the production of economical and high throughput methodologies to analyze and compare the molecular frameworks inside living organisms at large scale (Tulloch et al. 2018).

Proteomics

Proteins play prime roles in cellular functions. Total expressed proteins in a cell as well as an organism at any specified time can be studied using proteomics technology. Proteome analysis can provide complete information about the cellular processes and hence play an important role in the diagnosis and treatment of a wide spectrum of pathological conformations. The term "proteome" was introduced in the 1990s, since then, proteomics has been a principal tool in health sciences and clinics. Proteomics refer to the analysis of cell protein profiles. In the process of drug discovery, proteomics has played a valuable role because proteins are the main drug targets in disease conditions. Proteomics along with system biology plays a significant part in the selection of protein biomarkers and comprehensively understanding the disease associated pathways to design a compound that can mitigate or modulate a specific chemical pathway or cycle (Amiri-Dashatan et al. 2018). For example, in the case of high grade serous ovarian cancer patients, cancer/testis antigen 45 (CT45) was found to serve as a non-mutant tumor antigen as well as cell intrinsic enhancer of chemo sensitivity with the help of functional assays as well as quantitative proteomics. Moreover, CT45 was also linked to DNA damage signaling by acting as a regulator of protein phosphatase 4 (PP4) (Huang et al. 2012).

In recent times, both direct and indirect approaches have gained importance in drug target identification for small molecules. Direct approaches were introduced by Schreiber and coworkers, who worked on affinity matrices conjugated with FK506. Since then, remarkable progress has been made in the field of proteomics. Mass spectrometry (MS) has enabled us to identify voluminous data regarding the proteins present in a small volume of a given sample. Other techniques used for identification of target proteins include two dimensional electrophoresis and sodium docecylsulfate-polyacrylamide gel electrophoresis (SDS-PAGE) (Osada et al. 2014).

Advanced techniques such as ABPPs (Wiedl et al. 2011), SILAC (Mann 2006), ICAT (Tunon et al. 2010), iTRAQ (Ross et al. 2004), protein microarrays (Hall et al. 2007), and ChemProteoBase (Osada et al. 2014) can also be used in drug target identification.

Affinity-Based Protein Profiling (ABPPs)

Affinity-based protein profiling is a popular approach in the field chemical proteomics. In this technique, small molecular probes (soluble in a particular medium) are employed to capture the specific class of proteins that make covalent bonds with these probes through a reactive group. The reactive probe is usually incorporated with an affinity tag by means of a spacer (Bantscheff et al. 2009).

As an initial step, the biological sample is incorporated with the small molecule probe in order to allow it to bind with the protein for which it has affinity. After that, the probe-protein adducts are caught on the solid support having the affinity tag on it, e.g., in the case of P450 (cytochrome proteins), the reactive group is modulated so as to bind with the proteins and their active conformations. Hence, this particular strategy can be employed to functionally characterize the cytochrome P450 superfamily (Wright and Cravatt 2007).

SILAC

SILAC is the abbreviation of "stable isotope labeling by amino acids in cell culture". Along with LC-MS/MS, this technique is the most widely employed method to quantify protein abundance. In this technique, two populations of cells are grown; one with light or natural amino acids, while the second one is grown under exposure to heavy amino acids such as $^{13}C6$-arginine and $^{15}N2$-lysine. After that, the two populations are incorporated followed by a number of cell divisions ensuring the full labeling of the whole genome. The cells are subjected to mass spectrometry. The relative abundance of proteins can be assessed by comparison of the ratio of ion densities between SILAC peptide pairs (Wang et al. 2016).

This technique was employed by Voigt et al. (2013) in order to identify mitosis inhibitors. In their experiment, tetrahydropyran derivatives were synthesized using the process of Prins cyclization between polymer-bound aldehyde and homoallylic alcohol. The resulting compounds were treated with the HeLa and KB-V1 cells. It was noticed that tetrahydropyran derivatives target the CSE1L and tubulin proteins in a synergistic manner (Voigt et al. 2013).

Isotope Coded Affinity Tag (ICAT)

Isotope coded affinity tag (ICAT) makes possible the identification of a variety of peptides in two samples. The ICAT reagents comprising a thiol-specific reactive group, an isotopically light or heavy linker, and an affinity tag (such as biotin) are introduced in the samples (Adam et al. 2002). The presence of linkers (light or heavy) provides the probes that contain variable molecular masses but have identical chemical structure. The two protein samples can be mixed, digested, and purified by using techniques such as affinity chromatography, multidimensional liquid chromatography, and mass spectrometry. Afterward, the presence of light and heavy signals quantifies the relative protein abundance. This technique is an excellent non-gel-based method and is very useful for the evaluation of proteins in the presence of disease conditions and the effect of drugs on relative protein abundances. The major drawback associated with this technique is that it can be used only in case of peptides having cysteine as their essential constituent; hence, it can generate only limited data for protein quantification (Moseley et al. 2007).

ChemProteoBase

ChemProteoBase is a proteomic profiling system that is used for compound target analysis by integrating the proteome analysis acquired with the help of a two-dimensional gel electrophoresis system (Osada et al. 2014). The general work scheme for identification of targets by ChemProteoBase profiling is shown in Fig. 5a.

This particular database system was used by Futamura et al. (2013) to identify the molecular target of pyrrolizilactone (a fungal metabolite). The study proposed that pyrrolizilactone is a proteasome inhibitor, which is able to inhibit the trypsin-like activity of proteasome (Fig. 5b). It was analyzed with the help of two phenotypic profiling systems, namely ChemProteoBase and MorphoBase, used along with a two dimensional gel electrophoresis system (Futamura et al. 2013).

Morphology-Based Cell Assays

Visual analysis is an efficient way to screen bioactive molecules leading to the discovery of potential druggable targets followed by chemico-biological validation. MorphoBase, an encyclopedic cell morphology database, was established by Futamura et al. (2012) to discover small drug-like molecules. The roots of this method date back to 2004, when Perlman et al. examined 100 compounds at 13 threefold dilutions based on cellular components in HeLa cells. They classified the compounds with similar targets into the same groups (Futamura et al. 2012).

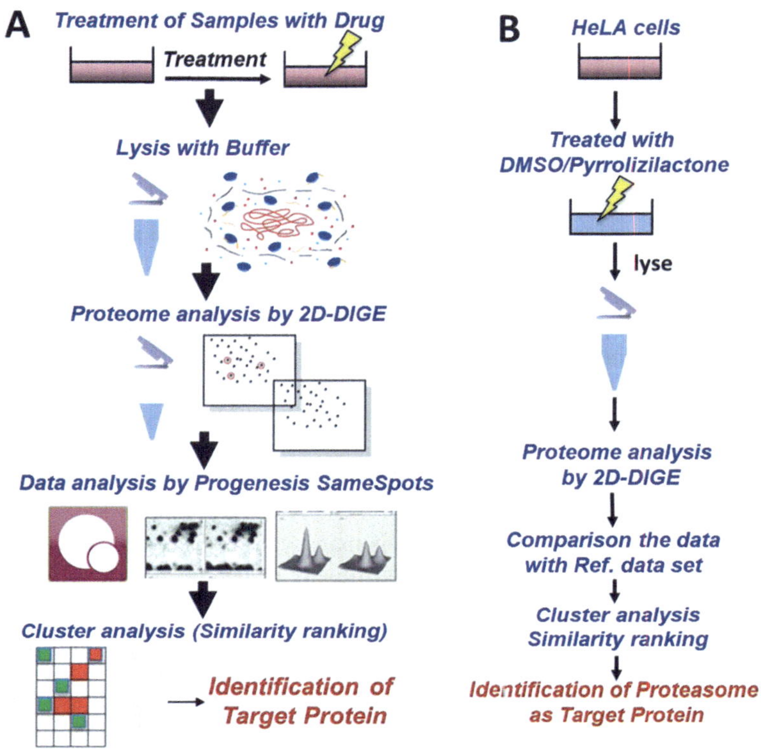

Fig. 5 Target Identification by phenotypic based approach (**a**) Work scheme for identification of molecular targets of small molecule by ChemProteoBase profiling (**b**) Identification of proteasome as molecular target of Pyrrolizilactone by ChemProteoBase profiling

MorphoBase consists in developing a high content image method followed by a phenotype profiling system with the help of software that can determine the similarities between compounds on the basis of multiparametric phenotypes responses based upon statistical analysis. This database system focuses on classifying the test compounds on the basis of their modes of action and predicting the side effects of these compounds (Fig. 6). Using this method, the molecular targets of compounds of interest can be determined in an unbiased manner. This method is a rather simple one consisting of a microscope, 96-well plate, and fluorescent nuclear staining. MorphoBase profiling is a rather new database system that can pinpoint the exact mechanism of action of a drug or compound of interest on the basis of morphological changes taking place in the treated cells (Muroi et al. 2016).

Futamura et al. (2012) used MorphoBase to examine the effects of 30 well referenced drugs on the basis of cell morphology of two mammalian cell lines; HeLa and rat kidney cells infected with ts25 (a T-class mutant of Rous sarcoma virus Prague stain), src^{ts}- NRK cells in a dose and time dependent manner. After that, these cells were treated with the aforementioned drugs followed by progressive alterations in

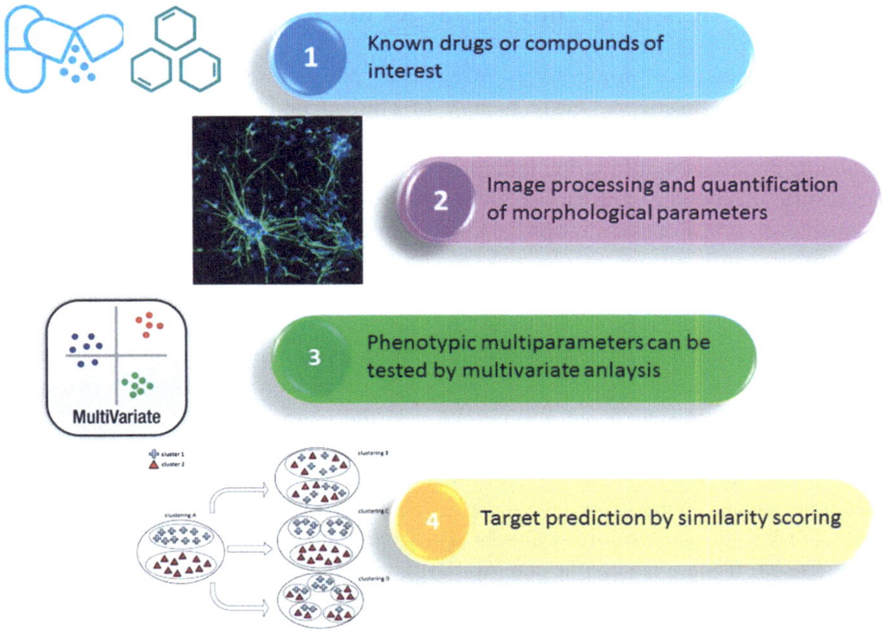

Fig. 6 Steps in MorphoBase profiling

cell morphology. The cell morphology was classified into three categories, namely flattened, polygonal, and rounded. These categories were further sub grouped based upon their variable size and presence of spikes, vacuoles, and granular structures. Contrary to that, HeLa cells showed rather simpler morphological alterations, which were classified into the categories, namely flattened, round up, and toxic or growth inhibition such as different phases of cell cycle (G1/S, G2/M, and flattened with embossed nucleus). The preliminary data emerging from these observations was insufficient to classify the drugs into therapeutic groups. It was, however, possible to discriminate the phenotypic responses of different cell lines against certain drugs on the basis of their mechanism of action. To minimize the human errors in recording phenotypic responses, cells were analyzed by an automatic system, i.e., IN Cell Analyzer, which performed high content image analysis and generated the quantitative morphological data by recognizing the cytoskeletal and subcellular components systematically. This analyzer collected 1500 cells from each well of treated cells and analyzed the cells by using the image segmentation procedures. The next procedure was to sort this multimetric data by using multivariate statistical tools in order to analyze, visualize, and rank the multiparametric high content phenotype results. Principal component analysis (PCA) was performed to visualize the phenotypic responses. The results of statistical analysis concluded that the drugs having similar mode of actions form a cluster. Using this system, the researchers (Futamura et al. 2012) identified NPD6689, NPD8617, and NPD 8969 as tubulin inhibitors (Futamura et al. 2012).

5 Genetic Approaches for Target Identification

Human genetics have progressively revealed its power to enhance the chances of successful drug discovery. Early history of clinical practice of translation from genetics began in 2003 with the discovery of the gene PCSK9 causing hypercholesterolemia due to gain of function mutations (Abifadel et al. 2003). A few years later, the results of targeted sequencing of PCSK9 exposed a considerable decrease in plasma level of low density lipoprotein that ultimately resulted in reduced incidence of coronary cardiac diseases (Cohen et al. 2005).

The use of genomic methods for drug discovery significantly improves current models in two aspects. First, the ideal drug mutation model states that single gene knockdown in a cell or an organism will result in dysfunction of only one protein, which may be an ideal target for a drug; second, the powerful aspect of performing a genetic screen to search out any mutation in the genome that is responsible for phenotypic effects allows revealing important functions (Cohen et al. 2006).

Because exploration of human genome for drug discovery is not always possible, model organisms such as *Caenorhabditis elegans* (nematode), *Saccharomyces cerevisae* and *Schizosaccharomyces pombe* (yeast) and *Drosophila melanogaster* (fruit fly) are used. If there is need of potential drug targets for DNA repair and cell division, then yeast is the organism of choice. In the case of multicellular genome conservation studies, the fruit fly is a potentially valuable model (Bosch and Rosich 2008).

CRISPR Cas System

Clustered regularly interspaced short palindromic repeats (CRISPRs) are repeated sequences of DNA obtained from prokaryotes, bacteria, and archaea. In 1987, they were discovered accidentally by a team of Japanese scientists (Ishino et al.), during an experiment in which they inserted an unusual series of repeated sequence interspaced with a specific sequence; spacer sequence in *E.coli,* while analyzing the responsible gene for alkaline phosphatase (Ishino et al. 2018). Furthermore, they elaborated that during an attack of virus on prokaryotes, CRISPR arrays are transcribed into short sequences in order to make small CRISPR RNA (crRNA). These arrays instruct CRISPR associated sequence proteins to cut complementary DNA or viral RNA sequences on the bases of CRISPR Cas system type. Functional role of Cas proteins as nucleases was discovered by Makarvoa et al. They conducted a relative genome investigation of CRISPR as well as Cas genes and projected the CRISPR Cas system function in resemblance to interference RNA in which protein complexes led to silencing the genes through mRNA cleavage. Some Cas proteins are responsible for DNA cleavage and others cleave RNA. For instance, Cas9 performs cleavage of DNA, whereas Cas13 cleaves RNA (Makarova et al. 2006).

6 Different CRISPR-Cas Systems and Their Applications

CRISPR-Cas9 is a general tool for genome engineering of eukaryotic cellular systems. It was first discovered from *Streptococcus pyogenes*. In this technique, segments of guide RNA (gRNA) are used to target nuclease proteins in the genomic site, which ultimately results in generation of a double strand break (DSB) in DNA tracked by the process of DNA repair (Fig. 7). This process of DNA repair adapts two processes; one is homology directed repair (HDR) and other is non-homologous end join (NHEJ) path in occurrence of a donor pattern of nucleotide sequence (Hsu et al. 2014).

Some researchers have reported that insertions and deletions may cause frame shift mutations and functional dysfunctioning of target genes while following the NHEJ path. In comparison, HDR is an error-free process and performs repairing of DSB in accordance with homologous template DNA donor sequences. CRISPR has been labeled as an efficient approach due to its adaptability, scalability, multiplex gene mutation, and ease of application (Jinek et al. 2012).

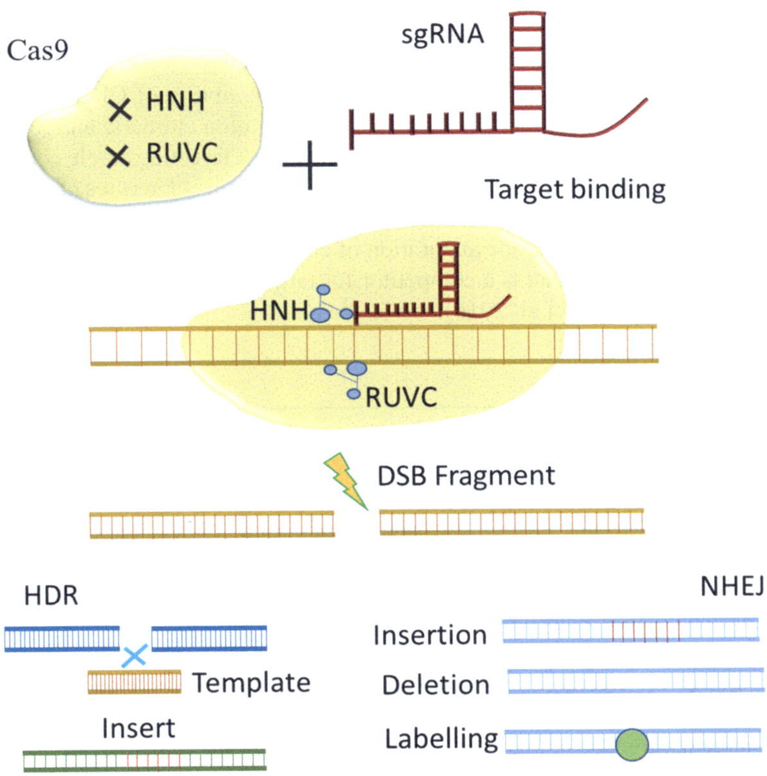

Fig. 7 Illustration of the steps involved in gene editing in CRIPSER-Cas9 technology

One study indicated CRISPR-Cas9 screening led toward identification of novel targets in therapy of acute myeloid leukemia by the use of AML cell lines. According to previous reports mRNA decapping enzyme scavenger (DCPS) is essential for survival. To check out the interaction among DCPS with the pre-mRNA mechanisms of enzymes of metabolic enzymes, spliceosomes, the technique of mass spectroscopy is used. CRIPSER-Cas9 technology assists in determining pre-mRNA metabolic pathways by categorizing DCPS as a target for AML therapy to treat muscular atrophy exhibited anti-leukemic activity (Yamauchi et al. 2018).

The field of drug discovery is becoming more efficient owing to new genetic discoveries, which integrate large scale genomics into target identification. Inherent drug resistance of an organism is generally categorized as an operative method in identification of particular cellular targets.

For example, a research study indicates leucyl-tRNA synthetase as a target of antifungal tavaborole obtained from a culture of *Saccharomyces cerevisiae* in occurrence of drug to create impulsive resistance mutants. DNA of the target was fragmented into 4 to 10 kb sized fragments and then inserted into a yeast vector to screen the aptitude to shield tavaborole-susceptible cells by tavaborole selection (Rock et al. 2007). Vector clone sequences of small persisting colonies exhibit those mutations in leucyl-tRNA synthetase that prevent the binding of a drug to its target (Ioerger et al. 2013).

Another study reported the competent gene editing ability of CRISPER in primary T-cell based immunotherapy such as next generation chimeric antigen receptor (CAR) T cells (Selvam et al. 2017). Cas13 is an RNA targeting single component enzyme that has nucleotide binding (HEPN) domains from eukaryotes and prokaryotes that directly target single stranded RNA. It also holds nucleases that differ in function and are responsible for maturation of crRNA that results in the creation of Cas13: crRNA complex that is a competitor for target RNA binding CRISPR-Cas9 technology (Abudayyeh et al. 2016).

7 Conclusions

Drug target identification acts as a keystone in drug discovery. With the advancement of knowledge and technology, phenotype-based models have replaced the affinity-based models. Identification of natural compounds to be used as drugs is an ordeal involving transcriptomics, metabolomics, genomics, and proteomics carefully imbricated in a centrally aligned framework of computational approaches. Every technique has some advantages and disadvantages. Affinity-based methods involve the direct interaction of drug targets with molecules of interest; hence, highly optimized experimental conditions driven by expert researchers are obligatory. A mere negligence in this regard may result in wastage of resources. Likewise, genomic-based approaches also depend upon the high state of the art level expertise in dealing with DNA and RNA modifications and measurement. Computational methods are based upon different tools of information technology coupled with statistics and chemoinformatics in order to detect the pathophysiological mechanisms of a wide spectrum of pathological conditions, and propose putative target(s) for drug molecules. However, the experimental support from proteomics and genetic techniques are required in order to validate the results from computational-based methods. Thus, there is a need to carefully analyze the pros and cons of all these techniques before planning the methodology required to carry out an experiment.

References

Abifadel M, et al. Mutations in PCSK9 cause autosomal dominant hypercholesterolemia. Nat Genet. 2003;34(2):154–6.

Abudayyeh OO, et al. C2c2 is a single-component programmable RNA-guided RNA-targeting CRISPR effector. Science. 2016;353(6299):aaf5573.

Adam GC, Sorensen EJ, Cravatt BF. Chemical strategies for functional proteomics. Mol Cell Proteomics. 2002;1(10):781–90.

Amiri-Dashatan N, et al. Proteomics applications in health: biomarker and drug discovery and food industry. Iran J Pharm Res. 2018;17(4):1523–36.

Ator MA, Mallamo JP, Williams M. Overview of drug discovery and development. Curr Protoc Pharmacol. 2006; Chapter 9: p. Unit9 9

Bantscheff M, Scholten A, Heck AJ. Revealing promiscuous drug-target interactions by chemical proteomics. Drug Discov Today. 2009;14(21-22):1021–9.

Bosch F, Rosich L. The contributions of Paul Ehrlich to pharmacology: a tribute on the occasion of the centenary of his Nobel Prize. Pharmacology. 2008;82(3):171–9.

Cohen J, et al. Low LDL cholesterol in individuals of African descent resulting from frequent nonsense mutations in PCSK9. Nat Genet. 2005;37(2):161–5.

Cohen JC, et al. Sequence variations in PCSK9, low LDL, and protection against coronary heart disease. N Engl J Med. 2006;354(12):1264–72.

Dart ML, et al. Homogeneous assay for target engagement utilizing bioluminescent thermal shift. ACS Med Chem Lett. 2018;9(6):546–51.

Dobson CM. Chemical space and biology. Nature. 2004;432(7019):824–8.

Drews J. Drug discovery: a historical perspective. Science. 2000;287(5460):1960–4.

Elgert C, et al. Thermal shift assay: Strengths and weaknesses of the method to investigate the ligand-induced thermostabilization of soluble guanylyl cyclase. J Pharm Biomed Anal. 2020;181:113065.

Floris M, et al. Genetic-driven druggable target identification and validation. Trends Genet. 2018;34(7):558–70.

Futamura Y, et al. Morphobase, an encyclopedic cell morphology database, and its use for drug target identification. Chem Biol. 2012;19(12):1620–30.

Futamura Y, et al. Identification of a molecular target of a novel fungal metabolite, pyrrolizilactone, by phenotypic profiling systems. Chembiochem. 2013;14(18):2456–63.

Gong F, et al. Dichloroacetate induces protective autophagy in LoVo cells: involvement of cathepsin D/thioredoxin-like protein 1 and Akt-mTOR-mediated signaling. Cell Death Dis. 2013;4:e913.

Hall DA, Ptacek J, Snyder M. Protein microarray technology. Mech Ageing Dev. 2007;128(1):161–7.

Hsu PD, Lander ES, d Zhang F. Development and applications of CRISPR-Cas9 for genome engineering. Cell. 2014;157(6):1262–78.

Huang F, et al. Chemical proteomics: terra incognita for novel drug target profiling. Chin J Cancer. 2012;31(11):507–18.

Hughes JP, et al. Principles of early drug discovery. Br J Pharmacol. 2011;162(6):1239–49.

Ioerger TR, et al. Identification of new drug targets and resistance mechanisms in Mycobacterium tuberculosis. PLoS One. 2013;8(9):e75245.

Ishino Y, Krupovic M, Forterre P. History of CRISPR-Cas from encounter with a mysterious repeated sequence to genome editing technology. J Bacteriol. 2018;200(7)
Jafari R, et al. The cellular thermal shift assay for evaluating drug target interactions in cells. Nat Protoc. 2014;9(9):2100–22.
Jessen KA, et al. The discovery and mechanism of action of novel tumor-selective and apoptosis-inducing 3,5-diaryl-1,2,4-oxadiazole series using a chemical genetics approach. Mol Cancer Ther. 2005;4(5):761–71.
Jinek M, et al. A programmable dual-RNA-guided DNA endonuclease in adaptive bacterial immunity. Science. 2012;337(6096):816–21.
Jung HJ, Kwon HJ. Target deconvolution of bioactive small molecules: the heart of chemical biology and drug discovery. Arch Pharm Res. 2015;38(9):1627–41.
Katsila T, et al. Computational approaches in target identification and drug discovery. Comput Struct Biotechnol J. 2016;14:177–84.
Kim Y, et al. Identification and validation of VEGFR2 kinase as a target of voacangine by a systematic combination of DARTS and MSI. Biomol Ther. 2020;10(4)
Koscielny G, et al. Open targets: a platform for therapeutic target identification and validation. Nucleic Acids Res. 2017;45(D1):D985–94.
Kotake Y, et al. Splicing factor SF3b as a target of the antitumor natural product pladienolide. Nat Chem Biol. 2007;3(9):570–5.
Lomenick B, et al. Target identification using drug affinity responsive target stability (DARTS). Proc Natl Acad Sci U S A. 2009;106(51):21984–9.
Lomenick B, Olsen RW, Huang J. Identification of direct protein targets of small molecules. ACS Chem Biol. 2011;6(1):34–46.
Macalino SJ, et al. Role of computer-aided drug design in modern drug discovery. Arch Pharm Res. 2015;38(9):1686–701.
Makarova KS, et al. A putative RNA-interference-based immune system in prokaryotes: computational analysis of the predicted enzymatic machinery, functional analogies with eukaryotic RNAi, and hypothetical mechanisms of action. Biol Direct. 2006;1:7.
Mann M. Functional and quantitative proteomics using SILAC. Nat Rev Mol Cell Biol. 2006;7(12):952–8.
Mesange F, et al. Identification of two tamoxifen target proteins by photolabeling with 4-(2-morpholinoethoxy)benzophenone. Bioconjug Chem. 2002;13(4):766–72.
Moller M, et al. Concordance between morphological and biochemical estimates of fluid spaces in rat brain cortex slices. Exp Brain Res. 1974;21(3):299–314.
Morretta E, et al. Discovering the biological target of 5-epi-sinuleptolide using a combination of proteomic approaches. Mar Drugs. 2017;15(10)
Moseley FL, et al. The use of proteomics to identify novel therapeutic targets for the treatment of disease. J Pharm Pharmacol. 2007;59(5):609–28.
Muroi M, Futamura Y, Osada H. Integrated profiling methods for identifying the targets of bioactive compounds: MorphoBase and ChemProteoBase. Nat Prod Rep. 2016;33(5):621–5.
Nagasawa I, et al. Identification of a small compound targeting PKM2-regulated signaling using 2D gel electrophoresis-based proteome-wide CETSA. Cell. Chem Biol. 2020;27(2):186–196 e4.
Osada H, et al. A Target identification system based on MorphoBase, ChemProteoBase, and photo-cross-linking beads. In: Waldmann H, Janning P, editors. Concepts and case studies in chemical biology. John Wiley & Sons, Inc.; 2014.
Pai MY, et al. Drug affinity responsive target stability (DARTS) for small-molecule target identification. Methods Mol Biol. 2015;1263:287–98.
Rock FL, et al. An antifungal agent inhibits an aminoacyl-tRNA synthetase by trapping tRNA in the editing site. Science. 2007;316(5832):1759–61.
Ross PL, et al. Multiplexed protein quantitation in Saccharomyces cerevisiae using amine-reactive isobaric tagging reagents. Mol Cell Proteomics. 2004;3(12):1154–69.
Santos R, et al. A comprehensive map of molecular drug targets. Nat Rev Drug Discov. 2017;16(1):19–34.

Schenone M, et al. Target identification and mechanism of action in chemical biology and drug discovery. Nat Chem Biol. 2013;9(4):232–40.
Selvam C, et al. Therapeutic potential of chemically modified siRNA: recent trends. Chem Biol Drug Des. 2017;90(5):665–78.
Sliwoski G, et al. Computational methods in drug discovery. Pharmacol Rev. 2014;66(1):334–95.
Smith E, Collins I. Photoaffinity labeling in target- and binding-site identification. Future Med Chem. 2015;7(2):159–83.
Swinney DC, Anthony J. How were new medicines discovered? Nat Rev Drug Discov. 2011;10(7):507–19.
Takeuchi T, Schumacker PT, Kozmin SA. Identification of fumarate hydratase inhibitors with nutrient-dependent cytotoxicity. J Am Chem Soc. 2015;137(2):564–7.
Tulloch LB, et al. Direct and indirect approaches to identify drug modes of action. IUBMB Life. 2018;70(1):9–22.
Tunon J, et al. Proteomic strategies in the search of new biomarkers in atherothrombosis. J Am Coll Cardiol. 2010;55(19):2009–16.
Voigt T, et al. A natural product inspired tetrahydropyran collection yields mitosis modulators that synergistically target CSE1L and tubulin. Angew Chem Int Ed Engl. 2013;52(1):410–4.
Wang J, et al. Target identification of natural and traditional medicines with quantitative chemical proteomics approaches. Pharmacol Ther. 2016;162:10–22.
Webb Y, et al. Photoaffinity labeling and mass spectrometry identify ribosomal protein S3 as a potential target for hybrid polar cytodifferentiation agents. J Biol Chem. 1999;274(20):14280–7.
Wiedl T, et al. Activity-based proteomics: identification of ABHD11 and ESD activities as potential biomarkers for human lung adenocarcinoma. J Proteome. 2011;74(10):1884–94.
Wright AT, Cravatt BF. Chemical proteomic probes for profiling cytochrome p450 activities and drug interactions in vivo. Chem Biol. 2007;14(9):1043–51.
Yamauchi T, et al. Genome-wide CRISPR-Cas9 screen identifies leukemia-specific dependence on a Pre-mRNA metabolic pathway regulated by DCPS. Cancer Cell. 2018;33(3):386–400 e5.
Yu H, et al. A systematic prediction of multiple drug-target interactions from chemical, genomic, and pharmacological data. PLoS One. 2012;7(5):e37608.
Zheng XS, Chan TF, Zhou HH. Genetic and genomic approaches to identify and study the targets of bioactive small molecules. Chem Biol. 2004;11(5):609–18.

Introduction to Target Validation

Cecilia Ortíz, Santiago Ruatta, and Marcelo Comini

Contents

Abstract

Although any macromolecule linked to a pathophysiological phenomenon is a natural candidate for therapeutic intervention, not all of them fulfil the criteria of drug target candidates. The selection of molecular targets with the highest potential for developing selective, effective and safe drugs depends, to a significant extent, on the correct assessment of their structural, biochemical and biological

C. Ortíz
Laboratory Redox Biology of Trypanosomes, Institut Pasteur de Montevideo, Montevideo, Uruguay

Laboratory of Biotechnology, Universidad ORT Uruguay, Montevideo, Uruguay

S. Ruatta
Laboratory Redox Biology of Trypanosomes, Institut Pasteur de Montevideo, Montevideo, Uruguay

Laboratory of Applied Chemistry, Facultad de Bioquímica y Ciencias Biológicas, Universidad Nacional del Litoral, Santa Fe, Argentina

M. Comini (✉)
Laboratory Redox Biology of Trypanosomes, Institut Pasteur de Montevideo, Montevideo, Uruguay
e-mail: mcomini@pasteur.edu.uy

M. T. Scotti, C. L. Bellera (eds.), *Drug Target Selection and Validation*, Computer–Aided Drug Discovery and Design,
https://doi.org/10.1007/978-3-030-95895-4_4

properties. The validation of a molecular target is the first experimental approach to be conducted in a target-based drug discovery campaign and is critical to diminish the high attrition rate and costs associated with the subsequent phases. In medicinal chemistry, target validation embraces the determination of the indispensability or physiological relevance of the molecular target, its druggability and assayability in vitro or in vivo. Different computational, genetic-, biochemical- and chemical-based approaches can be used complementary to address this goal. In this chapter, we provide an overview of the diverse strategies available to (bio)chemically and biologically validate a molecular target and discuss important concepts in this area.

Keywords

Forward genetics · Reverse genetics · RNA interference · Knock-out · CRISPR-Cas9 · Dominant negative · Chemogenomics · Chemical validation · Structure · activity relationship · Chemical probes

1 Definition and Requirements of a Molecular Target

In the field of medicinal chemistry, a molecular target is defined as a biological component of a living organism whose activity is associated with a patho-physiological effect and is suited to selective pharmacological intervention. The therapeutic strategy aims to restore homeostasis by affecting the activity of the miss-functioning molecular target/cells or by eliminating the pathogen (etiological approach) or by compensating the defects associated with the pathophysiological state (palliative approach). Although many macromolecules such as nucleic acids (e.g. DNA: promotors, regulators, genes, ARN: messenger, transfer, micro ARNs), proteins (e.g. receptors/ligands, enzymes, structural or regulatory proteins, lipoproteins), ribosomes (ribosomal RNA molecules and proteins) and lipids (membrane lipids and lipids second messengers) play important roles in cell and tissue physiology not all of them qualify as drug target candidates. In part, this is because an effective inhibition or modification of their function cannot be achieved by drug-like molecules due to molecular or biochemical/biological issues. For instance, macromolecules lacking structural pockets or elements suitable for binding small ligands (chemical or biological entities) in a specific and high affinity fashion do have a low prospect for becoming drug-target candidates. Even molecular targets shown to be indispensable for cell viability may not be suitable candidates for drug development because of their high turnover rate or poor metabolic control on particular pathway/process. Targeting these types of molecules will require long-term therapeutic regimens and/or drug administration at multiple doses or at high concentrations, which may compromise treatment safety (e.g. side effects).

Another important aspect to consider is the resiliency and functional redundancy present in most biological systems that allow organisms to maintain metastable conditions upon perturbations originating endogenously or exogenously. In this regard,

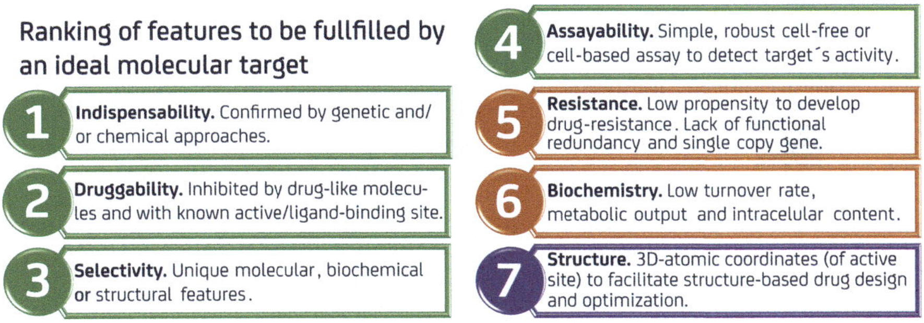

Fig. 1 Scheme outlining the main features of a molecular target for qualifying as a drug target candidate

even a unique and druggable molecule may not be a suitable drug target candidate if isoforms of it, alternative pathways or metabolites may take over the process where it participates. Ideally, the process/pathway where the target is functionally engaged should have a low or null propensity for metabolic rewiring, redundancy or development of drug resistance upon target inhibition. In line with this, molecules being critical and single nodes or bottlenecks of pathways or processes qualify as top target candidates from an 'omics' and physiological point of view. In contrast, for some pathological conditions or pathogens, the modification of a particular phenotype may require interfering with the activity of multiple targets. Such scenario demands the determination of the quantitative contribution of the pathway components and a polypharmacological approach (Morrow et al. 2010). Despite being challenging, a multi-target strategy does not demand a complete inhibition of the targets and the chances for developing multiple resistance is lower when compared to mono-target pharmacology. In this regard, the elucidation and analysis of protein–protein interaction networks (PINs) have become an extremely useful strategy to uncover the molecular mechanisms of diseases and drug discovery (Hao et al. 2018).

Therefore, the early assessment of molecular targets showing the best profile to deliver drug candidates (Fig. 1) is of paramount relevance for a target-based drug discovery campaign. In this respect, the validation of a molecule as a drug candidate is not only circumscribed to confirm its essentiality or biological role but, equally important, to determine its selective druggability, the feasibility for assaying its activity in vitro/in vivo and its potential for developing drug resistance. Although secondary, the availability of structural information is highly advantageous for the later stages of the drug development program because it will propel the rational (structure-based) design and optimization of the hit compounds.

2 Experimental-Based Approaches

The experimental assessment to determine if a macromolecule is involved in a disease process and if its modulation may have a therapeutic effect is the first step towards the validation of a drug target candidate. This can be achieved by performing

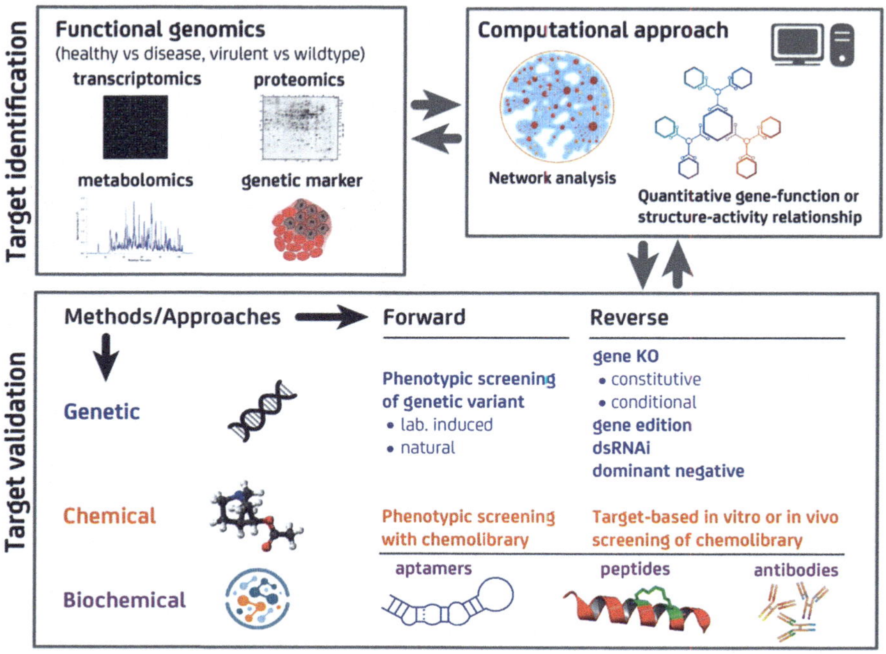

Fig. 2 Strategies used to identify and validate molecular targets. Experimental data from functional genomics and target validation approaches can be rationally analysed by computational approaches with large capabilities to integrate multiple information and descriptors. Reciprocally, the computational methods turn powerful tools to predict and suggest novel molecular targets that should be subjected to further identification and/or validation

the in vitro and/or in vivo phenotypic characterization of genetic variants of the organism of interest (Abuin et al. 2002). Complementary strategies to validate a molecular target include the modulation of protein expression or the (bio)chemical interference with target activity using small chemical compounds or ligands (i.e. antibodies, peptides or aptamers) (Devlin et al. 1990; Pini and Bracci 2000; Burgstaller et al. 2002). Below are briefly described the different genetic- and (bio) chemical-based approaches that can be used to validate and study drug target candidates, and the contribution of computational approaches to data analysis (Fig. 2).

Forward Genetics

The term *forward genetics* refers to the approach where a heritable condition, naturally occurring or laboratory-induced, is identified and associated with a particular disease-related phenotype (e.g. virulence, invasiveness, high proliferation, drug resistance) (Zhang 2002; Slonim 2002; Dedrick et al. 2019). However, though the effect (phenotype) is known and well characterized, its cause (genotype or biochemical) does not and may not necessarily be single but multifactorial. In *forward genetics*, one of the most important technical aspects is to have available or, if not, to design a specific and quantitative assay to discriminate robustly *wild-type* individuals from those bearing an abnormal phenotype. This requires detailed knowledge of the *wild-type* condition and the selection of measurable parameters to identify the aberrant phenotype. If not occurring naturally (e.g. patient sample, field isolation for infectious agents), the mutant lines should be generated in the laboratory with methods that may originate random or specific phenotypes. A broad spectrum of different methodologies, including radiation, chemical or biological mutagenic agents, can be used to induce mutations in the organism of interest (Lindsay 2003; Munroe and Schimenti 2009; Raphael and Talbot 2011; Bucan 2013; Ojelade and Rothenfluh 2013) (Fig. 3). For many decades, radiation and chemical-based approaches were the most robust and popular techniques employed to modify the genomic DNA with phenotypic consequences. Although easy to implement, the first one may induce massive and uncontrolled mutations such as deletions, inversions or translocations that frequently and simultaneously affect the molecular integrity of multiple genes. In contrast, the chemical-based technique mostly produces single-base mutations that are randomly distributed in the genome (Munroe et al. 2004). Given the randomness of these methodologies, their main limitation is that the location and extent of the genetic changes, and hence the identity of the target gene(s) linked to an adverse phenotype, may remain unknown or difficult to assess thoroughly. The discovery and development of biological agents able to perform 'insertional mutagenesis' has, in part, solved this problem. This technique consists in using

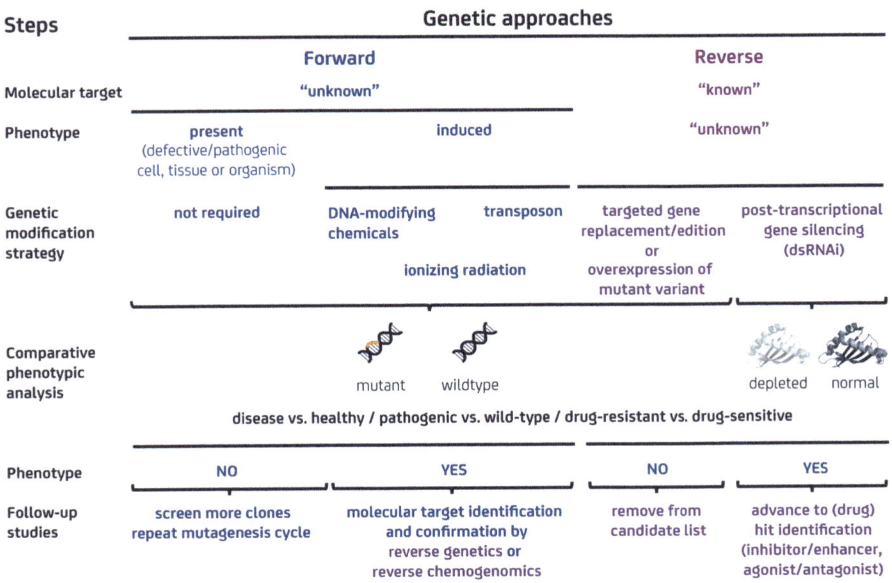

Fig. 3 Strategies for the validation of molecular targets based on genetics approaches

transposable genetic elements (transposons) that guide, in a region-selective manner, the generation of small or large genomic mutations (Hayes 2003; Reznikoff and Winterberg 2008; Largaespada 2009; Suster et al. 2009; Friedrich et al. 2017; Beckmann and Largaespada 2020). Taking advantage of this property, transposons harbouring elements with distinct insertion biases were developed and characterized with the aim to extend the number of locus/genes potentially affected by the insertion event (Lorens et al. 2001). The sequence-specific mode of action of transposons further facilitates the accurate identification of the targeted genomic sequences and, hence, to easily link phenotype to genotype. Regarding cell-based screenings, these are commonly accomplished by using libraries of chemical and biological agents (e.g., cDNA libraries, delivered and expressed from viral vectors) (Lorens et al. 2001; Lokey 2003). Animal screenings were historically performed in flies and worms and, more recently, extended to zebrafish, mice (Hrabé de Angelis et al. 2000; Nolan et al. 2000) and even in humans (e.g., by identifying naturally occurring genetic mutations and polymorphisms) (Moresco et al. 2013).

Independently of the mutagenic approach applied, the next step is to perform a phenotypic screening of the mutant organisms to identify abnormalities with respect to the behaviour of the *wild-type* condition. For approaches targeting multiple sequences/genes in a random fashion, the screening phase is usually the most time-consuming and laborious (Moresco et al. 2013; Schneeberger 2014). For this reason and to attain specificity, the conditions of the screening assay should be adapted to expose the desired phenotype (i.e. growth medium supplemented or deprived in special components, specific physicochemical or biological stress, etc.).

Once the mutants of interest are selected, the potential target gene(s) implied in the phenotype should be identified by standard DNA sequencing approaches. Finally, and in order to provide concluding evidence about the role of the gene in the disease or physiological phenotype, it is highly recommended to subject the identified molecular target to further characterization and validation studies either in vitro or in vivo (Carter and Shieh 2015a, b).

Although *forward genetics* is labour intensive, several consortia operating on an open-source basis undertook the systematic, genome-wide production, distribution and phenotypic analysis of mutant organisms (e.g. FlyBase.org, wormbase.org, MGD: mouse genome database, zfin.org). These initiatives contributed not only to speed-up the identification of both disease targets and models (Miyoshi et al. 2019) but also to 'popularize' the exploitation of the mutants by research laboratories.

Reverse Genetics

The validation of molecular targets by the *reverse genetics* approach consists in the deliberate modification of specific gene sequences with high 'surgical' precision, in order to alter their molecular nature or expression levels, and in analysing the phenotypic consequences thereof (Lindsay 2003). At variance with the *forward genetics* approach, the *reverse genetics* strategy demands prior knowledge of the sequence/gene to be targeted and, in some cases, even of the genomic context

(Takahashi et al. 1994). For simplicity, the techniques can be divided into those that fully abrogate (i.e. gene *knockout*: KO) and those that modulate the expression or function of the gene of interest. Gene KO, either by elimination or disruption of the corresponding DNA sequence, is a very robust technique that provides conclusive evidence about the indispensability or the contribution to a particular phenotype of the molecular target (Lindsay 2003; Carter and Shieh 2015a).

Targeted Gene Replacement and Editing

The classical method for knocking genes out and, hence, abrogate gene expression, consists in replacing or disrupting the coding sequence by homologous recombination techniques. For this purpose, a *knockout* (KO) DNA construct containing a reporter gene, usually a DNA sequence whose translation product confers resistance to a specific antibiotic, is flanked by sequences complementary to the gene of interest or adjacent to its 5′ and 3′ untranslated regions. The KO construct is prepared by standard molecular biology techniques and introduced into the organism by physicochemical (electroporation, nucleofection) or biological methods (e.g. lentivirus) (Deutscher et al. 2006). The reporter gene allows the isolation of a cell population or clones harbouring the genetic/phenotypic modification: defective on the gene of interest and resistant to the selection antibiotic. In the case of a diploid organism, a second recombination event must take place to force the modification of the second allele. This can be achieved by introducing a second DNA construct containing a different reporter/resistance gene or by inducing the duplication and recombination of the first KO construct into the second allele by increasing the concentration of selective pressure. This classical KO strategy is suitable to target single copy genes or multi-copy genes arranged in tandem (Hall et al. 2009). However, knocking out essential genes on an entire organism or in a germline can lead to a lethal phenotype (e.g., embryonic lethality), to compensatory mechanisms or to complex phenotypes that hinder establishing a clear genotype–phenotype relationship (Wang and Abate-Shen 2014; Deng 2014). In some cases, metabolic complementation, achieved by supplementing the growth media with the missing factor (metabolite produced by the targeted gene), can help bypass the lethal or compensatory phenotypes and, hence, to conduct a characterization of the KO-line under sub-lethal concentrations of the metabolite. In most cases, such a strategy cannot be implemented, but this limitation can be overcome by generating a conditional knockout (cKO) organism/cell line (Deng 2014; Lander et al. 2013). A cKO can be generated by co-expressing an inducible ectopic copy of the gene of interest in an organism defective in the corresponding genomic copy(ies) of this gene (i.e. a KO or mutant). Once the transgenic organism expressing the ectopic sequence is obtained, the next steps include the (stepwise for diploid organisms) deletion of the target gene under sustained expression of the transgene. Upon confirmation of the KO genotype, the expression of the ectopic gene can be interrupted and the consequent phenotype studied (Paape et al. 2020).

cKO can also be generated using the systems based on site/sequence specific recombinases and nucleases such as the Cre/loxP (Sauer and Henderson 1989) and CRISPR-Cas9 (Quadros et al. 2017). For the Cre/loxP method, a DNA construct (called *floxed* sequence) containing the sequence of the target gene flanked by a pair of short DNA sequences (*loxP* sites) is integrated into the genome by homologous recombination. The deletion of the target sequence is activated by inducing the expression of the Cre recombinase previously integrated into the genome of the organism or by performing a transient transfection of the cells with an episomal plasmid encoding for the recombinase. Cre recombinase will then drive the specific recognition and recombination at the loxP sites eliminating the target sequence from the genome (Carter and Shieh 2015a; Zhang et al. 1996; Feil et al. 1997, 2009; Metzger and Chambon 2001; Friedel et al. 2011). In the case of animal models, the production of cKO organisms requires the mating of a 'floxed' progeny with a transgenic one encoding for the inducible expression of Cre. By varying the promoter sequence that controls Cre expression, it was possible to exert temporal and cell/tissue specific control of the genetic KO (Patzke and Südhof 2016). The inducible expression of Cre can be achieved using the coupled estrogen receptor (ER)/tamoxifen system (Carter and Shieh 2015a; Metzger and Chambon 2001) or the tetracycline system (Bäckman et al. 2009). It is worth noting that the Cre/loxP technology can also be used to drive the inversion or translocation of the sequence of interest. Although highly specific and efficient, the Cre/loxP is a low throughput method for genetic manipulation, and compared to the CRISPR system, is more expensive and time consuming (e.g. 1–2 years for mice models). In fact, the generation of site-specific floxed mice is an important bottleneck of this method (Carter and Shieh 2015a).

The CRISPR-Cas9 technology uses a guiding RNA (gRNA) containing the sequence/site to be targeted and an endonuclease called Cas9 (Hsu et al. 2014). When these components are incorporated into a cell, Cas9 will bind to the single strand gRNA, which will guide the complex ARN-protein to a specific genomic site. The gRNA will bind the complementary genomic DNA sequence and Cas9 will catalyse the precise excision of the nearby double-stranded genomic DNA. At this point, the nicked DNA can be repaired, usually, by the non-homologous end joining (NHEJ) pathway (Cong et al. 2013; Mali et al. 2013) or by homologous recombination. The homologous recombination pathway offers the possibility to drive the locus-specific insertion of a sequence of interest (e.g. reporter-, resistance-, mutant-gene or regulatory elements: promoter/transactivator/enhancer/operon); therefore, this strategy is named *knock-in* or gene targeted replacement (Slaymaker et al. 2016; Kleinstiver et al. 2016; Havlicek et al. 2017). The gRNA and, eventually, the donor DNA can be incorporated to the cell by the transfection techniques mentioned above. With respect to Cas9, the protein or a vector encoding for it can be co-transfected along with the other components of the system. Under these conditions, Cas9 will exert its function in a transient (short-lived) fashion. Eventually, transgenic cell lines or organism encoding for the inducible expression of Cas9 can be generated. Efforts have been made towards robust CRISPR-Cas9 inducible systems because a leaky expression of the endonuclease may lead to unwanted off-target effects. In this regard, a CRISPR-Cas9 estrogen receptor chimera (Dow et al. 2015; Liu et al. 2016), similar to the one described above for classical cKO, has been developed. By selecting the proper genome locus and regulatory elements controlling Cas9 expression, it is possible to select the cell type and time point (e.g. developmental stage) at which a specific gene may be deleted or expressed. Although specific, the CRISPR-Cas9 technology currently has a low efficacy index, with only 1–10% of experiments reporting the successful deletion of the target gene (Lindsay 2003; Quadros et al. 2017). The most common drawbacks of this technique are associated with partial gene editing, duplicated or 'off target' insertions, as well as deletions effected by non-homologous end joining (Yang et al. 2013; Miano et al. 2016; Bishop et al. 2016). The recent finding that long single-stranded DNAs (ssDNAs) serve as very efficient donors for gene replacement has contributed to overcome some of these problems. The novel technology has been baptized: Easi-CRISPR. This novel approach has a higher efficiency for targeted DNA-cassette insertion at different loci (typically 30–60% and, in some cases, even 100%) and shorten the time (~2 months) to generate mice's founder strains (Quadros et al. 2017; Miura et al. 2018). The design of Easi-CRISPR donor is simple, and cassettes are commercially available, can be customized and rapidly synthesized (Quadros et al. 2017). The method proved suitable for the high and low-throughput generation of floxed animal models, which is a major limitation for the extensive use of the Cre/LoxP-system. An important limitation of all genome editing methods is that the effectiveness of the recombination event decreases with the size of the sequence to be inserted.

Post-transcriptional Gene Silencing

The post-transcriptional silencing of gene expression mediated by double stranded RNA, also known as RNA interference (RNAi), is an ancient biological mechanism that different eukaryotic organisms (Napoli et al. 1990; Romano and Macino 1992; Fire et al. 1998; Zhang et al. 2004; Cerutti and Casas-Mollano 2006; Setten et al. 2019) use to defend themselves against the intrusion of exogenous nucleic acids, usually of viral origin, and to regulate gene transcription (Bartel 2004). Briefly, the molecular mechanism involves the stepwise: (i) cleavage of double-stranded RNA into shorter molecules (siRNA or short interfering RNA: 21–25 nucleotides in length) by the ribonuclease DICER, (ii) the activation of the RNA-Induced Silencing Complex (RISC) by the sRNAs and (iii) the recognition and degradation of the mRNAs bearing sequences complementary to the siRNA by the RISC-siRNA complex (Hannon 2002; Stewart et al. 2003; Bantounas et al. 2004; Sen and Blau 2006; Rao et al. 2009; Wilson and Doudna 2013). The final output is a depletion of the target mRNA and its translation product. At variance with the KO approach, RNAi supresses but does not fully abrogate the expression of the target protein (Lindsay 2003; Lavery and King 2003). The magnitude of the downregulation very much depends on the level and half-life of the targeted transcript and protein. The depletion of highly abundant transcripts/proteins with high half-life demands a sustained activation of the RNAi system. The level (e.g., Western blot technique) and/or activity of the target protein must be determined to confirm the performance of the RNAi and correlate it with the phenotypic changes (Jackson et al. 2003). RNAi has gained an important popularity among the genetic reverse tools used to validate molecular targets (Dean 2001; Yu et al. 2002; Genovesio et al. 2011; Kolev et al. 2011; Moser et al. 2013; Mueller et al. 2014). In part, this is because most eukaryotic organisms possess the RNAi machinery, is far less laborious and costly than the KO or gene replacement/editing methods because it does not necessarily require the genetic manipulation of the host cell or organism (i.e. they can directly be transfected with exogenous siRNA or short hairpin RNA: shRNA) (Lindsay 2003; Rao et al. 2009). Furthermore, the technique proved highly specific and effective in downregulating gene expression (>50–90% mRNA and protein knockdown) and is amenable for high throughput applications (Premsrirut et al. 2011; Alsford et al. 2011; Glover et al. 2015). In the past, the potential off-target effects associated with this methodology were an issue because the selected siRNA sequences could also be present in other sites of the genome besides the gene of interest. Along with the availability of full genome sequences from different organisms, several RNAi libraries, databases and algorithms were created and are available to analyse and predict effective and selective siRNA sequences (Boutros et al. 2004; Arziman et al. 2005; Conte et al. 2015; Hu et al. 2021). The RNAi technology also admits the development of transgenic organisms where the expression of the double strand RNA is under the control of an inducible (tissue or cell cycle-specific) promoter (Sifuentes-Romero et al. 2011). At variance with the transient downregulation achieved by transfecting siRNA or shRNA, stable RNAi-cell lines or organisms allow for detailed functional studies of the molecular target. Compared to the KO approach, a remarkable advantage of the RNAi approach is that the phenotypic effect resembles a pharmacological scenario where full inhibition of a molecular target is never achieved (Sledz et al. 2003). Another comparative advantage of the RNAi technology with respect to the KO approach is the possibility to downregulate simultaneously the expression of distinct or multicopy genes (Stortz et al. 2017). However, in contrast to the KO approach, the RNAi strategy does not yield useful information for gene products that demand a high level of depletion to impair their function (e.g. >90%). The limited availability of effective si/sh-RNAi delivery systems for different cell types (i.e. primary or non-dividing cells, specific tissues) and, in particular, for whole organisms prevents a more extensive application of this method because the saturation of the RISC complex with si/shRNAi determines the effectiveness of the silencing. The delivery may rely on physicochemical (electroporation, nucleofection, lipofection) and biological (viral or plasmidic vectors) methods, and the optimization thereof is cell-line dependent (Sledz et al. 2003). For instance, the application of the virus-mediated delivery is limited to the target cells/tissues harbouring the corresponding viral receptors.

Modulation of Protein Function

Another strategy to reveal the critical role of a gene in a particular pathophysiological context consists in overexpressing transiently a mutant of the molecular target, a biochemical competitor or interactor. The overexpression of any of these variants will compete with the endogenous protein interfering with its function in the wild type cell/organism; therefore, the strategy has been named 'dominant-negative' (Herskowitz 1987; Sheppard 1994). Taking into account the physical basis of the molecular interference, the 'dominant negative' approach is suitable for oligomeric proteins, or proteins forming part of polypeptide complexes or playing regulatory roles by interacting with other cellular components (Lopes et al. 2016). Nonetheless, it can also be applied to unravel the role of monomeric proteins. For instance, overexpression of a protein competitor or interactor (not a mutant of the molecular target) that competes for the use of the cognate substrate or avoid accumulation of its product (i.e. by degrading or converting it into a non-biologically useful form or impair the functioning of the molecular target) may, in an indirect manner, disclose the relevance of the molecular target for disease (Frattali et al. 1992). Furthermore, the overexpression of point mutants allows dissecting structure–function relationships in the molecular target that can be pharmacologically addressed on a rational basis (Sheppard 1994). The dominant negative protein or gene can be incorporated to the cell or organism via standard molecular biology techniques (lipofection or nanoparticles containing the protein/mutein, or via the incorporation of an episomal or integrative plasmid or viral vectors). Stable transgenic cell lines/animals with inducible expression of the dominant negative molecule can also be generated and provide a useful platform for more exhaustive pathophysiological and pharmacological studies. The technique was employed to disclose the role and validate several protein targets such as integrins (Takada et al. 1992; Lee et al. 2012), tyrosine kinase receptors (Kashles et al. 1991; Sandhöfer et al. 2016), proteins related to Huntington's disease (Lopes et al. 2016), the thyroid hormone receptor (Yen et al. 1992; Nishiyama et al. 2003) or the tumour suppressor protein p53 (Xue et al. 2007; Bossi et al. 2008) and a component of the mitochondrial iron-sulfur cluster machinery (Manta et al. 2013). One of the limitations of this approach is that it requires a prior molecular/biochemical knowledge of the target to spot the residues/regions to be mutated to cause a biological impact. The approach is also prone to produce off-target effects as a consequence of a high intracellular content of the dominant negative species that may indirectly affect other cellular processes (e.g. by miss-localization or non-specific sequestration/interaction with other proteins) (Herskowitz 1987; Lagna and Hemmati-Brivanlou 1997; Cheng et al. 2003; Sledz and Williams 2005).

Other alternative strategies to interfere with protein function encompass the use of high affinity reagents such as antibodies, peptides and RNA aptamers (Devlin et al. 1990; Pini and Bracci 2000), but they are rarely used for target validation studies because their design is time consuming and expensive (Lindsay 2003).

Chemical-Based Approaches

In many cases, genetic validation of a target is not possible for technical or biological reasons (e.g. refractoriness or lack of genetic tools for manipulating specific organisms, multicopy genes, insufficient depletion by RNAi, etc.). The chemical validation of molecular targets is an alternative option to overcome these limitations and, at the same time, to identify ligands/inhibitors with pharmacological potential (Fig. 4). Chemogenomics is an interdisciplinary field that studies a biological, genomic or proteomic response of an cell/organism exposed to different chemical (drug-like) compounds (Bredel and Jacob 2004), attempting to fully match target and ligand space, and ultimately identify all ligands of all targets (Caron et al. 2001). In chemogenomics-based drug discovery, large collections of chemical products are screened for the parallel identification of biological targets and biologically active compounds. By integrating genomics, bioinformatics, combinatorial and synthetic chemistry, chemogenomics leads to the rational development and rapid screening of

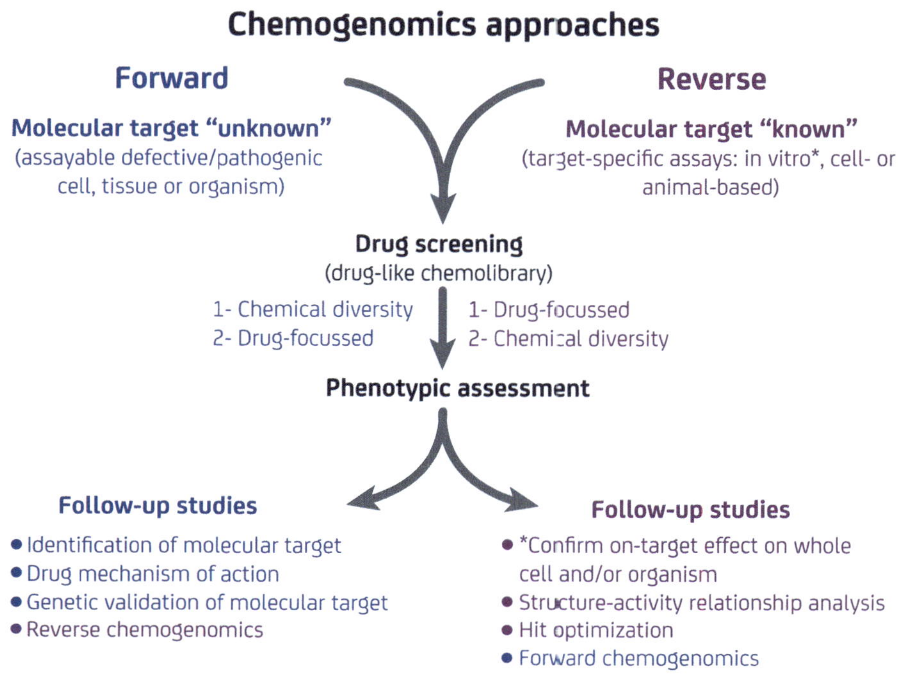

Fig. 4 Chemical-based approaches applied to the nomination and validation of molecular targets. Depending on the initial knowledge on the target/system under study and the bio-tools/equipment available, target nomination and/or validation can be approached applying one or both chemogenomics strategies. For instance, compound bioprofiling by forward chemogenomics is followed by the identification of the gene and protein involved in the adverse phenotype and, eventually, by investigating the biochemical/biological basis of the dysfunction. In contrast, reverse chemogenomics begin with the identification of the target's ligands/inhibitors and, next, by using them to validate the target and examine the phenotype. Asterisk: for molecular targets identified by means of cell-free assays, the on-target effect of the hit compounds should be confirmed on biological models.

target-specific chemical ligands. Among all biomedical applications, chemogenomics stands out in cancer research. Because human cancers show complex and multiple pathogenic aberrations, this strategy can help in designing (multi)target or patient-specific therapies, the so-called precision-medicine (Bredel and Jacob 2004). The experimental approaches relying on chemical compounds are complementarily used to validate drug target candidates and are powerful tools to determine compound mode of action and drug resistance (Collett et al. 2019), as well as for nominating new drug targets and genes involved in biological pathways.

Forward Chemogenomics

When the molecular basis of a phenotype is unknown, then a forward chemogenomics approach is mandatory. The strategy consists in performing a phenotypic screening on a panel of cells, tissues or complete organisms (e.g. eukaryotic or prokaryotic, unicellular or multicellular, physiological or pathological) exposed to a series of chemicals with the aim to exert a gain or loss of function, depending on the disease model. Upon detection of a phenotype, the next step is to identify the gene(s) whose function(s) was modulated by the compound. One important requirement of this approach is to have a robust and phenotype-specific cell/organism-based assay to increase the chances of detecting specific and potent modulators of the target or defective gene function. This also contributes to significantly shortening the time gap for identifying the gene responsible for the phenotype (uncontrolled proliferation, miss-functioning, drug resistance, virulence, metastasis, etc.). In this regard, any prior knowledge on the function or process targeted will narrow the set of genes to be studied as potential disease candidates. Not least, the integration of the information relative to the compound structural motifs associated with the emerging phenotype is extremely useful for the identification and classification of compound families and functional groups that may further be developed or optimized as drug candidates.

Reverse Chemogenomics

This approach consists in identifying inhibitors/ligands that interfere with the in vitro activity of the molecular target of interest and then on using the active compounds to validate in vivo the essentiality or function of the target gene. In this regard, the first stage of this approach is identical to the target-based drug discovery. A *sine qua non* condition is that the target of interest must be assayable via robust and, if possible, simple methodologies. The screening assays can be divided into cell-free, cell-based and organismal assays (Wallqvist et al. 2002). The cell-free methods rely on measuring the activity of the recombinant, synthetic or native form of the molecular target. Such assays may be adapted to high throughput screening setups and will render information about the target–ligand interaction. It is very important that the assay conditions (substrate concentrations, buffer system and physicochemical parameters) are close to the physiological ones in the system of interest. The cell/organism-based methods render additional information about the biochemical routes affected by the compounds and its on-target effect. Thus, although more laborious and costly, the target-specific cell/organism-based assays allow the direct validation of the molecular target. Depending on the information available about the target or related molecules, the chemolibrary to be screened may consist of random or selected (i.e. inhibitors/(ant)agonists reported for related proteins, drug-focussed libraries) compounds. The focussed libraries are usually associated with a high hit ratio and to a limited chemical novelty of the hits while the opposite applies for libraries presenting a large chemical diversity. The most promising hits arising from the in vitro screening assay not only serve to validate the molecular target but also as chemical warheads that may be subject to further hit-to-lead optimization.

Chemical Probes

The recent efforts in developing drug discovery technologies led researchers to generate small-molecule tools to help elucidate the roles of the targeted proteins in healthy and diseased cells and tissues. A chemical probe is a selective small-molecule modulator of a protein's function that allows the user to ask mechanistic and phenotypic questions about its molecular target in biochemical, cell-based or animal studies. They can rapidly and reversibly inhibit a protein or a protein domain in cells or animals, be used in almost any cell type and reveal temporal features of target inhibition (Arrowsmith et al. 2015). In this way, they allow the evaluation of a particular role a protein plays, whilst leaving other interactions intact. In combination with techniques such as RNAi, they draw a complete picture of a protein's function within a cell (Arrowsmith et al. 2015). Chemical probes are defined by four main criteria (Frye 2010; Workman and Collins 2010; Schwarz and Gestwicki 2018):

1. A minimal in vitro potency of less than 100 nM.
2. Greater than 30-fold selectivity over sequence-related proteins.
3. Profiled against an industry standard selection of pharmacologically relevant targets.
4. On-target cellular effects at greater than 1 mM.

Whilst not designed with drug-like characteristics in mind (for example absorption, distribution, metabolism and excretion (ADME) properties) (Arrowsmith et al. 2015), there is nevertheless a belief that chemical probes can provide a small-molecule starting point to accelerate the drug discovery process. Their utility in interrogating the function of a protein and thus its relevance (Sweis 2015) as a drug target has led a consortium of industrial and academic researchers to establish a collaboration for the development of probes for the entire proteome (Carter et al. 2019). Historically, demonstrating on-target activity in cells, and especially in animal models, has proven very challenging, particularly for those proteins with unknown cellular function. Although this task remains a challenge, the interdisciplinary approach of chemogenomics is a powerful way to link the target, the chemical probe and the bioactivity.

As for the state-of-the-art in chemical probes development, the data obtained by researchers is compiled at the collaborative database chemicalprobes.org, providing an invaluable tool for systematizing the progress made so far. For example, by filtering results in the database, one can obtain a list of model agonist or antagonist compounds for protein kinases, dehydrogenases, neurotransmitters receptors (α-1D adrenoceptor, serotonin receptor), various types of transferases (methyl, acetyl), apoptosis regulators, cytoskeletal proteins and transcription factors, to mention a few.

3 Computational-Based Approaches

Predictive Chemogenomics

The analysis of a matrix containing information about the structural/physicochemical features of the compounds vs their molecular target-binding affinity or the magnitude of the phenotypic changes exerted set the basis of a structure–activity relationship, and provides a rational frame from which the outcome of untested interactions can be predicted (Rognan 2007). Predictive chemogenomics base their logic in the following complementary statements: 'compounds sharing some chemical similarity should also share targets' and vice versa 'targets sharing similar ligands should share similar molecular patterns (binding sites)'.

Filling the full theoretical chemogenomic matrix implies that data on 'unliganded' targets should be gathered from the closest 'liganded' neighbouring targets, and that data on 'untargeted' ligands should be gathered from the closest 'targeted' ligands. To navigate the ligand space, structural and physicochemical properties, as well as geometrical parameters, functional groups and chemical families are defined as descriptors (Rognan 2007). The two-dimensional matrix analysis uses multivariate linear regression (Kauvar et al. 1995) to measure structure–activity relationship distances between two targets/compounds (Vieth et al. 2004) and predict a global pharmacological profile (Krejsa et al. 2003). Therefore, the larger the number of targets and chemical diversity explored, the more confident the prediction of the affinity/potency of a new compound to an existing target. Thus, predictive genomics is useful to rank potential new molecular targets of known compounds and, vice versa, novel ligands of known targets. In both cases, experimental confirmation of the in silico prediction is required.

Molecular Network Technology

Clearly, the experimental approaches described above are very potent in providing conclusive evidence about the involvement of a molecule in normal and abnormal physiological states, and their potential as molecular targets. However, compared to computational approaches, they are more laborious and, many times, give an atomistic view of the biological facts. Molecular network technology is an approach that embraces diverse fields such as systems biology, network biology, chemoinformatics and pharmacology, and allows inferring the relevance of a molecular target/pathway/bioactive compound. This technology emerged as a need for the analysis of the steadily increasing number of functional genomics (transcriptomics and proteomics) and chemogenomics data/studies for different pathophysiological states and pathogens. This approach relies on assessing, filtering, integrating and correlating multifactorial data available in the literature (e.g. PubMed) and genome/'omics'/gene-pathways/bio-databases of specific organisms by means of computational methods (Cheng et al. 2011). Software-based tools available from open (i.e. Cytoscape; https://cytoscape.org) or commercial (e.g. Ingenuity Pathway Analysis from Qiagen) sources offer the possibility to analyse and visualize molecular interactions between components of networks and biological pathways, and to discover regulatory molecules and molecular causal relationships based on experimental results. The analysis can be used to construct a ranking of potential molecular/drug candidates based on their contribution to a pathway or to the desired therapeutic effect, relationships and druggability, whenever information on chemical inhibition is available (Wren et al. 2004). Another advantage of this poly-functional analysis is that the information about implicit (drug) targets relationships or central network hubs or bottlenecks is highly valuable for suggesting polypharmacological approaches against complex diseases (Wessel et al. 2015; Barrangou et al. 2015; Chen and Butte 2016). As for any computational approach, the experimental confirmation of the predicted targets and drug interactions via the functional or drug screening techniques described in previous sections is a must and may require iterative cycles that will contribute to invigorate the searching and analytic software algorithms.

4 General Remarks and Prospective

The huge scientific and technological advances, and the many breakthroughs in the understanding of disease biology boosted by the 'omics' revolution and the post-genome era (Lander et al. 2001; Ward 2001; Chen et al. 2016) does not yet keep pace with the discovery and approval of novel drugs and therapies. In a drug discovery campaign, selecting the right target is crucial but it is just the beginning of a long journey where target validation is the first and most relevant step. In fact, a recent study estimated that drug discovery projects showing a greater confidence in target validation, in linking target to disease or in understanding the role of the target in the pathology were significantly more successful in clinical Phase II than projects without such scientific background (Cook et al. 2014). Genetics and chemical approaches can be used in a complementary fashion not only with the purpose to validate the molecular target but also to assess its druggabilitty and to deliver preliminary hit compounds. The discovery and further development of gene editing and silencing tools introduced molecular precision (CRISPR-Cas9) and pharmacological perspective (RNAi or dominant negative) to the validation process (Deveraux et al. 2003; Meissner et al. 2017; Rautela et al. 2020). The in vitro or in vivo (cell or organism) assayability of the molecular target or abnormal phenotype is a determinant factor for allowing the implementation of drug screening strategies (chemogenomics) and the rational optimization of the hits. For the chemical-based approaches, it is highly recommended to perform the experiments with drug-like molecules under (nearly) physiological conditions because this will shorten the time and investment required for turning a hit into a lead. In this regard, the possibility for drug repositioning should be taken into consideration by using libraries consisting of clinical drugs for distinct medical applications. Large chemical libraries can also be screened in the search of novel molecular targets. Although this strategy imposes the subsequent identification of the target, it already reveals hit candidates. The significant advance in 'omics' technologies applied to study the basis of different pathologies (transmissible and non-transmissible diseases and/or pathogens) turned them into a valuable source of data for identifying molecules, processes and pathways playing critical roles in abnormal phenotypes. The computational approaches are potent tools for integrating and mining the multiple data and descriptors arising from functional genomics or large chemogenomics studies. The predictive capacity of the in silico methods has been improved and depends to a great extent on the iteration with wet-laboratory experiments (Duarte et al. 2019). Thus, hit candidates (molecular target or ligand) highlighted by this technology should be experimentally confirmed before advancing on the drug discovery campaign. Precision medicine will require an intensive support of the experimental approaches outlined in this chapter to validate novel targets. Undoubtedly, these approaches set the basis for a robust and promising target product profile drug discovery campaign and contribute to a more holistic comprehension of pathophysiological processes.

Acknowledgements S.R. is a postdoctoral fellow of the Consejo Nacional de Investigaciones Científicas y Técnicas (CONICET, Argentina). M.A.C. is thankful for the support from FOCEM (Fondo para la Convergencia Estructural del Mercosur, COF 03/11) and from the Programa de Desarrollo de Ciencias Básicas (PEDECIBA, Uruguay). C.O. and M.A.C. belong to the Sistema Nacional de Investigadores from the Agencia Nacional de Investigación e Innovación, Uruguay.

References

Abuin A, Holt KH, Platt KA, et al. Full-speed mammalian genetics: in vivo target validation in the drug discovery process. Trends Biotechnol. 2002;20:36–42. https://doi.org/10.1016/S0167-7799(01)01843-1.

Alsford S, Turner DJ, Obado SO, et al. High-throughput phenotyping using parallel sequencing of RNA interference targets in the African trypanosome. Genome Res. 2011;21:915–24. https://doi.org/10.1101/gr.115089.110.

Arrowsmith CH, Audia JE, Austin C, Baell J, Bennett J, Blagg J, Bountra C, et al. The promise and peril of chemical probes. Nat Chem Biol. 2015;11(8):536–41. https://doi.org/10.1038/nchembio.1867.

Arziman Z, Horn T, Boutros M. E-RNAi: a web application to design optimized RNAi constructs. Nucleic Acids Res. 2005;33:W582–8. https://doi.org/10.1093/nar/gki468.

Bäckman CM, Zhang Y, Malik N, et al. Generalized tetracycline induced Cre recombinase expression through the ROSA26 locus of recombinant mice. J Neurosci Methods. 2009;176:16–23. https://doi.org/10.1016/j.jneumeth.2008.08.024.

Bantounas I, Phylactou LA, Uney JB. RNA interference and the use of small interfering RNA to study gene function in mammalian systems. J Mol Endocrinol. 2004;33:545–57. https://doi.org/10.1677/jme.1.01582.

Barrangou R, Birmingham A, Wiemann S, et al. Advances in CRISPR-Cas9 genome engineering: lessons learned from RNA interference. Nucleic Acids Res. 2015;43:3407–19. https://doi.org/10.1093/nar/gkv226.

Bartel DP. MicroRNAs: genomics, biogenesis, mechanism, and function. Cell. 2004;116:281–97. https://doi.org/10.1016/s0092-8674(04)00045-5.

Beckmann PJ, Largaespada DA. Transposon insertion mutagenesis in mice for modeling human cancers: critical insights gained and new opportunities. Int J Mol Sci. 2020;21:E1172. https://doi.org/10.3390/ijms21031172.

Bishop KA, Harrington A, Kouranova E, et al. CRISPR/Cas9-mediated insertion of loxP sites in the mouse Dock7 gene provides an effective alternative to use of targeted embryonic stem cells. G3 (Bethesda). 2016;6:2051–61. https://doi.org/10.1534/g3.116.030601.

Bossi G, Marampon F, Maor-Aloni R, et al. Conditional RNA interference *in vivo* to study mutant p53 oncogenic gain of function on tumor malignancy. Cell Cycle. 2008;7:1870–9. https://doi.org/10.4161/cc.7.12.6161.

Boutros M, Kiger AA, Armknecht S, et al. Genome-wide RNAi analysis of growth and viability in *Drosophila* cells. Science. 2004;303:832–5. https://doi.org/10.1126/science.1091266.

Bredel M, Jacob E. Chemogenomics: an emerging strategy for rapid target and drug discovery. Nat Rev Genet. 2004;5(4):262–75. https://doi.org/10.1038/nrg1317.

Bucan M. Mouse genetics. In: Maloy S, Hughes K, editors. Brenner's encyclopedia of genetics. 2nd ed. San Diego: Academic Press; 2013. p. 486–8.

Burgstaller P, Girod A, Blind M. Aptamers as tools for target prioritization and lead identification. Drug Discov Today. 2002;7:1221–8. https://doi.org/10.1016/s1359-6446(02)02522-9.

Caron PR, Mullican MD, Mashal RD, Wilson KP, Su MS, Murcko MA. Chemogenomic approaches to drug discovery. Curr Opin Chem Biol. 2001;5(4):464–70. https://doi.org/10.1016/s1367-5931(00)00229-5.

Carter M, Shieh J. Manipulating endogenous genes. In: Guide to research techniques in neuroscience. Elsevier; 2015a. p. 273–93.

Carter M, Shieh J. Identifying genes and proteins of interest. In: Guide to research techniques in neuroscience. Elsevier; 2015b. p. 203–18.

Carter AJ, Kraemer O, Zwick M, Mueller-Fahrnow A, Arrowsmith CH, Edwards AM. Target 2035: probing the human proteome. Drug Discov Today. 2019;24(11):2111–5. https://doi.org/10.1016/j.drudis.2019.06.020.

Cerutti H, Casas-Mollano JA. On the origin and functions of RNA-mediated silencing: from protists to man. Curr Genet. 2006;50:81–99. https://doi.org/10.1007/s00294-006-0078-x.

Chen B, Butte AJ. Leveraging big data to transform target selection and drug discovery. Clin Pharmacol Ther. 2016;99:285–97. https://doi.org/10.1002/cpt.318.

Chen X, Yan CC, Zhang X, et al. Drug-target interaction prediction: databases, web servers and computational models. Brief Bioinform. 2016;17:696–712. https://doi.org/10.1093/bib/bbv066.

Cheng JC, Moore TB, Sakamoto KM. RNA interference and human disease. Mol Genet Metab. 2003;80:121–8. https://doi.org/10.1016/j.ymgme.2003.08.011.

Cheng T, Li Q, Wang Y, Bryant SH. Identifying compound-target associations by combining bioactivity profile similarity search and public databases mining. J Chem Inf Model. 2011;51:2440–8. https://doi.org/10.1021/ci200192v.

Collett CF, Kitson C, Baker N, et al. Chemogenomic profiling of antileishmanial efficacy and resistance in the related Kinetoplastid parasite *Trypanosoma brucei*. Antimicrob Agents Chemother. 2019;63:e00795–19. https://doi.org/10.1128/AAC.00795-19.

Cong L, Ran FA, Cox D, et al. Multiplex genome engineering using CRISPR/Cas systems. Science. 2013;339:819–23. https://doi.org/10.1126/science.1231143.

Conte D, MacNeil LT, Walhout AJM, Mello CC. RNA Interference in *Caenorhabditis Elegans*. Curr Protoc Mol Biol. 2015;109:26.3.1–30. https://doi.org/10.1002/0471142727.mb2603s109.

Cook D, Brown D, Alexander R, et al. Lessons learned from the fate of AstraZeneca's drug pipeline: a five-dimensional framework. Nat Rev Drug Discov. 2014;13:419–31. https://doi.org/10.1038/nrd4309.

Dean NM. Functional genomics and target validation approaches using antisense oligonucleotide technology. Curr Opin Biotechnol. 2001;12:622–5. https://doi.org/10.1016/s0958-1669(01)00270-1.

Dedrick RM, Guerrero-Bustamante CA, Garlena RA, et al. Engineered bacteriophages for treatment of a patient with a disseminated drug-resistant *Mycobacterium abscessus*. Nat Med. 2019;25:730–3. https://doi.org/10.1038/s41591-019-0437-z.

Deng C-X. Conditional *Knockout* mouse models of cancer. Cold Spring Harb Protoc. 2014:pdb.top074393. 2014; https://doi.org/10.1101/pdb.top074393.

Deutscher D, Meilijson I, Kupiec M, Ruppin E. Multiple *knockout* analysis of genetic robustness in the yeast metabolic network. Nat Genet. 2006;38:993–8. https://doi.org/10.1038/ng1856.

Deveraux QL, Aza-Blanc P, Wagner KW, et al. Exposing oncogenic dependencies for cancer drug target discovery and validation using RNAi. Semin Cancer Biol. 2003;13:293–300. https://doi.org/10.1016/s1044-579x(03)00043-9.

Devlin JJ, Panganiban LC, Devlin PE. Random peptide libraries: a source of specific protein binding molecules. Science. 1990;249:404–6. https://doi.org/10.1126/science.2143033.

Dow LE, Fisher J, O'Rourke KP, et al. Inducible in vivo genome editing with CRISPR-Cas9. Nat Biotechnol. 2015;33(4):390–4. https://doi.org/10.1038/nbt.3155.

Duarte Y, Márquez-Miranda V, Miossec MJ, González-Nilo F. Integration of target discovery, drug discovery and drug delivery: a review on computational strategies. WIREs Nanomed Nanobiotechnol. 2019;11 https://doi.org/10.1002/wnan.1554.

Feil R, Wagner J, Metzger D, Chambon P. Regulation of Cre recombinase activity by mutated estrogen receptor ligand-binding domains. Biochem Biophys Res Commun. 1997;237:752–7. https://doi.org/10.1006/bbrc.1997.7124.

Feil S, Valtcheva N, Feil R. Inducible Cre mice. Methods Mol Biol. 2009;530:343–63. https://doi.org/10.1007/978-1-59745-471-1_18.

Fire A, Xu S, Montgomery MK, et al. Potent and specific genetic interference by double-stranded RNA in *Caenorhabditis elegans*. Nature. 1998;391:806–11. https://doi.org/10.1038/35888.

Frattali AL, Treadway JL, Pessin JE. Insulin/IGF-1 hybrid receptors: implications for the dominant-negative phenotype in syndromes of insulin resistance. J Cell Biochem. 1992;48:43–50. https://doi.org/10.1002/jcb.240480108.

Friedel RH, Wurst W, Wefers B, Kühn R. Generating conditional knockout mice. Methods Mol Biol. 2011;693:205–31. https://doi.org/10.1007/978-1-60761-974-1_12.

Friedrich MJ, Rad L, Bronner IF, et al. Genome-wide transposon screening and quantitative insertion site sequencing for cancer gene discovery in mice. Nat Protoc. 2017;12:289–309. https://doi.org/10.1038/nprot.2016.164.

Frye SV. The art of the chemical probe. Nat Chem Biol. 2010;6(3):159–61. https://doi.org/10.1038/nchembio.296.

Genovesio A, Giardini MA, Kwon Y-J, et al. Visual genome-wide RNAi screening to identify human host factors required for *Trypanosoma cruzi* infection. PLoS One. 2011;6:e19733. https://doi.org/10.1371/journal.pone.0019733.

Glover L, Alsford S, Baker N, et al. Genome-scale RNAi screens for high-throughput phenotyping in bloodstream-form African trypanosomes. Nat Protoc. 2015;10:106–33. https://doi.org/10.1038/nprot.2015.005.

Hall B, Limaye A, Kulkarni AB. Overview: generation of gene knockout mice. Curr Protoc Cell Biol. Chapter 19:Unit 19.12 19.12.1-17. 2009; https://doi.org/10.1002/0471143030.cb1912s44.

Hannon GJ. RNA interference. Nature. 2002;418:244–51. https://doi.org/10.1038/418244a.

Hao T, Wang Q, Zhao L, et al. Analyzing of molecular networks for human diseases and drug discovery. Curr Top Med Chem. 2018;18:1007–14. https://doi.org/10.2174/1568026618666180813143408.

Havlicek S, Shen Y, Alpagu Y, et al. Re-engineered RNA-guided FokI-nucleases for improved genome editing in human cells. Mol Ther. 2017;25:342–55. https://doi.org/10.1016/j.ymthe.2016.11.007.

Hayes F. Transposon-based strategies for microbial functional genomics and proteomics. Annu Rev Genet. 2003;37:3–29. https://doi.org/10.1146/annurev.genet.37.110801.142807.

Herskowitz I. Functional inactivation of genes by dominant negative mutations. Nature. 1987;329:219–22. https://doi.org/10.1038/329219a0.

Hrabé de Angelis MH, Flaswinkel H, Fuchs H, et al. Genome-wide, large-scale production of mutant mice by ENU mutagenesis. Nat Genet. 2000;25:444–7. https://doi.org/10.1038/78146.

Hsu PD, Lander ES, Zhang F. Development and applications of CRISPR-Cas9 for genome engineering. Cell. 2014;157:1262–78. https://doi.org/10.1016/j.cell.2014.05.010.

Hu Y, Comjean A, Rodiger J, et al. FlyRNAi.org—the database of the Drosophila RNAi screening center and transgenic RNAi project: 2021 update. Nucleic Acids Res. 2021;49:D908–15. https://doi.org/10.1093/nar/gkaa936.

Jackson AL, Bartz SR, Schelter J, et al. Expression profiling reveals off-target gene regulation by RNAi. Nat Biotechnol. 2003;21:635–7. https://doi.org/10.1038/nbt831.

Kashles O, Yarden Y, Fischer R, et al. A dominant negative mutation suppresses the function of normal epidermal growth factor receptors by heterodimerization. Mol Cell Biol. 1991;11:1454–63. https://doi.org/10.1128/mcb.11.3.1454-1463.1991.

Kauvar LM, Higgins DL, Villar HO, Sportsman JR, Engqvist-Goldstein A, Bukar R, Bauer KE, Dilley H, Rocke DM. Predicting ligand binding to proteins by affinity fingerprinting. Chem Biol. 1995;2(2):107–18. https://doi.org/10.1016/1074-5521(95)90283-x.

Kleinstiver BP, Pattanayak V, Prew MS, et al. High-fidelity CRISPR-Cas9 nucleases with no detectable genome-wide off-target effects. Nature. 2016;529:490–5. https://doi.org/10.1038/nature16526.

Kolev NG, Tschudi C, Ullu E. RNA Interference in protozoan parasites: achievements and challenges. Eukaryot Cell. 2011;10:1156–63. https://doi.org/10.1128/EC.05114-11.

Krejsa CM, Horvath D, Rogalski SL, Penzotti JE, Mao B, Barbosa F, Migeon JC. Predicting ADME properties and side effects: the BioPrint approach. Curr Opin Drug Discov Devel. 2003;6(4):470–80.

Lagna G, Hemmati-Brivanlou A. 4 Use of dominant negative constructs to modulate gene expression. Curr Top Dev Biol. 1997;36:75–98. Elsevier

Lander ES, Linton LM, Birren B, et al. Initial sequencing and analysis of the human genome. Nature. 2001;409:860–921. https://doi.org/10.1038/35057062.

Lander N, Ulrich PN, Docampo R. *Trypanosoma brucei* vacuolar transporter chaperone 4 (TbVtc4) is an acidocalcisome polyphosphate kinase required for in vivo infection. J Biol Chem. 2013;288:34205–16. https://doi.org/10.1074/jbc.M113.518993.

Largaespada DA. Transposon mutagenesis in mice. Methods Mol Biol. 2009;530:379–90. https://doi.org/10.1007/978-1-59745-471-1_20.

Lavery KS, King TH. Antisense and RNAi: powerful tools in drug target discovery and validation. Curr Opin Drug Discov Devel. 2003;6:561–9.

Lee CN, Heidbrink JL, McKinnon K, et al. RNA interference characterization of proteins discovered by proteomic analysis of pancreatic cancer reveals function in cell growth and survival. Pancreas. 2012;41:84–94. https://doi.org/10.1097/MPA.0b013e3182236385.

Lindsay MA. Target discovery. Nat Rev Drug Discov. 2003;2:831–8. https://doi.org/10.1038/nrd1202.

Liu KI, Ramli MN, Woo CW, et al. A chemical-inducible CRISPR-Cas9 system for rapid control of genome editing. Nat Chem Biol. 2016;12(11):980–7. https://doi.org/10.1038/nchembio.2179.

Lokey RS. Forward chemical genetics: progress and obstacles on the path to a new pharmacopoeia. Curr Opin Chem Biol. 2003;7:91–6. https://doi.org/10.1016/S1367-5931(02)00002-9.

Lopes C, Aubert S, Bourgois-Rocha F, et al. Dominant-negative effects of adult-onset huntingtin mutations alter the division of human embryonic stem cells-derived neural cells. PLoS One. 2016;11:e0148680. https://doi.org/10.1371/journal.pone.0148680.

Lorens JB, Sousa C, Bennett MK, et al. The use of retroviruses as pharmaceutical tools for target discovery and validation in the field of functional genomics. Curr Opin Biotechnol. 2001;12:613–21. https://doi.org/10.1016/s0958-1669(01)00269-5.

Mali P, Yang L, Esvelt KM, et al. RNA-guided human genome engineering via Cas9. Science. 2013;339:823–6. https://doi.org/10.1126/science.1232033.

Manta B, Pavan C, Sturlese M, et al. Iron-sulfur cluster binding by mitochondrial monothiol glutaredoxin-1 of *Trypanosoma brucei*: molecular basis of iron-sulfur cluster coordination and relevance for parasite infectivity. Antioxid Redox Signal. 2013;19:665–82. https://doi.org/10.1089/ars.2012.4859.

Meissner KA, Lunev S, Wang Y-Z, et al. Drug target validation methods in malaria—protein interference assay (PIA) as a tool for highly specific drug target validation. Curr Drug Targets. 2017;18:1069–85. https://doi.org/10.2174/1389450117666160201115003.

Metzger D, Chambon P. Site- and time-specific gene targeting in the mouse. Methods. 2001;24:71–80. https://doi.org/10.1006/meth.2001.1159.

Miano JM, Zhu QM, Lowenstein CJ. A CRISPR path to engineering new genetic mouse models for cardiovascular research. Arterioscler Thromb Vasc Biol. 2016;36:1058–75. https://doi.org/10.1161/ATVBAHA.116.304790.

Miura H, Quadros RM, Gurumurthy CB, Ohtsuka M. Easi-CRISPR for creating knock-in and conditional knockout mouse models using long ssDNA donors. Nat Protoc. 2018;13:195–215. https://doi.org/10.1038/nprot.2017.153.

Miyoshi C, Kim SJ, Ezaki T, et al. Methodology and theoretical basis of forward genetic screening for sleep/wakefulness in mice. Proc Natl Acad Sci U S A. 2019;116:16062–7. https://doi.org/10.1073/pnas.1906774116.

Moresco EMY, Li X, Beutler B. Going forward with genetics. Am J Pathol. 2013;182:1462–73. https://doi.org/10.1016/j.ajpath.2013.02.002.

Morrow JK, Tian L, Zhang S. Molecular networks in drug discovery. Crit Rev Biomed Eng. 2010;38:143–56. https://doi.org/10.1615/critrevbiomedeng.v38.i2.30.

Moser LA, Pollard AM, Knoll LJ. A genome-wide siRNA screen to identify host factors necessary for growth of the parasite *Toxoplasma gondii*. PLoS One. 2013;8:e68129. https://doi.org/10.1371/journal.pone.0068129.

Mueller A-K, Hammerschmidt-Kamper C, Kaiser A. RNAi in *Plasmodium* CPD. Curr Pharm Des. 2014;20:278–83. https://doi.org/10.2174/13816128113199990027.

Munroe R, Schimenti J. Mutagenesis of mouse embryonic stem cells with ethylmethanesulfonate. Methods Mol Biol. 2009;530:131–8. https://doi.org/10.1007/978-1-59745-471-1_7.

Munroe RJ, Ackerman SL, Schimenti JC. Genome wide two-generation screens for recessive mutations by ES cell mutagenesis. Mamm Genome. 2004;15:960–5. https://doi.org/10.1007/s00335-004-2406-0.

Napoli C, Lemieux C, Jorgensen R. Introduction of a Chimeric Chalcone synthase gene into petunia results in reversible co-suppression of homologous genes in trans. Plant Cell. 1990;2:279–89. https://doi.org/10.1105/tpc.2.4.279.

Nishiyama K, Baba S, Yamada T, et al. Embryonic lethal effect of expressing a dominant negative mutant human thyroid hormone receptor alpha1 in mice. Endocr J. 2003;50:561–70. https://doi.org/10.1507/endocrj.50.561.

Nolan PM, Peters J, Strivens M, et al. A systematic, genome-wide, phenotype-driven mutagenesis programme for gene function studies in the mouse. Nat Genet. 2000;25:440–3. https://doi.org/10.1038/78140.

Ojelade SA, Rothenfluh A. Chapter 6: Alcohol and *Drosophila melanogaster*. In: Miller PM, editor. Biological Research on Addiction. San Diego: Academic Press; 2013. p. 51–9.
Paape D, Prendergast CT, Price HP, et al. Genetic validation of *Leishmania* genes essential for amastigote survival *in vivo* using N-myristoyltransferase as a model. Parasit Vectors. 2020;13:132. https://doi.org/10.1186/s13071-020-3999-1.
Patzke C, Südhof TC. The conditional KO approach: Cre/Lox technology in human neurons. Rare Dis. 2016;4:e1131884. https://doi.org/10.1080/21675511.2015.1131884.
Pini A, Bracci L. Phage display of antibody fragments. Curr Protein Pept Sci. 2000;1:155–69. https://doi.org/10.2174/1389203003381397.
Premsrirut PK, Dow LE, Kim SY, et al. A rapid and scalable system for studying gene function in mice using conditional RNA interference. Cell. 2011;145:145–58. https://doi.org/10.1016/j.cell.2011.03.012.
Quadros RM, Miura H, Harms DW, et al. Easi-CRISPR: a robust method for one-step generation of mice carrying conditional and insertion alleles using long ssDNA donors and CRISPR ribonucleoproteins. Genome Biol. 2017;18:92. https://doi.org/10.1186/s13059-017-1220-4.
Rao DD, Vorhies JS, Senzer N, Nemunaitis J. siRNA vs. shRNA: similarities and differences. Adv Drug Deliv Rev. 2009;61:746–59. https://doi.org/10.1016/j.addr.2009.04.004.
Raphael AR, Talbot WS. Chapter 1: New insights into signaling during myelination in *Zebrafish*. In: Birchmeier C, editor. Current topics in developmental biology. Academic Press; 2011. p. 1–19.
Rautela J, Surgenor E, Huntington ND. Drug target validation in primary human natural killer cells using CRISPR RNP. J Leukoc Biol. 2020;108:1397–408. https://doi.org/10.1002/JLB.2MA0620-074R.
Reznikoff WS, Winterberg KM. Transposon-based strategies for the identification of essential bacterial genes. Methods Mol Biol. 2008;416:13–26. https://doi.org/10.1007/978-1-59745-321-9_2.
Rognan D. Chemogenomic approaches to rational drug design. Br J Pharmacol. 2007;152(1):38–52. https://doi.org/10.1038/sj.bjp.0707307.
Romano N, Macino G. Quelling: transient inactivation of gene expression in *Neurospora crassa* by transformation with homologous sequences. Mol Microbiol. 1992;6:3343–53. https://doi.org/10.1111/j.1365-2958.1992.tb02202.x.
Sandhöfer N, Bauer J, Reiter K, et al. The new and recurrent FLT3 juxtamembrane deletion mutation shows a dominant negative effect on the wild-type FLT3 receptor. Sci Rep. 2016;6:28032. https://doi.org/10.1038/srep28032.
Sauer B, Henderson N. Cre-stimulated recombination at loxP-containing DNA sequences placed into the mammalian genome. Nucleic Acids Res. 1989;17:147–61. https://doi.org/10.1093/nar/17.1.147.
Schneeberger K. Using next-generation sequencing to isolate mutant genes from forward genetic screens. Nat Rev Genet. 2014;15:662–76. https://doi.org/10.1038/nrg3745.
Schwarz D, Gestwicki JE. Revisiting the "Art of the Chemical Probe". ACS Chem Biol. 2018;13(5):1109–10. https://doi.org/10.1021/acschembio.8b00390.
Sen GL, Blau HM. A brief history of RNAi: the silence of the genes. FASEB J. 2006;20:1293–9. https://doi.org/10.1096/fj.06-6014rev.
Setten RL, Rossi JJ, Han S-P. The current state and future directions of RNAi-based therapeutics. Nat Rev Drug Discov. 2019;18:421–46. https://doi.org/10.1038/s41573-019-0017-4.
Sheppard D. Dominant negative mutants: tools for the study of protein function *in vitro and in vivo*. Am J Respir Cell Mol Biol. 1994;11:1–6. https://doi.org/10.1165/ajrcmb.11.1.8018332.
Sifuentes-Romero I, Milton SL, García-Gasca A. Post-transcriptional gene silencing by RNA interference in non-mammalian vertebrate systems: where do we stand? Mutat Res. 2011;728:158–71. https://doi.org/10.1016/j.mrrev.2011.09.001.
Slaymaker IM, Gao L, Zetsche B, et al. Rationally engineered Cas9 nucleases with improved specificity. Science. 2016;351:84–8. https://doi.org/10.1126/science.aad5227.
Sledz CA, Williams BRG. RNA interference in biology and disease. Blood. 2005;106:787–94. https://doi.org/10.1182/blood-2004-12-4643.
Sledz CA, Holko M, de Veer MJ, et al. Activation of the interferon system by short-interfering RNAs. Nat Cell Biol. 2003;5:834–9. https://doi.org/10.1038/ncb1038.

Slonim DK. From patterns to pathways: gene expression data analysis comes of age. Nat Genet. 2002;32(4):502–8. https://doi.org/10.1038/ng1033.

Stewart SA, Dykxhoorn DM, Palliser D, et al. Lentivirus-delivered stable gene silencing by RNAi in primary cells. RNA. 2003;9:493–501. https://doi.org/10.1261/rna.2192803.

Stortz JA, Serafim TD, Alsford S, et al. Genome-wide and protein kinase-focused RNAi screens reveal conserved and novel damage response pathways in *Trypanosoma brucei*. PLoS Pathog. 2017;13:e1006477. https://doi.org/10.1371/journal.ppat.1006477.

Suster ML, Sumiyama K, Kawakami K. Transposon-mediated BAC transgenesis in zebrafish and mice. BMC Genomics. 2009;10:477. https://doi.org/10.1186/1471-2164-10-477.

Sweis RF. Target (In)validation: a critical, sometimes unheralded, role of modern medicinal chemistry. ACS Med Chem Lett. 2015;6(6):618–21. https://doi.org/10.1021/acsmedchemlett.5b00183.

Takada Y, Ylänne J, Mandelman D, et al. A point mutation of integrin beta 1 subunit blocks binding of alpha 5 beta 1 to fibronectin and invasin but not recruitment to adhesion plaques. J Cell Biol. 1992;119:913–21. https://doi.org/10.1083/jcb.119.4.913.

Takahashi JS, Pinto LH, Vitaterna MH. Forward and reverse genetic approaches to behavior in the mouse. Science. 1994;264:1724–33.

Vieth M, Higgs RE, Robertson DH, Shapiro M, Gragg EA, Hemmerle H. Kinomics-structural biology and chemogenomics of kinase inhibitors and targets. Biochim Biophys Acta. 2004;1697(1–2):243–57. https://doi.org/10.1016/j.bbapap.2003.11.028.

Wallqvist A, Rabow AA, Shoemaker RH, Sausville EA, Covell DG. Establishing connections between microarray expression data and chemotherapeutic cancer pharmacology. Mol Cancer Ther. 2002;1(5):311–20.

Wang J, Abate-Shen C. Analyses of tumor-suppressor genes in germline mouse models of cancer. Cold Spring Harb Protoc. 2014;2014:807–12. https://doi.org/10.1101/pdb.top069773.

Ward SJ. Impact of genomics in drug discovery. Biotechniques. 2001;31(3):626–34. https://doi.org/10.2144/01313dd01.

Wessel J, Chu AY, Willems SM, et al. Low-frequency and rare exome chip variants associate with fasting glucose and type 2 diabetes susceptibility. Nat Commun. 2015;6:5897. https://doi.org/10.1038/ncomms6897.

Wilson RC, Doudna JA. Molecular mechanisms of RNA interference. Annu Rev Biophys. 2013;42:217–39. https://doi.org/10.1146/annurev-biophys-083012-130404.

Workman P, Collins I. Probing the probes: fitness factors for small molecule tools. Chem Biol. 2010;17(6):561–77. https://doi.org/10.1016/j.chembiol.2010.05.013.

Wren JD, Bekeredjian R, Stewart JA, et al. Knowledge discovery by automated identification and ranking of implicit relationships. Bioinformatics. 2004;20:389–98. https://doi.org/10.1093/bioinformatics/btg421.

Xue W, Zender L, Miething C, et al. Senescence and tumour clearance is triggered by p53 restoration in murine liver carcinomas. Nature. 2007;445:656–60. https://doi.org/10.1038/nature05529.

Yang H, Wang H, Shivalila CS, et al. One-step generation of mice carrying reporter and conditional alleles by CRISPR/Cas-mediated genome engineering. Cell. 2013;154:1370–9. https://doi.org/10.1016/j.cell.2013.08.022.

Yen PM, Sugawara A, Refetoff S, Chin WW. New insights on the mechanism(s) of the dominant negative effect of mutant thyroid hormone receptor in generalized resistance to thyroid hormone. J Clin Invest. 1992;90:1825–31. https://doi.org/10.1172/JCI116058.

Yu J-Y, DeRuiter SL, Turner DL. RNA interference by expression of short-interfering RNAs and hairpin RNAs in mammalian cells. Proc Natl Acad Sci U S A. 2002;99:6047–52. https://doi.org/10.1073/pnas.092143499.

Zhang MQ. Extracting functional information from microarrays: a challenge for functional genomics. Proc Natl Acad Sci U S A. 2002;99:12509–11. https://doi.org/10.1073/pnas.212532499.

Zhang Y, Riesterer C, Ayrall AM, et al. Inducible site-directed recombination in mouse embryonic stem cells. Nucleic Acids Res. 1996;24:543–8. https://doi.org/10.1093/nar/24.4.543.

Zhang H, Kolb FA, Jaskiewicz L, et al. Single processing center models for human Dicer and bacterial RNase III. Cell. 2004;118:57–68. https://doi.org/10.1016/j.cell.2004.06.017.

Structure-Based Binding Pocket Detection and Druggability Assessment

Santiago Rodríguez, Juan I. Alice, Carolina L. Bellera, and Alan Talevi

Contents

Abstract

Target-focused (often described as "rational") drug discovery starts with the identification and validation of one or more target biomolecules whose modulation will interfere with disease progression, or at least alleviate disease symptoms. In addition to verifying the association between such biomolecules and the disease, target validation often involves the assessment of some relevant properties, *druggability* among them.

Druggability is often defined as the ability of a target to bind a small, drug-like molecule with high affinity and specificity. If a small molecule has already been reported to bind to a target in a specific manner, the target is traditionally regarded as druggable (recently, thus, it has been proposed that different targets might present different degrees of druggability, with some of them being "easy

S. Rodríguez · J. I. Alice · C. L. Bellera · A. Talevi (✉)
Laboratory of Bioactive Compounds Research and Development (LIDeB), Department of Biological Sciences, Faculty of Exact Sciences, University of La Plata (UNLP), La Plata, Buenos Aires, Argentina

Argentinean National Council of Scientific and Technical Research (CONICET)—CCT La Plata, La Plata, Buenos Aires, Argentina
e-mail: atalevi@biol.unlp.edu.ar

M. T. Scotti, C. L. Bellera (eds.), *Drug Target Selection and Validation*, Computer-Aided Drug Discovery and Design,
https://doi.org/10.1007/978-3-030-95895-4_5

to drug" and some of them being "difficult to drug"). Druggability has also been implied based on evolutionary relationships (using a sort of transitivity or "guilt by association" principle), or by predicting druggable pockets from different levels of structural information, without comparison with any template. The latter approximations, which will invariably depart from the detection of binding pockets and later assign a score to differentiate druggable from non-druggable proteins, are the scope of the present chapter.

Importantly, current methods to assess target druggability from structural data might be challenged by the occurrence of hidden or cryptic binding sites, which could more appropriately be assessed by studying the target dynamically, rather than from static representations of the biomolecule of interest.

Keywords
Druggability · Target druggability · Druggability prediction · Drug target · Target-focused drug discovery · Binding site · Binding site identification · Cryptic binding site · Hidden binding site · Drug discovery · Target validation · Functional assessment · Binding site detection · Function modeling · Proteins

1 Introduction

During the second half of the twentieth century, target-focused drug discovery became the dominant paradigm in the drug discovery field, although the extent of its validity and its productivity has recently been called into question (Swinney and Anthony 2011; Margineanu 2016). The target-centered approximation relies on the search of exquisitely selective drugs able to modulate a drug target (usually, a protein) that has been associated with disease progression or, at least, with disease symptomatology. As a hypothesis-driven approach, it requires the pre-definition of the targeted molecule.

The assessment of target validation may subtly vary across different domains of pharmacology. Whereas association of the target with the pathophysiology of disease is always a must, other properties depend on the therapeutic goal. For instance, essentiality and selectivity are relevant properties when developing anti-infective agents (Talevi et al. 2019). Low impact on the modulation of physiological conditions and differential expression across the body for specific targeting are, in contrast, desired when the target corresponds to the organism that is intended to be cured (Gashaw et al. 2011). High-throughput assayability, vulnerability, liability to drug resistance, and availability of an experimental structure of the target are perhaps not fundamental but desirable properties in a target (Talevi et al. 2019; Gashaw et al. 2011; Tonge 2018).

No matter the therapeutic goal, *druggability* is an essential property to take into consideration when assessing a potential target for a drug discovery project. It refers to the likelihood of being able to bind and to modulate a target with a small molecule (Owens 2007). In general, if a small molecule has already been reported to specifically bind with the target of interest, in a specific manner, such target is regarded to be druggable. This does not mean that absence of a known high-affinity ligand is necessarily equivalent to lack of druggability, as it is not feasible to confront the entire drug-like space with a given protein. Accordingly, druggability prediction tools are required to assess druggability. A "guilt by association" principle has often been used as evidence of draggability, as evolutionary relationships of the target of interest with known druggable targets have been used as hints of druggability (Owens 2007; Hajduk et al. 2005; Wang et al. 2013).

Analysis of protein–protein interaction (PPI) networks has also been proposed as a suitable approach to identify druggable proteins from system-level properties: some studies suggest that, similar to essential and disease proteins, druggable proteins preferentially occupy specific regions within PPI networks (Yildirim et al. 2007; Yao and Rzhetsky 2008; Dezső and Ceccarelli 2020). For instance, it was observed that the average connectivity of successful drug targets is significantly higher than the average connectivity within the network but still relatively small compared to the maximum connectivity observed in the network (Yao and Rzhetsky 2008; Dezső and Ceccarelli 2020) (Fig. 1). Only drug targets associated with severe, life-threatening conditions (for which the cost-benefit balance is altered) usually afford a higher connectivity, which explains the usually severe side-effects of, let us say, treatments for oncologic or autoimmune disorders. In general, successful drug targets tend to act as bridge nodes between two or more clusters of closely interacting molecules. Another system-level property that could be used as a predictor of druggability is the ratio of the number of nonsynonymous to synonymous single-nucleotide polymorphisms (SNPs): successful drug targets have a considerably smaller ratio than human genes on average (Yao and Rzhetsky 2008). System-level properties can be coupled with machine

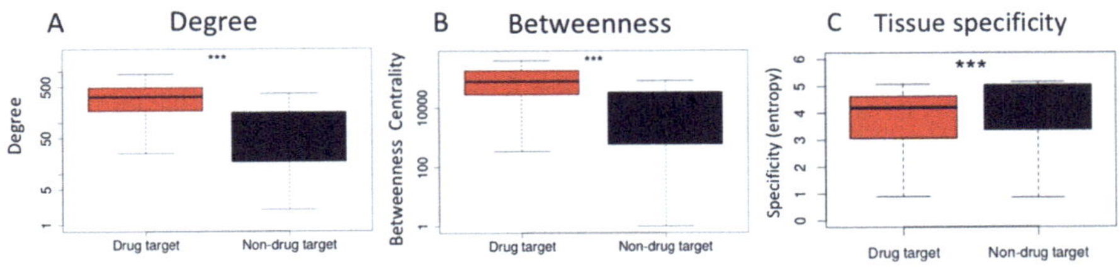

Fig. 1 Some system-level features seemingly associated with druggability. Topological analysis of the PPI network was realized from data downloaded from the STRING database. Tissue specificity was assessed from RNAseq tissue expression data downloaded from the Genotype-Tissue Expression and the Human Protein Atlas and characterized by Shannon's entropy (the higher the entropy, the wider the distribution of the protein). The red and black bars correspond to drug and non-drug targets, in that order, and $p < 0.001$ is indicated as ***. It can be appreciated that, on average, drug targets seem to be more highly connected (higher degree) than non-drug targets, and that the betweenness centrality of drug targets surpasses that of non-targets. Drug targets also have narrower distribution (i.e., tissue-specificity) in comparison with non-targets

learning approximations to differentiate druggable from undruggable (or scarcely druggable) targets (Kandoi et al. 2015).

Alternatively, druggable pockets can be identified in a protein from structural information (from primary to tertiary structure), which constitutes the scope of this chapter; such approximation can also be combined with machine learning tools to provide multi-variate algorithms for druggability assessment.

2 Binding Pocket Prediction

Structure-based druggability assessment typically involves a combination of automated binding pocket detection tools, and an algorithm to quantify the druggability of the so-identified putative ligand-binding sites (providing a score for each detected pocket). A binding pocket is usually defined as a concave region, cleft or groove on a protein surface (Kawabata and Go 2007). Protein cavities have been classified in different manners by different authors resorting to (sometimes subtle) topological distinctions (Simões et al. 2017; Kawabata 2019). In some cases, the differences between different types of cavities are visually and conceptually obvious, though. For instance, the term *void* has been used to denote a buried, closed cavity, whereas *channel* refers to a cavity with double entry. Remarkably, binding pockets should be distinguished from "*hot spots,*" which are (sub) regions of the binding surface that contribute a disproportionate amount to the binding free energy (Brenke et al. 2009).

Depending on the software used, the input for binding site detection tools may be the protein sequence or the three-dimensional structure of the query protein. Moreover, some tools can only inspect monomers, whereas others can be used to assess oligomeric proteins.

Most sequence-based approaches (also sometimes called "evolutionary algorithms") rely on the analysis of residue conservation, by assuming that ligand-binding residues are functionally important and thus likely to be conserved through evolution (Yang et al. 2013a; Capra and Singh 2007). These methods can be advantageous in the sense that they make a prediction from sequence alone, but their accuracy tends to be low because non-binding residues can also be highly conserved due to the other roles, such as fold stabilization. Other tools use the sequence as input to model the 3D structure though three-dimensional structure prediction such as I-TASSER, and only then putative binding pockets are detected (Wu et al. 2018).

On the other hand, methods based on the three-dimensional structure of the protein of interest have roughly been classified into those that identify the pockets by recognizing the surface cavities on the 3D structural model of the target protein (without using any template), and those that infer binding pocket information from known template proteins with established binding sites and which have global and/or local structure similarity to the query (Yang et al. 2013a). According to large-scale benchmarking exercises, this type of template-based approach represents an accurate option when templates with close homology are found within ligand–protein complex structure databases (Schmidt et al. 2011; Gallo Cassarino et al. 2014). Interestingly, some recent template-based tools employ, instead of sequence alignment, matching algorithms that align similar microenvironments or physicochemical properties between sites or even entire proteins. This allows not only detecting similar sites between homologs but also between unrelated proteins (Liu et al. 2018).

After examination of the existing template-free, geometry-based approximations, Kawabata and Go classified the wide spectrum of pocket detection algorithms into three categories, namely *grid-based*, *sphere-based*, and *α-shape-based* (Kawabata and Go 2007). They also concluded that all the pocket-finding programs made their decisions based on two properties of the pocket: size and depth (which can be associated with pocket surface and volume). A binding pocket should have a space into which a binding ligand enters (the "entry" to the pocket should have a minimal size to allow ligand entrance); depth, on the other hand, denotes how deep a pocket is buried from the protein surface (which has been allegorically called "sea-level" (Laskowski et al. 1996): a large binding surface area is expected for buried pockets, which would provide the binding affinity to specific ligands, as the protein might enclose many ligand atoms using geometric and chemical complementarities (Kawabata and Go 2007). *Grid-based methods* cover the proteins in a three-dimensional grid, where empty grid points are regarded as pockets if they satisfy some geometric or energetic conditions. In *sphere-based methods*, pockets are estimated using spherical probes; these are subsequently clustered or analyzed to yield cavities assumed to be binding sites. The *α-shape-based method*, an alpha shape is defined as a subset of Delaunay tessellations of protein atoms, disregarding edges longer than the sum of the radii of two atoms; a pocket is then defined as an empty Delaunay tetrahedron.

More recently, other categories of binding pocket predictive tools have been considered besides template- and geometry-based methods (Clark et al. 2020). For instance, *energy-based methods* use probe molecules or chemical moieties to locate regions of the protein where intermolecular interactions such as hydrogen bonding or pi-stacking are likely to be formed. *Machine-learning methods*, which are the key to many recent developments, incorporate previously established physicochemical predictors into a machine-learning context, training prediction algorithms by using methods such as ensemble learning or deep learning. Other approximations rely on consensus of previously developed tools. As it happens with druggability assessment tools (see next section), the accessibility to pocket detection methods varies depending on the program, from tools available through web servers or freely downloadable open-source applications, to tools included within commercial packages.

Table 1 classifies a non-exhaustive list of binding pocket prediction tools, depending on their underlying principle. Figure 2 shows the putative binding pockets predicted by different binding pocket detection/druggability prediction tools for HIV-1 protease and trypanothione synthetase.

Several databases dedicated to the collection of binding data of known protein–ligand complexes have been developed, which could be useful for training, validation and/or comparison of binding pocket detection tools; noteworthy, among them,

Table 1 An arbitrary, non-exhaustive selection of binding pocket detection tools with different underlying principles

Class	Software package or tool	Reference
Template-based	3DLigandSite	Wass et al. (2010)
	FINDSITE	Brylinski and Skolnick (2008)
	firestar	López et al. (2007)
	I-TASSER Suite	Yang et al. (2015)
	IntFOLD	Roche et al. (2011)
	ProBis	Konc and Janezic (2010)
Geometry-based	CAVITATOR	Gao and Skolnick (2013)
	CASTp	Binkowski et al. (2003)
	Depth	Tan et al. (2013)
	LIGSITE	Hendlich et al. (1997)
	Fpocket	Le Guilloux et al. (2009)
	Ghecom	Kawabata (2010)
	SURFNET	Laskowski (1995)
Energy-based	AutoSite	Ravindranath and Sanner (2016)
	FTSite	Ngan et al. (2012)
	PocketFinder	An et al. (2005)
	Q-SiteFinder	Laurie and Jackson (2005)
	SITEHOUND	Hernández et al. (2009)
Others (machine learning-based, consensus approaches)	COACH-D	Wu et al. (2018)
	DeepCSeqSite	Cui et al. (2019)
	DeepSite	Jiménez et al. (2017)
	Kalasanty	Stepniewska-Dziubinska et al. (2020)
	LigandRFs	Chen et al. (2014)
	LIGSITEcsc	Huang and Schroeder (2006)

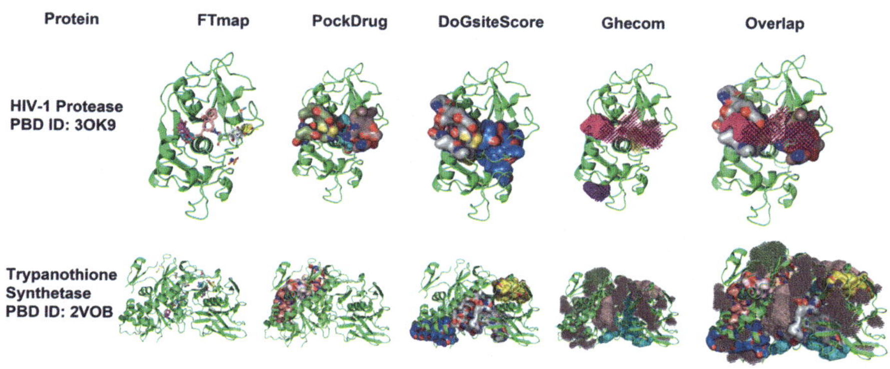

Fig. 2 Binding pockets predicted by different binding pocket detection/druggability assessment tools for HIV protease (up) and trypanothione synthetase (down). The overlap of all putative pockets is shown on the right, revealing high consensus in the predictions for HIV protease and low consensus in the case of trypanothione synthetase

we can mention MPStruc (http://blanco.biomol.uci.edu/mpstruc/), PDBbind (Liu et al. 2015), Relibase (Hendlich et al. 2003), LigASite (Dessailly et al. 2008), and sc-PDB (Kellenberger et al. 2006). Dias et al. (Dias et al. 2019) recently introduced CavBench, a benchmarking software to compare binding pocket detection tools using established, ground-truth datasets of binding sites and statistical classification metrics (e.g., recall, precision, etc.). CavBench is XML-extensible so that new cavity/binding-site datasets, and new cavity detection methods can be accommodated. It also includes a ground-truth dataset of cavities from 660 apo proteins and 1633 holo proteins.

Exhaustively reviewing the long and ever-growing list of existing binding site detection tools is beyond the scope of this chapter; however, the next subsections include a representative selection of some of the available tools. At least one program from each of the categories included in Table 1 has been overviewed.

(a) **Firestar**
Firestar (López et al. 2007, 2011) estimates the probability that a residue is involved in ligand binding based on pairwise sequence alignments between the query sequence and PSI-BLAST profiles pre-generated for FireDB (Lopez et al. 2007) consensus sequences. Briefly, FireDB is a databank of known functionally important residues, integrating residues involved in known ligand-binding pockets and catalytic sites. The data comes from close atomic contacts (closer than 1.0 Å) in ligand–protein complexes in Protein Data Bank crystal structures and catalytic residues reliably annotated in the Catalytic Site Atlas (Furnham et al. 2014). The sequences in FireDB have been clustered at 97% sequence identity, and the consensus sequence used for the PSI-Blast run has been built for each cluster. Firestar can be publicly accessed at http://firedb.bioinfo.cnio.es/Php/FireStar.php (last accessed August 2021). Its output consists of the predicted binding sites ranked by number of binding sites ordered by number of homologs found, a reliability score, a list of the source alignments, and the list of residues that take part in the pocket. This is also highlighted in a visual summary.

(b) **I-TASSER Suite**
The I-TASSER Suite comprises a diversity of functionalities and is available both from an on-line server (https://zhanglab.ccmb.med.umich.edu/I-TASSER/, last accessed August 2021) and as a stand-alone package. Possibly best known by its protein structure modeling pipeline, which has been consistently selected as the best for protein structure prediction community-wide experiments, it also includes function modeling methods. The I-TASSER Suite pipeline involves four general steps: threading template identification, iterative structure assembly simulation, model selection and refinement, and structure-based function annotation (Yang et al. 2015). First, the query is threaded by LOMETS (a local meta-threading application containing eight fold-recognition programs) (Wu and Zhang 2007) through a nonredundant structure library, and structural templates are identified. The topology of the modeled protein is constructed by assembling aligned fragments excised from different threading-aligned

templates and the structure of unaligned regions is obtained by ab initio folding. The structure folding and reassembly is then obtained through Monte Carlo simulations under an optimized knowledge-based force field. The lowest free-energy conformations are identified by structure clustering and a new round of assembly simulation is performed to remove steric clashes and refine the protein topology. For function annotation, the structure models with the highest confidence are matched against a database of ligand–protein interactions, BioLiP5 (Yang et al. 2013b) to identify homologous function templates, from which ligand-binding site information is deducted. The suite includes three complementary algorithms for functional annotation: COFACTOR, TM-SITE, and S-SITE, which are integrated by COACH (see further information later in this chapter).

(c) **Fpocket**

Fpocket (Le Guilloux et al. 2009) is an open-source pocket detection software (https://github.com/Discngine/fpocket, last accessed August 2021) based on the theory of alpha spheres (Liang et al. 1998). An alpha sphere is a sphere that contains no internal atom and contacts four atoms on its boundary. Consequently, the four atoms are at an equal distance to the sphere center: this distance corresponds to the sphere radius. It follows that the sphere radii reflect the local curvature defined by the four atoms: if the atoms were in a plane, this would correspond to a sphere of infinite radius; contrariwise, four atoms at the apex of a tetrahedron would lead to a radius close to that of the van der Waals radius. In a protein, small spheres will be located within the protein, large spheres would be at the exterior, and cavities will correspond to intermediate spheres. It is possible to filter the ensemble of alpha spheres (according to some maximal and minimal radii values) to address pocket detection. Fpocket comprises three essential steps: (a) the whole ensemble of spheres is determined from the protein structure; (b) clusters of spheres close together are identified to detect pockets; (c) properties from the atoms in the pocket are analyzed to score each pocket; for such purpose, the spheres are classified as polar or apolar depending on the number of polar atoms with which they enter into contact.

(d) **PocketFinder and Q-siteFinder**

PocketFinder (An et al. 2005) detects binding pockets based on a grid potential map of van der Waals interactions of the protein. The map is used to find regions of consistently high van der Waals attraction. The first step to do so is the creation of the grid potential map using a probe atom (e.g., an aliphatic carbon) in orthogonal parallelepiped surrounding protein atoms. Then the potential map is smoothed, and putative ligand envelopes are built by contouring the resulting map: envelopes smaller than 100 $Å^3$ are subsequently disregarded.

The underlying principle to PocketFinder is somehow similar to that of Q-siteFinder (Laurie and Jackson 2005), where the interaction between the protein and a $-CH_3$ probe is quantified at a grid of 0.9 Å resolution enclosing the entire protein. Those probes with the most favorable binding energy are kept and close individual probe coordinates are subsequently clustered, computing the total interaction energies of probes within each cluster.

(e) **COACH-D**

COACH-D is an enhanced, recently reported version of the COACH algorithm (Yang et al. 2013a). Five independent methods are used to predict the protein–ligand binding pockets in the query protein. Four of them correspond to template-based methods that predict the binding pockets by matching the query structure with ligand-binding templates from BioLiP. The fifth is a mixed structure-based method (ConCavity) that, for the binding pocket prediction, considers both the structure geometry and the sequence conservation and structure geometry for the binding sites. The results from these individual approaches are combined with consensus predictions by the COACH algorithm. COACH-D is freely available through its webserver (https://yanglab.nankai.edu.cn/COACH-D/, last accessed August 2021).

(f) **Kalasanty**

Kalasanty (Stepniewska-Dziubinska et al. 2020) is a good example of the recent trend in the field, where state-of-the-art machine learning approaches are used to improve the accuracy of pocket detection (and druggability assessment) tools. In this case, a fully convolutional neural network is implemented, the kind of which is typically used for image processing and, in particular, in image segmentation. Image segmentation tries to locate an object, or many objects, within an image. The output of such models is a set of scores assigned to each pixel; the score relates to the probability that a given pixel belongs to the sought object. In Kalasanty, the input is a grid representation of the 3D structure of a protein, and the pursued object is the binding pocket. The model output is a probability density: each point in the space is given a probability of being part of a pocket.

The model was trained and validated using the sc-PDB21 dataset (Desaphy et al. 2015), which consists of known binding sites, accompanied with prepared protein structures (17,594 binding sites in total). The model correctly predicted the center location of 44.6% of the pockets in the test set, and only 5% of the correctly located pockets had wrongly predicted shape. The source code is publicly accessible at http://gitlab.com/cheminfIBB/kalasanty (last accessed August 2021).

3 Druggability Prediction

Druggability predictors are mostly structure-based approaches that rely on empirical structure–property relationships. To provide wide coverage of the protein space, a substantial dataset is required to train and validate the model, assuring generalizability. Typically, the model yields a score value for each pocket, which can be used to classify them as druggable or non-druggable; sometimes, a gray zone (borderline behavior) is admitted. As described in the case of pocket detection tools, druggability prediction has recently benefited from the application of state-of-the-art machine learning methods, as well as consensus scoring functions. Below the reader may find a far-from-exhaustive list of some relevant druggability assessment tools.

(a) **Dscore**
Dscore is the druggability scoring function in Schrödinger's commercial software for locating and characterizing binding sites, SiteMap (Halgren 2009). SiteMap site-finding algorithm places a 1-Å grid around the entire protein (or, optionally, within a box centered on a co-crystallized or docked ligand). Each grid point is classified as being "inside" or "outside" the protein by comparing the distance to neighboring protein atoms to the van der Waals radius of such atoms. The outside points are then examined to define which are in good van der Waals contact with the protein and reasonably well enclosed by it. Candidate site points are grouped together if they lie within a given threshold distance (by default, 1.76 Å), and site-point groups with a gap between them that occurs in an exposed region and is less than or equal to a specified distance (by default, 6.5 Å) are merged and constitute a predicted site. A scoring function called SiteScore is used to quantitatively assess and compare the predicted binding sites. It is a weighted sum of three properties: the square root of the number of site points, the enclosure score, and the hydrophilic score. The same features, weighted differently, are used to assess druggability through Dscore. For an extensive set of 538 proteins taken from the PDBbind database (Wang et al. 2005), SiteScore ranks the co-crystallized binding site as the top scoring predicted site in 98.5 and 87.5% of the cases for those co-crystallized ligand-protein complexes of subnM or nM to subμM affinity, respectively.

(b) **DrugPred**
According to DrugPred developers (Krasowski et al. 2011), a binding site was defined to be druggable if it can bind small orally available, non-prodrug drug-like ligands noncovalently. All dockable approved drugs in DrugBank at that moment were docked into a pocket to establish its boundaries. All the ligands for which a docking pose was obtained and for which the ratio of van der Waals score to the number of heavy atoms was below or equal −1.2 were merged into a "superligand," which was considered as a negative print of the binding site. A number of descriptors of the binding site were computed, including volume, surface area, compactness, and amino acid composition (fraction of hydrophobic, charged, polar, and "multifunctional" amino acids). A discriminant function to differentiate druggable from non-druggable proteins was then derived using partial least squares. A non-redundant set of druggable and less-druggable binding sites (the NRDLD), was compiled ad hoc to infer and validate the druggability predictor. The model achieved an accuracy of approximately 90% in such dataset, and approximately 80% for other datasets previously compiled by other authors.

(c) **DoGSiteScorer**
DoGSiteScorer (Volkamer et al. 2012) initially predicts potential pockets on the protein surface. For this purpose, a grid is spanned around the protein and the grid points are flagged depending on their spatial overlap with any protein

atom. Later, a difference of Gaussian (DoG) filter is applied, identifying where the location of a sphere-like object is favorable. These positions are clustered to potential subpockets; neighboring subpockets are ultimately merged to pockets. Once the pockets are identified, several geometric and physico-chemical properties are computed, including pocket volume, pocket surface, pocket depth, pocket shape, pocket enclosure, pocket atom counts, functional group and amino acid compositions, lipophilic surface, and overall hydrophobicity. If a ligand is provided with the protein, the overlap between ligand and pocket volume is also computed. A Support Vector Machine model based on a subset of such descriptors then assigns a druggability score between 0 and 1: the higher the score, the more druggable the pocket is predicted to be. The model was trained and validated using the nonredundant druggable dataset (NRDD) (Schmidtke and Barril 2010), with cross-validation showing a mean accuracy of 90%. It is available through webserver at https://bio.tools/dogsitescorer (last accessed August 2021).

(d) **PockDrug**

PockDrug druggability model (Hussein et al. 2015) was built and validated using the NRDLD set, whereas Druggable Cavity Directory database (http://fpocket.sourceforge.net/dcd/, last accessed August 2021) was used to collect a second test set for further external validation. PockDrug is a consensus druggability model emerging from a combination of linear discriminant models based on a selection of 52 physicochemical and geometrical pocket descriptors by cross-validation using part of the NRDLD set estimated by fpocket as the training set. From the so-obtained models, the seven most robust and efficient ones were chosen using the NRDLD independent test set estimated using four different pocket detection methods (among them fpocket and DoGSite). The consensus druggability model yields a druggability probability that corresponds to the mean of the probabilities of such seven best models, plus their associated standard deviation. PockDrug assigns an estimated average druggability probability of 0.87 ± 0.15 for druggable pockets, in contrast with an average of 0.18 ± 0.15 for less druggable pockets (using the NRDLD set and the pockets estimated from it by the four estimation methods). It is freely available online at http://pockdrug.rpbs.univ-paris-diderot.fr/cgi-bin/index.py?page=home (last accessed August 2021).

4 Additional Challenges: Hidden/Cryptic Binding Sites

Cryptic binding sites are not apparent in proteins (and thus, non-visible crystallographically) if not in the presence of ligands that upon the binding event induce remarkable protein rearrangement (Clark et al. 2020; Kuzmanic et al. 2020). They are thus intrinsically related to the notions of induced-fit binding and conformational selection binding (Paul and Weikl 2016). Whereas these sites can be drugged and might be key to modulate apparently undruggable targets, they are understandably more difficult to identify by pocket detection algorithms, especially if using their apo structures. Molecular dynamics, i.e., computational methods based on atomistic molecular simulations, possibly represent the best approach to identify and characterize cryptic binding sites (and to eventually sample conformational states to feed druggability prediction tools) (Vajda et al. 2018; Zheng 2021). For instance, CrypticScout is a recently developed tool to reveal cryptic binding sites by means of mixed-solvent molecular dynamics simulations using benzene as a hydrophobic probe to detect cryptic pockets (Martinez-Rosell et al. 2020). Other specialized computational tools that have been developed combine molecular dynamics with fragment docking and machine learning approaches (Oleinikovas et al. 2016; Cimermancic et al. 2016).

Due to the existence of cryptic sites, the druggable proteome might be significantly larger than previously believed, as it has been estimated that it could increase from ~40% to ~78% of disease-associated proteins (Cimermancic et al. 2016).

5 Final Remarks

The drug discovery community has realized, in the past two decades, that druggability assessment is a vital component of drug target selection and validation to undertake new "rational," target-focused drug discovery projects. This might be performed through very different approximations, from system-level data to guilt-by-association (template-based) strategies, and also from template-independent structure-based approximations, which have been the focus of this chapter. Among these, we might mention two sub-classes, geometry- and energy-based methods.

Druggablity assessment through structure-based approximations consists, basically, of two steps: first, putative binding pockets are identified in the protein; second, a druggability score is computed for each pocket based on empirical functions whose independent variable are pocket descriptors.

Lately, as in other fields of bioinformatics and cheminformatics, the performance of druggability scoring functions has been boosted by the use of state-of-the-art machine learning methods, such as deep learning and consensus scoring.

Refinement of druggability tools currently implies studying the putative drug targets in a dynamic manner, to reveal cryptic pockets. This trend will possibly expand the predicted druggable proteome considerably.

Acknowledgements SR and JIA are fellows of the Argentinean National Council of Scientific and Technical Research (CONICET); CLB and AT occupy permanent positions at CONICET. JIA, CLB, and AT occupy permanent positions at UNLP.

References

An J, Totrov M, Abagyan R. Pocketome via comprehensive identification and classification of ligand binding envelopes. Mol Cell Proteomics. 2005;4:752–61.

Binkowski TA, Naghibzadeh S, Liang J. CASTp: computed atlas of surface topography of proteins. Nucleic Acids Res. 2003;31:3352–5.

Brenke R, Kozakov D, Chuang GY, Beglov D, Hall D, Landon MR, et al. Fragment-based identification of druggable 'hot spots' of proteins using Fourier domain correlation techniques. Bioinformatics. 2009;25:621–7.
Brylinski M, Skolnick J. A threading-based method (FINDSITE) for ligand-binding site prediction and functional annotation. Proc Natl Acad Sci U S A. 2008;105:129–34.
Capra JA, Singh M. Predicting functionally important residues from sequence conservation. Bioinformatics. 2007;23:1875–82.
Chen P, Huang JZ, Gao X. LigandRFs: random forest ensemble to identify ligand-binding residues from sequence information alone. BMC Bioinformatics. 2014;15:S4.
Cimermancic P, Weinkam P, Rettenmaier TJ, Bichmann L, Keedy DA, Woldeyes RA, et al. CryptoSite: expanding the druggable proteome by characterization and prediction of cryptic binding sites. J Mol Biol. 2016;428:709–19.
Clark JJ, Orban ZJ, Carlson HA. Predicting binding sites from unbound versus bound protein structures. Sci Rep. 2020;10:15856.
Cui Y, Dong Q, Hong D, Wang X. Predicting protein-ligand binding residues with deep convolutional neural networks. BMC Bioinformatics. 2019;20:93.
Desaphy J, Bret G, Rognan D, Kellenberger E. sc-PDB: a 3D-database of ligandable binding sites--10 years on. Nucleic Acids Res. 2015;43:D399–404.
Dessailly BH, Lensink MF, Orengo CA, Wodak SJ. LigASite--a database of biologically relevant binding sites in proteins with known apo-structures. Nucleic Acids Res. 2008;36:D667–73.
Dezső Z, Ceccarelli M. Machine learning prediction of oncology drug targets based on protein and network properties. BMC Bioinformatics. 2020;21:104.
Dias S, Simões T, Fernandes F, Martins AM, Ferreira A, Jorge J, et al. CavBench: a benchmark for protein cavity detection methods. PLoS One. 2019;14:e0223596.
Furnham N, Holliday GL, de Beer TA, Jacobsen JO, Pearson WR, Thornton JM. The catalytic site Atlas 2.0: cataloging catalytic sites and residues identified in enzymes. Nucleic Acids Res. 2014;42:D485–9.
Gallo Cassarino T, Bordoli L, Schwede T. Assessment of ligand binding site predictions in CASP10. Proteins. 2014;82(Suppl. 2):154–63.
Gao M, Skolnick J. APoc: large-scale identification of similar protein pockets. Bioinformatics. 2013;29:597–604.
Gashaw I, Ellinghaus P, Sommer A, Asadullah K. What makes a good drug target? Drug Discov Today. 2011;16:1037–43.
Hajduk PJ, Huth JR, Tse C. Predicting protein druggability. Drug Discov Today. 2005;10:1675–82.
Halgren TA. Identifying and characterizing binding sites and assessing druggability. J Chem Inf Model. 2009;49:377–89.
Hendlich M, Rippmann F, Barnickel G. LIGSITE: automatic and efficient detection of potential small molecule-binding sites in proteins. J Mol Graph Model. 1997;15:359–63.
Hendlich M, Bergner A, Günther J, Klebe G. Relibase: design and development of a database for comprehensive analysis of protein-ligand interactions. J Mol Biol. 2003;326:607–20.
Hernández M, Ghersi D, Sanchez R. SITEHOUND-web: a server for ligand binding site identification in protein structures. Nucleic Acids Res. 2009;37:W413–6.
Huang B, Schroeder M. LIGSITEcsc: predicting ligand binding sites using the Connolly surface and degree of conservation. BMC Struct Biol. 2006;6:19.
Hussein HA, Borrel A, Geneix C, Petitjean M, Regad L, Camproux AC. PockDrug-server: a new web server for predicting pocket druggability on holo and apo proteins. Nucleic Acids Res. 2015;43:W436–42.
Jiménez J, Doerr S, Martínez-Rosell G, Rose AS, De Fabritiis G. DeepSite: protein-binding site predictor using 3D-convolutional neural networks. Bioinformatics. 2017;33:3036–42.
Kandoi G, Acencio ML, Lemke N. Prediction of druggable proteins using machine learning and systems biology: a mini-review. Front Physiol. 2015;6:366.
Kawabata T. Detection of multiscale pockets on protein surfaces using mathematical morphology. Proteins. 2010;78:1195–211.

Kawabata T. Detection of cave pockets in large molecules: spaces into which internal probes can enter, but external probes from outside cannot. Biophys Physicobiol. 2019;16:391–406.
Kawabata T, Go N. Detection of pockets on protein surfaces using small and large probe spheres to find putative ligand binding sites. Proteins. 2007;68:516–29.
Kellenberger E, Muller P, Schalon C, Bret G, Foata N, Rognan D. sc-PDB: an annotated database of druggable binding sites from the Protein Data Bank. J Chem Inf Model. 2006;46:717–27.
Konc J, Janezic D. ProBiS algorithm for detection of structurally similar protein binding sites by local structural alignment. Bioinformatics. 2010;26:1160–8.
Krasowski A, Muthas D, Sarkar A, Schmitt S, Brenk R. DrugPred: a structure-based approach to predict protein druggability developed using an extensive nonredundant data set. J Chem Inf Model. 2011;51:2829–42.
Kuzmanic A, Bowman GR, Juarez-Jimenez J, Michel J, Gervasio FL. Investigating cryptic binding sites by molecular dynamics simulations. Acc Chem Res. 2020;53:654–61.
Laskowski RA. SURFNET: a program for visualizing molecular surfaces, cavities, and intermolecular interactions. J Mol Graph. 1995;13:323–30.
Laskowski RA, Luscombe NM, Swindells MB, Thornton JM. Protein clefts in molecular recognition and function. Protein Sci. 1996;5:2438–52.
Laurie AT, Jackson RM. Q-SiteFinder: an energy-based method for the prediction of protein-ligand binding sites. Bioinformatics. 2005;21:1908–16.
Le Guilloux V, Schmidtke P, Tuffery P. Fpocket: an open source platform for ligand pocket detection. BMC Bioinformatics. 2009;10:168.
Liang J, Edelsbrunner H, Woodward C. Anatomy of protein pockets and cavities: measurement of binding site geometry and implications for ligand design. Protein Sci. 1998;7:1884–7.
Liu Z, Li Y, Han L, Li J, Liu J, Zhao Z, et al. PDB-wide collection of binding data: current status of the PDBbind database. Bioinformatics. 2015;31:405–12.
Liu T, Ish-Shalom S, Torng W, Lafita A, Bock C, Mort M, et al. Biological and functional relevance of CASP predictions. Proteins. 2018;86(Suppl. 1):374–86.
López G, Valencia A, Tress ML. firestar--prediction of functionally important residues using structural templates and alignment reliability. Nucleic Acids Res. 2007;35:W573–7.
Lopez G, Valencia A, Tress M. FireDB--a database of functionally important residues from proteins of known structure. Nucleic Acids Res. 2007;35:D219–23.
Lopez G, Maietta P, Rodriguez JM, Valencia A, Tress ML. firestar--advances in the prediction of functionally important residues. Nucleic Acids Res. 2011;39:W235–41.
Margineanu DG. Neuropharmacology beyond reductionism—a likely prospect. Biosystems. 2016;141:1–9.
Martinez-Rosell G, Lovera S, Sands ZA, De Fabritiis G. PlayMolecule CrypticScout: predicting protein cryptic sites using mixed-solvent molecular simulations. J Chem Inf Model. 2020;60:2314–24.
Ngan CH, Hall DR, Zerbe B, Grove LE, Kozakov D, Vajda S. FTSite: high accuracy detection of ligand binding sites on unbound protein structures. Bioinformatics. 2012;28:286–7.
Oleinikovas V, Saladino G, Cossins BP, Gervasio FL. Understanding cryptic pocket formation in protein targets by enhanced sampling simulations. J Am Chem Soc. 2016;138:14257–63.
Owens J. Determining druggability. Nat Rev Drug Discov. 2007;6:187.
Paul F, Weikl TR. How to distinguish conformational selection and induced fit based on chemical relaxation rates. PLoS Comput Biol. 2016;12:e1005067.
Ravindranath PA, Sanner MF. AutoSite: an automated approach for pseudo-ligands prediction-from ligand-binding sites identification to predicting key ligand atoms. Bioinformatics. 2016;32:3142–9.
Roche DB, Buenavista MT, Tetchner SJ, McGuffin LJ. The IntFOLD server: an integrated web resource for protein fold recognition, 3D model quality assessment, intrinsic disorder prediction, domain prediction and ligand binding site prediction. Nucleic Acids Res. 2011;39:W171–6.
Schmidt T, Haas J, Gallo Cassarino T, Schwede T. Assessment of ligand-binding residue predictions in CASP9. Proteins. 2011;79(Suppl. 10):126–36.

Schmidtke P, Barril X. Understanding and predicting druggability. A high throughput method for detection of drug binding sites. J Med Chem. 2010;53:5858–67.

Simões T, Lopes D, Dias S, Fernandes F, Pereira J, Jorge J, et al. Geometric detection algorithms for cavities on protein surfaces in molecular graphics: a survey. Comput Graph Forum. 2017;36:643–83.

Stepniewska-Dziubinska MM, Zielenkiewicz P, Siedlecki P. Improving detection of protein-ligand binding sites with 3D segmentation. Sci Rep. 2020;10:5035.

Swinney DC, Anthony J. How were new medicines discovered? Nat Rev Drug Discov. 2011;10:507–19.

Talevi A, Carrillo C, Comini M. The Thiol-polyamine metabolism of *Trypanosoma cruzi*: molecular targets and drug repurposing strategies. Curr Med Chem. 2019;26:6614–35.

Tan KP, Nguyen TB, Patel S, Varadarajan R, Madhusudhan MS. Depth: a web server to compute depth, cavity sizes, detect potential small-molecule ligand-binding cavities and predict the pKa of ionizable residues in proteins. Nucleic Acids Res. 2013;41:W314–21.

Tonge PJ. Drug-target kinetics in drug discovery. ACS Chem Neurosci. 2018;9:29–39.

Vajda S, Beglov D, Wakefield AE, Egbert M, Whitty A. Cryptic binding sites on proteins: definition, detection, and druggability. Curr Opin Chem Biol. 2018;44:1–8.

Volkamer A, Kuhn D, Rippmann F, Rarey M. DoGSiteScorer: a web server for automatic binding site prediction, analysis and druggability assessment. Bioinformatics. 2012;28:2074–5.

Wang R, Fang X, Lu Y, Yang CY, Wang S. The PDBbind database: methodologies and updates. J Med Chem. 2005;48:4111–9.

Wang X, Wang R, Zhang Y, Zhang H. Evolutionary survey of druggable protein targets with respect to their subcellular localizations. Genome Biol Evol. 2013;5:1291–7.

Wass MN, Kelley LA, Sternberg MJ. 3DLigandSite: predicting ligand-binding sites using similar structures. Nucleic Acids Res. 2010;38:W469–73.

Wu S, Zhang Y. LOMETS: a local meta-threading-server for protein structure prediction. Nucleic Acids Res. 2007;35:3375–82.

Wu Q, Peng Z, Zhang Y, Yang J. COACH-D: improved protein-ligand binding sites prediction with refined ligand-binding poses through molecular docking. Nucleic Acids Res. 2018;46:W438–42.

Yang J, Roy A, Zhang Y. Protein-ligand binding site recognition using complementary binding-specific substructure comparison and sequence profile alignment. Bioinformatics. 2013a;29:2588–95.

Yang J, Roy A, Zhang Y. BioLiP: a semi-manually curated database for biologically relevant ligand-protein interactions. Nucleic Acids Res. 2013b;41:D1096–103.

Yang J, Yan R, Roy A, Xu D, Poisson J, Zhang Y. The I-TASSER Suite: protein structure and function prediction. Nat Methods. 2015;12:7–8.

Yao L, Rzhetsky A. Quantitative systems-level determinants of human genes targeted by successful drugs. Genome Res. 2008;18:206–13.

Yildirim MA, Goh KI, Cusick ME, Barabási AL, Vidal M. Drug-target network. Nat Biotechnol. 2007;25(10):1119–26.

Zheng W. Predicting cryptic ligand binding sites based on normal modes guided conformational sampling. Proteins. 2021;89:416–26.

Network-Based Target Identification

Zengrui Wu and Yun Tang

Contents

Abstract

Target identification may help to find potential therapeutic or adverse effects for active compounds and further optimize the structures of these compounds. However, identification of drug targets by experiments is usually expensive and time-consuming. Therefore, various types of computational methods have been developed for prediction of drug targets over the past decade, such as molecular docking-based, pharmacophore-based, similarity-based, machine learning-based, and network-based methods. Among them, network-based methods have two obvious advantages, namely independence of three-dimensional structures of targets and negative samples. In this chapter, we mainly focused on network-based methods for prediction of drug targets. At first, we introduced several representative types of network-based methods. Subsequently, web servers for network-based prediction of targets were introduced briefly. Then, the emphasis was put on the practical applications of network-based methods, including discovery of new targets and elucidation of the molecular mechanisms of

Z. Wu · Y. Tang (✉)
Shanghai Key Laboratory of New Drug Design, School of Pharmacy, East China University of Science and Technology, Shanghai, China
e-mail: ytang234@ecust.edu.cn

M. T. Scotti, C. L. Bellera (eds.), *Drug Target Selection and Validation*, Computer–Aided Drug Discovery and Design,
https://doi.org/10.1007/978-3-030-95895-4_6

therapeutic and adverse effects. Finally, the advantages and limitations of network-based methods were discussed. In summary, network-based methods provide useful tools for drug repurposing, de novo drug discovery, systems pharmacology, and systems toxicology.

Keywords

Network-based method · Recommendation algorithm · Target identification · Drug-target interaction · Drug repositioning · Systems pharmacology · Systems toxicology

1 Introduction

With the rapid advances in systems biology and network pharmacology, the drug discovery paradigm has been shifting from the linear mode of "one drug → one target → one disease" to the network mode of "multi-drugs → multi-targets → multi-diseases" (Hopkins 2008; Medina-Franco et al. 2013; Anighoro et al. 2014). Under the new paradigm, a drug usually interacts with multiple targets, rather than one target. The intricate interactions between drugs and targets can be described as drug–target interaction (DTI) networks (Yildirim et al. 2007), also known as polypharmacology networks (Hopkins 2008). In this type of network, nodes represent drugs and targets, and edges or links represent DTIs. A drug node and a target node are connected by an edge or a link if the drug can interact with the target, indicated by certain k_i, k_d, IC_{50}, or EC_{50} values under a threshold. Interactions of a drug with multiple targets can lead to both desired therapeutic and undesired adverse effects (Anighoro et al. 2014; Roth et al. 2004; Wu et al. 2018a). Hence, identification of targets for drugs can be converted into prediction of new edges or links, namely new DTIs, within a DTI network, which may help to find potential therapeutic or adverse effects for active compounds, in addition to virtual screening for therapeutic targets.

However, identification of DTIs using only biological and pharmacological experiments is usually expensive and time-consuming. Over the past decade, various types of computational methods have been developed for prediction of DTIs, such as molecular docking-based (Kharkar et al. 2014; Patel et al. 2015), pharmacophore-based (Wang et al. 2017; Yang 2010), similarity-based (Keiser et al. 2007; Gong et al. 2013), machine learning-based (Ding et al. 2014; D'Souza et al. 2020; Chen et al. 2016), and network-based (Wu et al. 2018a). The development of computational methods has made it possible to identify DTIs cheaply and efficiently.

Different methods have different characteristics. Molecular docking-based methods rely on three-dimensional (3D) structures of targets, but the structures of many targets have not been determined yet. For example, only a few members of the human G protein-coupled receptor (GPCR) superfamily, the largest protein family in the human proteome, have experimentally determined 3D structures (Stevens et al. 2013; Xiang et al. 2016). Pharmacophore-based methods rely on pharmacophore models. In pharmacophore modeling, how to select training set compounds is a problem that often confuses users, even experienced ones (Yang 2010). Machine learning-based methods rely on machine learning models. In general, building a supervised machine learning model for DTI prediction requires both active and inactive DTIs determined by experiments, namely positive and negative samples (Ding et al. 2014; Chen et al. 2016; Cheng et al. 2012a; Yu et al. 2012). However, it is often difficult to find enough high-quality negative samples from public databases and literature (Cheng et al. 2012a; Yu et al. 2012). In contrast to these methods, network-based methods for prediction of DTIs are derived from algorithms for recommendation (Lu et al. 2012) and link prediction (Lu and Zhou 2011), and hence have two obvious advantages, namely independence of 3D structures of targets and negative samples (Wu et al. 2018a).

In this chapter, we mainly focused on network-based methods for prediction of drug targets. At first, we introduced several representative types of network-based methods, including network-based inference (NBI) methods, similarity-based inference methods, random walk-based methods, local community-based methods, and path-based methods. Subsequently, web servers for network-based prediction of targets were introduced briefly. Then, the emphasis was put on the practical applications of network-based methods, including discovery of new targets and elucidation of the molecular mechanisms of therapeutic and adverse effects. These application studies indicate that network-based methods provide useful tools for drug repurposing, de novo drug discovery, systems pharmacology, and systems toxicology.

2 Methodologies

Network-Based Inference Methods

NBI

More than ten years ago, Zhou et al. proposed a recommendation algorithm named NBI (Zhou et al. 2007), also known as probabilistic spreading (ProbS) (Zhou et al. 2010), aiming to predict possible future likes (called objects) for users based on their past preferences. We successfully applied this recommendation algorithm to the prediction of DTIs by treating drugs and targets as users and objects, respectively (Cheng et al. 2012b). In brief, NBI predicts potential DTIs by performing resource-spreading processes on a known DTI network (Cheng et al. 2012b).

Herein, we used an example to illustrate the processes. As shown in Fig. 1, if we have a DTI network consisting of three drugs and five targets, the following steps can be taken to predict potential targets for the drug highlighted with shadow. At first, initial resources are allocated to the target nodes connected with the highlighted drug node, while no resource is allocated to other unconnected target nodes. Then, two resource-spreading processes are performed. In the first resource-spreading process, each target node equally spreads its resources to the drug nodes connected with it. In the second resource-spreading process, each drug node equally spreads its resources to the target nodes connected with it. After the two resource-spreading processes, the amount of resources located in each target node can be

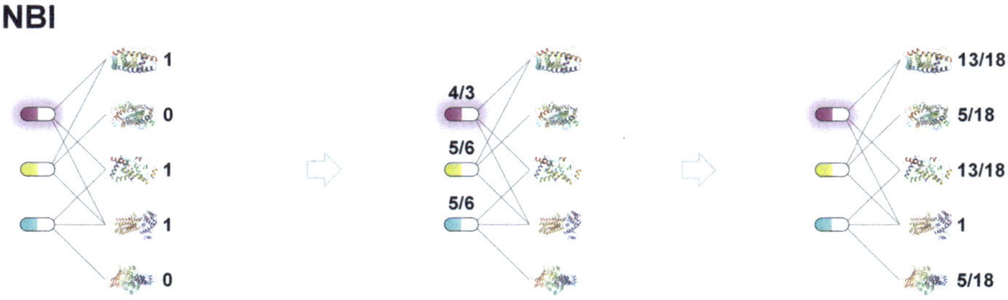

Fig. 1 The schematic diagram of NBI

recognized as the prediction score of the interaction between the highlighted drug and the target. Once the prediction score of an unconnected target node is high after two or more resource-spreading processes, which means that the probability of the target node connecting to the highlighted drug node is high, the target will be recommended to the highlighted drug.

More mathematically speaking, assuming that we have a DTI network consisting of N_D drugs ($D_1, D_2, \ldots, D_{N_D}$) and N_T targets ($T_1, T_2, \ldots, T_{N_T}$), the DTI network can be represented as an $N_D \times N_T$ adjacency matrix A_{DTI}. $A_{DTI}(i, j) = 1$ if drug D_i and target T_j are connected in the DTI network, otherwise = 0. Based on A_{DTI}, a matrix containing initial resources is defined as:

$$A_{NBI} = \begin{bmatrix} O & A_{DTI} \\ (A_{DTI})^T & O \end{bmatrix} \tag{1}$$

Based on the initial resource matrix, a transfer matrix is defined as:

$$W_{NBI}(i,j) = \frac{A_{NBI}(i,j)}{\sum_{l=1}^{N_D+N_T} A_{NBI}(i,l)} \tag{2}$$

Let k be the number of resource-spreading processes (usually set to 2), the resource-spreading processes can be described by the following equation:

$$F_{NBI} = A_{NBI} \times (W_{NBI})^k \tag{3}$$

The value of $F_{NBI}(i, N_D + j)$ is the prediction score of the interaction between drug D_i and target T_j.

Using the above-mentioned equations, the two resource-spreading processes shown in Fig. 1 can be described in a more mathematical way:

$$F_{NBI} = A_{NBI} \times \left(W_{NBI}\right)^k$$

$$= \left[\begin{array}{ccc|ccccc} 0 & 0 & 0 & 1 & 0 & 1 & 1 & 0 \\ 0 & 0 & 0 & 1 & 1 & 0 & 1 & 0 \\ 0 & 0 & 0 & 0 & 0 & 1 & 1 & 1 \\ \hline 1 & 1 & 0 & 0 & 0 & 0 & 0 & 0 \\ 0 & 1 & 0 & 0 & 0 & 0 & 0 & 0 \\ 1 & 0 & 1 & 0 & 0 & 0 & 0 & 0 \\ 1 & 1 & 1 & 0 & 0 & 0 & 0 & 0 \\ 0 & 0 & 1 & 0 & 0 & 0 & 0 & 0 \end{array}\right] \times \left[\begin{array}{ccc|ccccc} 0 & 0 & 0 & 1/3 & 0 & 1/3 & 1/3 & 0 \\ 0 & 0 & 0 & 1/3 & 1/3 & 0 & 1/3 & 0 \\ 0 & 0 & 0 & 0 & 0 & 1/3 & 1/3 & 1/3 \\ \hline 1/2 & 1/2 & 0 & 0 & 0 & 0 & 0 & 0 \\ 0 & 1 & 0 & 0 & 0 & 0 & 0 & 0 \\ 1/2 & 0 & 1/2 & 0 & 0 & 0 & 0 & 0 \\ 1/3 & 1/3 & 1/3 & 0 & 0 & 0 & 0 & 0 \\ 0 & 0 & 1 & 0 & 0 & 0 & 0 & 0 \end{array}\right]^2$$

$$= \left[\begin{array}{ccc|ccccc} 0 & 0 & 0 & 13/18 & 5/18 & 13/18 & 1 & 5/18 \\ 0 & 0 & 0 & 8/9 & 11/18 & 7/18 & 1 & 1/9 \\ 0 & 0 & 0 & 7/18 & 1/9 & 8/9 & 1 & 11/18 \\ \hline 13/18 & 8/9 & 7/18 & 0 & 0 & 0 & 0 & 0 \\ 5/18 & 11/18 & 1/9 & 0 & 0 & 0 & 0 & 0 \\ 13/18 & 7/18 & 8/9 & 0 & 0 & 0 & 0 & 0 \\ 1 & 1 & 1 & 0 & 0 & 0 & 0 & 0 \\ 5/18 & 1/9 & 11/18 & 0 & 0 & 0 & 0 & 0 \end{array}\right] \quad (4)$$

Obviously, $[1\ 0\ 1\ 1\ 0]$ in the first row of A_{NBI} are the initial resources allocated to the target nodes connected with the highlighted drug, and $\left[13/18\ \ 5/18\ \ 13/18\ \ 1\ \ 5/18\right]$ in the first row of F_{NBI} are the prediction scores of the interactions between the highlighted drug and the five targets, which are consistent with Fig. 1.

As the simplest of NBI methods, NBI only uses a known DTI network as input. This characteristic makes NBI run fast, but limits its application range. Specifically, NBI can only predict potential targets for the drugs within the known DTI network and cannot predict potential targets for any novel compounds outside of the DTI network, such as newly extracted natural products and newly synthesized compounds. The reason for this limitation is that novel compounds have no connections with the known DTI network.

Substructure-Drug-Target NBI (SDTNBI)

To overcome the limitation of NBI, we proposed a new method named SDTNBI (Wu et al. 2017). In brief, SDTNBI uses chemical substructures to bridge the gap between a known DTI network and new chemical entities. Specifically, the following steps are taken. At first, the chemical structures of the drugs in the DTI network and new chemical entities outside the DTI network are broken into fragments, namely substructures. After treating new chemical entities as special drugs, a drug-substructure association (DSA) network was constructed. In this network, nodes represent drugs and substructures, and edges represent DSAs. A drug and a substructure are connected by an edge if the chemical substructure of the drug contains the substructure. Then, a substructure-drug-target network is constructed by integrating the DTI and DSA networks. Obviously, this network consists of three types of nodes (drugs, substructures, and targets) and two types of edges (DTIs and DSAs), which is more heterogeneous than the DTI network. Finally, potential DTIs can be predicted by performing resource-spreading processes in the substructure-drug-target network.

Herein, we used an example to illustrate the processes. As shown in Fig. 2, if we have a substructure-drug-target network consisting of three drugs, four substructures, and five targets, the following steps can be taken to predict potential targets for the drug highlighted with shadow. At first, initial resources are allocated to the two types of nodes (substructure and target nodes) connected with the highlighted drug node, while no resources are allocated to other unconnected nodes. Then, two resource-spreading processes similar to those in NBI are performed. After the two resource-spreading processes, the amount of resources located in each target node can be recognized as the prediction score of the interaction between the highlighted drug and the target.

More mathematically speaking, assuming that there are a total number of N_D drugs and new chemical entities ($D_1, D_2, \ldots, D_{N_D}$) and a total number of N_T targets ($T_1, T_2, \ldots, T_{N_T}$), the DTI network can be represented as an $N_D \times N_T$ adjacency matrix A_{DTI}. $A_{DTI}(i, j) = 1$ if drug D_i and target T_j are connected in the DTI network, otherwise = 0. Assuming that there are a total number of N_S substructures ($S_1, S_2, \ldots, S_{N_S}$), the DSA network can be represented as an $N_D \times N_S$ adjacency matrix A_{DSA}.$A_{DSA}(i, j) = 1$ if drug D_i and substructure S_j are connected in the DSA network, otherwise = 0. Based on A_{DTI} and A_{DSA}, a matrix containing initial resources is defined as:

$$A_{SDTNBI} = \begin{bmatrix} O & A_{DSA} & A_{DTI} \\ (A_{DSA})^T & O & O \\ (A_{DTI})^T & O & O \end{bmatrix} \tag{5}$$

Meanwhile, two matrices similar to A_{DTI} and A_{DSA}, but that eliminated the influence of new chemical entities, are defined as:

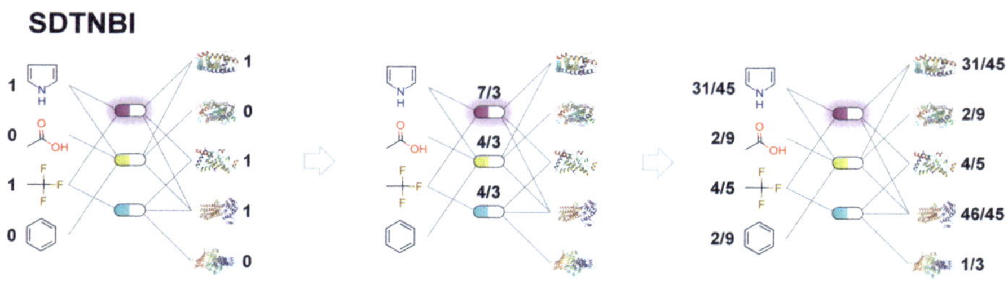

Fig. 2 The schematic diagram of SDTNBI

$$B_{DTI}(i,j) = \begin{cases} A_{DTI}(i,j) & \text{if } \sum_{N_T}^{l=1} A_{DTI}(i,l) \neq 0 \\ 0 & \text{otherwise} \end{cases} \tag{6}$$

$$B_{DSA}(i,j) = \begin{cases} A_{DSA}(i,j) & \text{if } \sum_{N_T}^{l=1} A_{DTI}(i,l) \neq 0 \\ 0 & \text{otherwise} \end{cases} \tag{7}$$

Normally, B_{DTI} is just equal to A_{DTI}. Based on these two matrices, a transfer matrix is defined as:

$$B_{SDTNBI} = \begin{bmatrix} O & B_{DSA} & B_{DTI} \\ (B_{DSA})^T & O & O \\ (B_{DTI})^T & O & O \end{bmatrix} \tag{8}$$

Then, the transfer matrix is normalized by the following equation:

$$W_{SDTNBI}(i,j) = \begin{cases} \frac{B_{SDTNBI}(i,j)}{\sum_{l=1}^{N_D+N_S+N_T} B_{SDTNBI}(i,l)} & \text{if } B_{SDTNBI}(i,j) \neq 0 \\ 0 & \text{otherwise} \end{cases} \tag{9}$$

Let k be the number of resource-spreading processes (usually set to 2), the resource-spreading processes can be described by the following equation:

$$F_{SDTNBI} = A_{SDTNBI} \times (W_{SDTNBI})^k \tag{10}$$

The value of $F_{SDTNBI}(i, N_D + N_S + j)$ is the prediction score of the interaction between drug D_i and target T_j.

Although SDTNBI can predict potential targets for various types of compounds, the performance of SDTNBI models is often worse than those of NBI models (Wu et al. 2018a). This may result from the potential imbalances. As mentioned above, there are different types of nodes and edges in a substructure–drug–target network. However, different types of nodes are treated equally in the allocation of initial resources, and the same weights are assigned to different types of edges.

Balanced SDTNBI (bSDTNBI)

To address the potential imbalances, we introduced three tunable parameters into SDTNBI, and proposed a new method named bSDTNBI (Wu et al. 2016). As shown in Fig. 3, the first parameter $\alpha \in [0, 1)$ is used to adjust the initial resources allocated to different types of nodes. A smaller α value means that fewer resources will be allocated to substructure nodes, while a larger α value means that more resources will be allocated to substructure nodes. Using this parameter, two matrices are defined based on the aforementioned A_{DTI} and A_{DSA}:

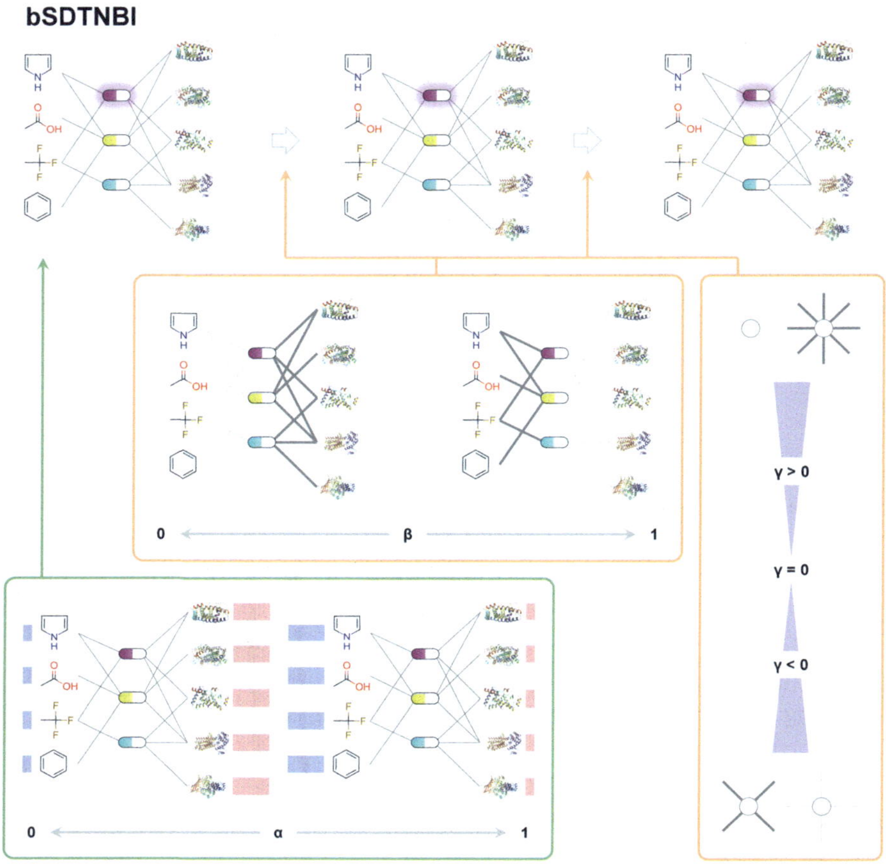

Fig. 3 The schematic diagram of bSDTNBI

$$A'_{DSA}(i,j) = \alpha \cdot \frac{A_{DSA}(i,j)}{\sum_{l=1}^{N_S} A_{DSA}(i,i)} \tag{11}$$

$$A'_{DTI}(i,j) = (1-\alpha) \cdot \frac{A_{DTI}(i,j)}{\sum_{l=1}^{N_T} A_{DTI}(i,l)} \tag{12}$$

Based on these two matrices, a matrix containing initial resources is defined as:

$$A_{bSDTNBI} = \begin{bmatrix} O & A'_{DSA} & A'_{DTI} \\ \left(A'_{DSA}\right)^T & O & O \\ \left(A'_{DTI}\right)^T & O & O \end{bmatrix} \tag{13}$$

The second parameter $\beta \in [0, 1)$ is used to adjust the weights of different types of edges. A smaller β value means that smaller weights will be assigned to the edges representing DSAs, while a larger β value means that larger weights will be assigned to the edges representing DSAs. Using this parameter and aforementioned B_{DTI} and B_{DSA}, a transfer matrix is defined as:

$$B_{bSDTNBI} = \begin{bmatrix} O & \beta \cdot B_{DSA} & (1-\beta)\cdot B_{DTI} \\ \beta \cdot (B_{DSA})^T & O & O \\ (1-\beta)\cdot (B_{DTI})^T & O & O \end{bmatrix} \quad (14)$$

The third parameter $\gamma \in (-\infty, +\infty)$ is used to adjust the influence of hub nodes. $\gamma > 0$ means that the influence of hub nodes will be strengthened, while $\gamma < 0$ means that the influence of hub nodes will be weakened. Using this parameter, the transfer matrix is normalized by the following equations:

$$C_{bSDTNBI}(i,j) = B_{bSDTNBI}(i,j) \cdot \left[\sum_{N_D+N_S+N_T}^{l=1} B_{bSDTNBI}(l,j) \right]^{\gamma} \quad (15)$$

$$W_{bSDTNBI}(i,j) = \begin{cases} \dfrac{C_{bSDTNBI}(i,j)}{\sum_{l=1}^{N_D+N_S+N_T} C_{bSDTNBI}(i,l)} & \text{if } C_{bSDTNBI}(i,j) \neq 0 \\ 0 & \text{otherwise} \end{cases} \quad (16)$$

Let k be the number of resource-spreading processes (usually set to 2), the resource-spreading processes can be described by the following equation:

$$F_{bSDTNBI} = A_{bSDTNBI} \times (W_{bSDTNBI})^k \quad (17)$$

The value of $F_{bSDTNBI}(i, N_D + N_S + j)$ is the prediction score of the interaction between drug D_i and target T_j.

Our previous study has shown that when the three parameters were optimized properly, the performance of bSDTNBI models was comparable to that of NBI models, and better than that of SDTNBI models (Wu et al. 2016).

Similarity-Based Inference Methods

Similarity inference methods are derived from another type of recommendation algorithm, named collaborative filtering algorithm (Huang et al. 2007). These methods use both the known DTI network and similarity data as inputs to predict potential DTIs. For example, in our previous article about NBI (Cheng et al. 2012b), we proposed two similarity inference methods. The first one is drug-based similarity inference (DBSI), based on the hypothesis that drugs with similar chemical structures tend to interact with similar targets. The chemical structure similarity between two drugs is measured by the Tanimoto coefficient (Willett et al. 1998). The other one is target-based similarity inference (TBSI), based on the hypothesis that proteins with similar sequences tend to be targeted by similar drugs. The sequence similarity between two target proteins is measured by the normalized Smith–Waterman score (Yamanishi et al. 2008). We compared these two methods with NBI and found that both of them underperformed NBI (Cheng et al. 2012b).

In addition to chemical structure similarity and protein sequence similarity, similarity-based inference methods can also use phenotype similarity to predict potential DTIs. For example, in a previous study (Cheng et al. 2013a), we proposed two similarity-based inference methods, named drug side-effect similarity inference (DSESI) and drug therapeutic similarity inference (DTSI). Among them, DSESI is based on the hypothesis that drugs with similar side effects tend to interact with similar targets, while DTSI is based on the hypothesis that drugs with similar Anatomical Therapeutic Chemical (ATC) classification codes tend to interact with similar targets. We compared these two methods with DBSI, and found that the performance of the three methods is: DTSI > DBSI > DSESI (Cheng et al. 2013a).

Random Walk-Based Methods

Random walk is a classical concept that has been used in link prediction (Liu and Lu 2010) and recommender systems (Lu et al. 2012). Over the past decade, random walk-based methods were also applied to the prediction of DTIs. The most representative of random walk-based methods is Network-based Random Walk with Restart on the Heterogeneous network (NRWRH) (Chen et al. 2012). In brief, NRWRH predicts potential DTIs by performing random walk with restart on a heterogeneous network. The heterogeneous network is constructed by integrating the known DTI network and two types of similarity, namely chemical structure similarity and protein sequence similarity. In the article about NRWRH (Chen et al. 2012), the authors declared that NRWRH outperformed not only several machine learning-based methods but also other random walk-based methods, including Random Walk with Restart on the Heterogeneous network (RWRH), Network-based Random Walk with Restart (NRWR), and Random Walk with Restart (RWR).

Local Community-Based Methods

Initially, the local-community-paradigm (LCP) theory was used for linking prediction in monopartite networks, such as brain networks and protein-protein interaction networks (Cannistraci et al. 2013). Then, local community-based methods were improved to make them suitable for use in bipartite networks (Daminelli et al. 2015). Recently, five local community-based methods, including Cannistraci-Alanis-Ravasi (CAR), Cannistraci-Jaccard (CJC), Cannistraci preferential attachment (CPA), Cannistraci-Adamic-Adar (CAA), and Cannistraci resource allocation (CRA), were applied in the prediction of DTIs and showed high performance (Duran et al. 2018).

Path-Based Methods

As the name implies, path-based methods use paths between nodes to predict new edges. For example, DASPfind uses simple paths of particular lengths on a heterogeneous network to predict potential DTIs (Ba-Alawi et al. 2016). The heterogeneous network is similar to that in NRWRH (Chen et al. 2012), constructed by integrating the DTI network and similarity data. The authors of DASPfind claimed that DASPfind outperformed several previous network-based methods, such as NRWRH (Chen et al. 2012) and DT-Hybrid (Alaimo et al. 2013).

3 Web Servers

Up to now, many web servers have been developed for prediction of drug targets (Chen et al. 2016). Different web servers use different methods. For example, TarFisDock uses molecular docking (Li et al. 2006), SEA uses 2D similarity searching (Keiser et al. 2007), ChemMapper uses 3D similarity searching (Gong et al. 2013), SwissTargetPrediction uses both 2D and 3D similarity (Daina et al. 2019), and TargetNet uses machine learning (Yao et al. 2016). However, there are only a few web servers to use network-based methods. In this section, two of them are introduced.

NetInfer

NetInfer (http://lmmd.ecust.edu.cn/netinfer) was developed based on our previous studies of NBI methods (Wu et al. 2020). This web server provides a user-friendly interface. Users can easily predict target proteins, microRNAs, pathways, ATC codes, or adverse drug events (ADEs) for the small molecules of their interests in a few steps. As shown in Fig. 4a, either inputting SMILES strings into the text box or drawing a chemical structure in the molecular editor is allowed. After selecting a prediction type, a set of default settings will be provided automatically. Novice users can directly use the default settings, while expert users can adjust the settings according to their experience. In general, it will not take more than one minute to obtain prediction results. As shown in Fig. 4b, the prediction results can be downloaded as comma-separated values (CSV) or tab-separated values (TSV) files. From April 2020 to February 2022, more than three thousand tasks have been submitted to NetInfer. We hope that our web server can help users to find new targets, therapeutic effects or adverse effects for their submitted molecules.

DASPfind

DASPfind (https://www.cbrc.kaust.edu.sa/daspfind/) was developed based on the aforementioned DASPfind method (Ba-Alawi et al. 2016). Compared with NetInfer,

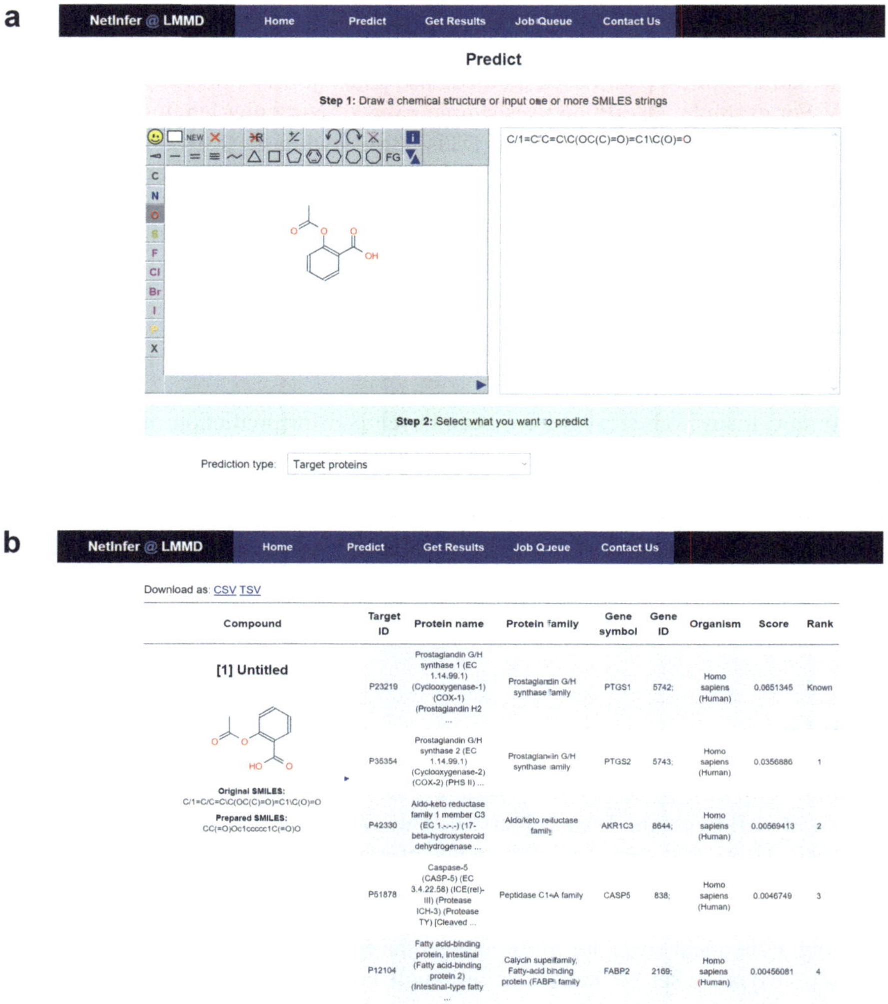

Fig. 4 The user interface of NetInfer, including (**a**) inputting chemical structures and (**b**) getting prediction results

this web server is not very user-friendly. Users need to upload three text files in specific formats, which contain DTIs, drug similarity data, and target similarity data, respectively. It may be difficult for novice users to prepare these files (Ba-Alawi et al. 2016).

4 Applications

Currently, most of the studies about network-based methods focus primarily on the evaluation indicators of methods. Only a few researchers used biological and pharmacological experiments to validate their prediction results. Over the past decade, we tried to use our NBI methods in practical applications, including discovery of new targets and elucidation of the molecular mechanisms of therapeutic and adverse effects. In this section, we describe some typical applications of network-based target identification methods.

Discovery of New Targets for Known Compounds

Studies on GPCRs

GPCRs are targeted by over 30% of the approved drugs (Stevens et al. 2013; Santos et al. 2017), representing the largest class of druggable targets for treatment of various human diseases, such as psychiatric (Roth et al. 2004), neurodegenerative (Schapira et al. 2006), and metabolic diseases (Oh and Olefsky 2016). Recently, we used bSDTNBI to investigate the polypharmacology of GPCR ligands (Wu et al. 2018b). At first, we constructed local and global DTI networks for human GPCRs. Based on the DTI networks, bSDTNBI models were constructed using different types of molecular fingerprints. After systematic evaluation, the best model was used to predict new GPCR targets for known GPCR ligands. According to the prediction results, 20 compounds predicted to target a GPCR named prostaglandin E2 receptor EP4 subtype (abbreviated as EP4) were purchased for experimental validation. Two of the purchased compounds, named AM966 and Ki16425, showed antagonistic effects on EP4 with IC_{50} values <10 μM in calcium flux and cAMP assays. Interestingly, both AM966 and Ki16425 are known antagonists for lysophosphatidic acid receptors. Considering that EP4 is associated with colon cancer, lung cancer, osteoporosis, and rheumatoid arthritis, our newly discovered EP4 antagonists may serve as lead compounds for the treatment of these diseases.

Studies on Nuclear Receptors

Nuclear receptors are involved in the pathology of various human diseases, such as cancer, diabetes, obesity, and autoimmune diseases (Gronemeyer et al. 2004; Kojetin and Burris 2014), and have been targeted by approximately 16% of the approved small-molecule drugs (Santos et al. 2017). In a previous study, we used NBI to predict potential targets for approved drugs (Cheng et al. 2012b). According to the prediction results, 31 drugs predicted to target estrogen receptor α (ERα) or estrogen receptor β (ERβ) were purchased for experimental validation. Four of the purchased drugs (diclofenac, simvastatin, ketoconazole, and itraconazole) showed antagonistic effects on ERα or ERβ with IC_{50} values <10 μM. Two of them (simvastatin and ketoconazole) further showed anti-proliferative effect on MDA-MB-231 human breast cancer cell line with IC_{50} values <10 μM, suggesting that these two drugs may have therapeutic effects on breast cancer.

In a more recent study, we used bSDTNBI to predict potential targets for the compounds from Enamine (Wu et al. 2016). According to the prediction results, 56 compounds predicted to target ERα were purchased for experimental validation. Twenty-seven of the purchased compounds showed antagonistic or agonistic effects on ERα with IC_{50} or EC_{50} values <10 μM. This high success rate indicates the high performance of bSDTNBI. The newly discovered ERα antagonists may serve as lead compounds for the treatment of ERα-positive breast cancer.

Elucidation of the Molecular Mechanisms of Therapeutic Effects

Studies on Approved Drugs

Recent studies have demonstrated that the use of nonsteroidal anti-inflammatory drugs (NSAIDs) is associated with lower risk of cancer (Nan et al. 2015). However, molecular mechanisms of the chemopreventive effects of NSAIDs remain unclear. In a previous study, we used SDTNBI to investigate the possible molecular mechanisms of the chemopreventive effects of NSAIDs (Wu et al. 2017). After predicting potential targets for NSAIDs, we constructed a drug–gene–disease network consisting of 21 NSAIDs and 29 cancer types by integrating DTIs with gene-disease associations. Some of the predicted DTIs were validated by previously published pharmacological studies or co-crystal structure data, suggesting the high performance of SDTNBI. Then, we found that NSAIDs may exert anticancer effects by inhibiting several enzymes associated with cancer, including prostaglandin G/H synthase 2 (PTGS2), aldo-keto reductase family 1 member C3 (AKR1C3), carbonic anhydrase 9 (CA9), carbonic anhydrase 12 (CA12), and cyclin-dependent kinase 2 (CDK2).

In a more recent study, we used bSDTNBI to investigate the possible anticancer mechanisms of antidepressant and anti-diabetic drugs (Wu et al. 2016). After predicting potential targets for approved drugs, we constructed a comprehensive drug–gene–disease network consisting of 666 approved drugs and 15 cancer types by integrating DTIs with gene–disease associations. According to the drug–gene–disease network and previously published studies, we found that tricyclic antidepressant drugs may exert anticancer effects by targeting serotonin receptors, and anti-diabetic drugs may exert anticancer effects by targeting cancer cell metabolism.

These two studies showed the practical value of our NBI methods in elucidation of the therapeutic mechanisms of approved drugs.

Studies on Natural Products

Natural products have been important sources of new drugs in the past decades (Newman and Cragg 2020). In a recent study, we used bSDTNBI to investigate the possible anticancer mechanisms of natural products (Fang et al. 2017a). At first, we collected natural products from several traditional Chinese medicine databases, and constructed a global DTI network consisting of 2388 natural products. Based on the DTI network, bSDTNBI models were constructed using different types of fingerprints. After systematic evaluation, the best model was used to predict potential targets for the natural products in the DTI network. Then, the associations between these natural products and 13 cancer types were obtained using a statistical approach. Taking three natural products (kaempferol, resveratrol, and genistein) as examples, we found that these natural products may exert anticancer effects by inhibiting several cancer-associated proteins and pathways.

In addition to anticancer effects, bSDTNBI was also used to investigate the possible molecular mechanisms of anti-aging effects of natural products (Fang et al. 2017b). These studies showed the practical value of our NBI methods in elucidation of the molecular mechanisms of natural products.

Elucidation of the Molecular Mechanisms of Adverse Effects

Drug-induced liver injury (DILI) is one of the leading causes of drug failure. In a recent study, we used bSDTNBI to investigate the possible molecular mechanisms of DILI (Peng et al. 2019). At first, we constructed a DTI network consisting of 896 DILI compounds and 1067 targets. After predicting potential targets for these DILI compounds, we found that 26 of the predicted target proteins were associated with hepatotoxicity and highly expressed in the liver, including CYP2E1, GSTA1, EPHX1, ADH1B, ADH1C, ALDH2, F7, and IL2. Then, a scoring function named DILI-Score was developed to further assess the hepatotoxicity of compounds. Taking tyrosine kinase inhibitors and TAK-875 as examples, we found these compounds may cause liver injury by targeting some of the 26 proteins. These findings may help to better understand the molecular mechanisms of DILI and reduce the risk of hepatotoxicity in drug discovery and development.

In addition to hepatotoxicity, bSDTNBI was also used to investigate the possible molecular mechanisms of skin sensitization (Di et al. 2019). These studies showed the practical value of our NBI methods in elucidation of the molecular mechanisms of adverse effects.

5 Advantages and Limitations

Since the beginning of the new century, various types of computational methods have been developed for prediction of targets (Chen et al. 2016; Lavecchia and Cerchia 2016). Among them, network-based methods have several advantages. Compared with molecular docking-based methods, network-based methods are independent of 3D structures of targets. Compared with supervised machine learning-based methods, network-based methods are independent of negative samples. These good characteristics make network-based methods able to predict potential targets for compounds on a larger scale, especially those targets lacking 3D structures or negative samples. Moreover, network-based methods run fast. Because networks can be represented as matrices, network-based methods can usually be described by simple matrix operations such as matrix multiplication (Cheng et al. 2012b; Wu et al. 2016, 2017; Chen et al. 2012). These operations can be easily parallelized. For example, with the support of GPU, bSDTNBI only takes tens of seconds to calculate prediction scores for all possible interactions between more than ten thousand compounds and more than one thousand targets. This high speed allows making large-scale predictions.

Although network-based methods achieved success in the prediction of targets, there remains room for improvement. First, they cannot make predictions for those targets without any known ligand. Second, they cannot output prediction scores correlated with binding affinities. Moreover, interaction types (e.g., agonism/antagonism, activation/inhibition) are not considered yet. In our view, there are two possible ways to overcome these limitations. One way is to apply more link prediction algorithms to the prediction of DTIs, such as hierarchical structure (Clauset et al. 2008), stochastic block (Guimera and Sales-Pardo 2009), and likelihood analysis (Pan et al. 2016). The other way is to integrate more types of biomedical data, such as drug-ADE associations (Cheng et al. 2013b), drug-induced gene expression profiles (Lamb et al. 2006; Subramanian et al. 2007), chemical ADMET properties (Yang et al. 2019), protein–protein interactions (Li et al. 2017), and protein structure data (Burley et al. 2021).

Nevertheless, network-based target identification methods provide useful tools for drug repurposing, de novo drug discovery, systems pharmacology, and systems toxicology. We hope that they will play greater roles in the future.

Acknowledgements This work was supported by the National Key Research and Development Program of China (Grants 2016YFA0502304 and 2019YFA0904800), the National Natural Science Foundation of China (Grants 81673356 and 81872800), the Shanghai Post-doctoral Excellence Program (Grant 2018199), the Shanghai Sailing Program (Grant 19YF1412700), the China Postdoctoral Science Foundation (Grant 2019 M661413), and the 111 Project (Grant BP0719034).

Conflicts of Interest The authors declare that they have no conflict of interest.

References

Alaimo S, Pulvirenti A, Giugno R, Ferro A. Drug-target interaction prediction through domain-tuned network-based inference. Bioinformatics. 2013;29(16):2004–8. https://doi.org/10.1093/bioinformatics/btt307.

Anighoro A, Bajorath J, Rastelli G. Polypharmacology: challenges and opportunities in drug discovery. J Med Chem. 2014;57(19):7874–87. https://doi.org/10.1021/jm5006463.

Ba-Alawi W, Soufan O, Essack M, Kalnis P, Bajic VB. DASPfind: new efficient method to predict drug-target interactions. J Cheminformatics. 2016;8:15. https://doi.org/10.1186/s13321-016-0128-4.

Burley SK, Bhikadiya C, Bi CX, Bittrich S, Chen L, Crichlow GV, Christie CH, et al. RCSB Protein Data Bank: powerful new tools for exploring 3D structures of biological macromolecules for basic and applied research and education in fundamental biology, biomedicine, biotechnology, bioengineering and energy sciences. Nucleic Acids Res. 2021;49(D1):D437–51. https://doi.org/10.1093/nar/gkaa1038.

Cannistraci CV, Alanis-Lobato G, Ravasi T. From link-prediction in brain connectomes and protein interactomes to the local-community-paradigm in complex networks. Sci Rep. 2013;3:1613. https://doi.org/10.1038/srep01613.

Chen X, Liu MX, Yan GY. Drug-target interaction prediction by random walk on the heterogeneous network. Mol BioSyst. 2012;8(7):1970–8. https://doi.org/10.1039/c2mb00002d.

Chen X, Yan CC, Zhang XT, Zhang X, Dai F, Yin J, Zhang YD. Drug-target interaction prediction: databases, web servers and computational models. Brief Bioinform. 2016;17(4):696–712. https://doi.org/10.1093/bib/bbv066.

Cheng FX, Zhou YD, Li J, Li WH, Liu GX, Tang Y. Prediction of chemical-protein interactions: multitarget-QSAR versus computational chemogenomic methods. Mol BioSyst. 2012a;8(9):2373–84. https://doi.org/10.1039/c2mb25110h.

Cheng FX, Liu C, Jiang J, Lu WQ, Li WH, Liu GX, Zhou WX, Huang J, Tang Y. Prediction of drug-target interactions and drug repositioning via network-based inference. PLoS Comput Biol. 2012b;8(5):e1002503. https://doi.org/10.1371/journal.pcbi.1002503.

Cheng FX, Li WH, Wu ZR, Wang XC, Zhang C, Li J, Liu GX, Tang Y. Prediction of polypharmacological profiles of drugs by the integration of chemical, side effect, and therapeutic space. J Chem Inf Model. 2013a;53(4):753–62. https://doi.org/10.1021/ci400010x.

Cheng FX, Li WH, Wang XC, Zhou YD, Wu ZR, Shen J, Tang Y. Adverse drug events: database construction and *in silico* prediction. J Chem Inf Model. 2013b;53(4):744–52. https://doi.org/10.1021/ci4000079.

Clauset A, Moore C, Newman MEJ. Hierarchical structure and the prediction of missing links in networks. Nature. 2008;453(7191):98–101. https://doi.org/10.1038/nature06830.

Daina A, Michielin O, Zoete V. SwissTargetPrediction: updated data and new features for efficient prediction of protein targets of small molecules. Nucleic Acids Res. 2019;47(W1):W357–64. https://doi.org/10.1093/nar/gkz382.

Daminelli S, Thomas JM, Duran C, Cannistraci CV. Common neighbours and the local-community-paradigm for topological link prediction in bipartite networks. New J Phys. 2015;17:113037. https://doi.org/10.1088/1367-2630/17/11/113037.

Di PW, Wu ZR, Yang HB, Li WH, Tang Y, Liu GX. Prediction of the allergic mechanism of haptens via a reaction-substructure-compound-target-pathway network system. Toxicol Lett. 2019;317:68–81. https://doi.org/10.1016/j.toxlet.2019.09.019.

Ding H, Takigawa I, Mamitsuka H, Zhu SF. Similarity-based machine learning methods for predicting drug-target interactions: a brief review. Brief Bioinform. 2014;15(5):734–47. https://doi.org/10.1093/bib/bbt056.

D'Souza S, Prema KV, Seetharaman B. Machine learning models for drug-target interactions: current knowledge and future directions. Drug Discov Today. 2020;25(4):748–56. https://doi.org/10.1016/j.drudis.2020.03.003.

Duran C, Daminelli S, Thomas JM, Haupt VJ, Schroeder M, Cannistraci CV. Pioneering topological methods for network-based drug-target prediction by exploiting a brain-network self-organization theory. Brief Bioinform. 2018;19(6):1183–202. https://doi.org/10.1093/bib/bbx041.

Fang JS, Wu ZR, Cai CP, Wang Q, Tang Y, Cheng FX. Quantitative and systems pharmacology. 1. *in silico* prediction of drug-target interaction of natural products enables new targeted cancer therapy. J Chem Inf Model. 2017a;57(11):2657–71. https://doi.org/10.1021/acs.jcim.7b00216.

Fang JS, Gao L, Ma HL, Wu QH, Wu T, Wu J, Wang Q, Cheng FX. Quantitative and systems pharmacology 3. Network-based identification of new targets for natural products enables potential uses in aging-associated disorders. Front Pharmacol. 2017b;8:747. https://doi.org/10.3389/fphar.2017.00747.

Gong JY, Cai CQ, Liu XF, Ku X, Jiang HL, Gao DQ, Li HL. ChemMapper: a versatile web server for exploring pharmacology and chemical structure association based on molecular 3D similarity method. Bioinformatics. 2013;29(14):1827–9. https://doi.org/10.1093/bioinformatics/btt270.

Gronemeyer H, Gustafsson JA, Laudet V. Principles for modulation of the nuclear receptor superfamily. Nat Rev Drug Discov. 2004;3(11):950–64. https://doi.org/10.1038/nrd1551.

Guimera R, Sales-Pardo M. Missing and spurious interactions and the reconstruction of complex networks. Proc Natl Acad Sci U S A. 2009;106(52):22073–8. https://doi.org/10.1073/pnas.0908366106.

Hopkins AL. Network pharmacology: the next paradigm in drug discovery. Nat Chem Biol. 2008;4(11):682–90. https://doi.org/10.1038/nchembio.118.

Huang Z, Zeng D, Chen HC. A comparison of collaborative-filtering recommendation algorithms for e-commerce. IEEE Intell Syst. 2007;22(5):68–78. https://doi.org/10.1109/mis.2007.4338497.

Keiser MJ, Roth BL, Armbruster BN, Ernsberger P, Irwin JJ, Shoichet BK. Relating protein pharmacology by ligand chemistry. Nat Biotechnol. 2007;25(2):197–206. https://doi.org/10.1038/nbt1284.

Kharkar PS, Warrier S, Gaud RS. Reverse docking: a powerful tool for drug repositioning and drug rescue. Future Med Chem. 2014;6(3):333–42. https://doi.org/10.4155/fmc.13.207.

Kojetin DJ, Burris TP. REV-ERB and ROR nuclear receptors as drug targets. Nat Rev Drug Discov. 2014;13(3):197–216. https://doi.org/10.1038/nrd4100.

Lamb J, Crawford ED, Peck D, Modell JW, Blat IC, Wrobel MJ, Lerner J, et al. The connectivity map: using gene-expression signatures to connect small molecules, genes, and disease. Science. 2006;313(5795):1929–35. https://doi.org/10.1126/science.1132939.

Lavecchia A, Cerchia C. *In silico* methods to address polypharmacology: current status, applications and future perspectives. Drug Discov Today. 2016;21(2):288–98. https://doi.org/10.1016/j.drudis.2015.12.007.

Li HL, Gao ZT, Kang L, Zhang HL, Yang K, Yu KQ, Luo XM, Zhu WL, Chen KX, Shen JH, Wang XC, Jiang HL. TarFisDock: a web server for identifying drug targets with docking approach. Nucleic Acids Res. 2006;34:W219–24. https://doi.org/10.1093/nar/gkl114.

Li TB, Wernersson R, Hansen RB, Horn H, Mercer J, Slodkowicz G, Workman CT, et al. A scored human protein-protein interaction network to catalyze genomic interpretation. Nat Methods. 2017;14(1):61–4. https://doi.org/10.1038/nm.4083.

Liu WP, Lu LY. Link prediction based on local random walk. Epl. 2010;89(5):58007. https://doi.org/10.1209/0295-5075/89/58007.

Lu LY, Zhou T. Link prediction in complex networks: a survey. Physica A. 2011;390(6):1150–70. https://doi.org/10.1016/j.physa.2010.11.027.

Lu LY, Medo M, Yeung CH, Zhang YC, Zhang ZK, Zhou T. Recommender systems. Phys Rep. 2012;519(1):1–49. https://doi.org/10.1016/j.physrep.2012.02.006.

Medina-Franco JL, Giulianotti MA, Welmaker GS, Houghten RA. Shifting from the single to the multitarget paradigm in drug discovery. Drug Discov Today. 2013;18(9-10):495–501. https://doi.org/10.1016/j.drudis.2013.01.008.

Nan HM, Hutter CM, Lin Y, Jacobs EJ, Ulrich CM, White E, Baron JA, et al. Association of aspirin and NSAID use with risk of colorectal cancer according to genetic variants. J Am Med Assoc. 2015;313(11):1133–42. https://doi.org/10.1001/jama.2015.1815.

Newman DJ, Cragg GM. Natural products as sources of new drugs over the nearly four decades from 01/1981 to 09/2019. J Nat Prod. 2020;83(3):770–803. https://doi.org/10.1021/acs.jnatprod.9b01285.

Oh DY, Olefsky JM. G protein-coupled receptors as targets for anti-diabetic therapeutics. Nat Rev Drug Discov. 2016;15(3):161–72. https://doi.org/10.1038/nrd.2015.4.

Pan LM, Zhou T, Lu LY, Hu CK. Predicting missing links and identifying spurious links via likelihood analysis. Sci Rep. 2016;6:22955. https://doi.org/10.1038/srep22955.

Patel H, Lucas X, Bendik I, Gunther S, Merfort I. Target fishing by cross-docking to explain polypharmacological effects. ChemMedChem. 2015;10(7):1209–17. https://doi.org/10.1002/cmdc.201500123.

Peng YY, Wu ZR, Yang HB, Cai YC, Liu GX, Li WH, Tang Y. Insights into mechanisms and severity of drug-induced liver injury via computational systems toxicology approach. Toxicol Lett. 2019;312:22–33. https://doi.org/10.1016/j.toxlet.2019.05.005.

Roth BL, Sheffler DJ, Kroeze WK. Magic shotguns versus magic bullets: selectively non-selective drugs for mood disorders and schizophrenia. Nat Rev Drug Discov. 2004;3(4):353–9. https://doi.org/10.1038/nrd1346.

Santos R, Ursu O, Gaulton A, Bento AP, Donadi RS, Bologa CG, Karlsson A, Al-Lazikani B, Hersey A, Oprea TI, Overington JP. A comprehensive map of molecular drug targets. Nat Rev Drug Discov. 2017;16(1):19–34. https://doi.org/10.1038/nrd.2016.230.

Schapira AHV, Bezard E, Brotchie J, Calon F, Collingridge GL, Ferger B, Hengerer B, Hirsch E, Jenner P, Le Novere N, Obeso JA, Schwarzschild MA, Spampinato U, Davidai G. Novel pharmacological targets for the treatment of Parkinson's disease. Nat Rev Drug Discov. 2006;5(10):845–54. https://doi.org/10.1038/nrd2087.

Stevens RC, Cherezov V, Katritch V, Abagyan R, Kuhn P, Rosen H, Wuthrich K. The GPCR network: a large-scale collaboration to determine human GPCR structure and function. Nat Rev Drug Discov. 2013;12(1):25–34. https://doi.org/10.1038/nrd3859.

Subramanian A, Narayan R, Corsello SM, Peck DD, Natoli TE, Lu XD, Gould J, et al. A next generation connectivity map: L1000 platform and the first 1,000,000 profiles. Cell. 2007;171(6):1437–1452.e1417. https://doi.org/10.1016/j.cell.2017.10.049.

Wang X, Shen YH, Wang SW, Li SL, Zhang WL, Liu XF, Lai LH, Pei JF, Li HL. PharmMapper 2017 update: a web server for potential drug target identification with a comprehensive target pharmacophore database. Nucleic Acids Res. 2017;45(W1):W356–60. https://doi.org/10.1093/nar/gkx374.

Willett P, Barnard JM, Downs GM. Chemical similarity searching. J Chem Inf Model. 1998;38(6):983–96. https://doi.org/10.1021/ci9800211.

Wu ZR, Lu WQ, Wu D, Luo AQ, Bian HP, Li J, Li WH, Liu GX, Huang J, Cheng FX, Tang Y. *In silico* prediction of chemical mechanism of action via an improved network-based inference method. Br J Pharmacol. 2016;173(23):3372–85. https://doi.org/10.1111/bph.13629.

Wu ZR, Cheng FX, Li J, Li WH, Liu GX, Tang Y. SDTNBI: an integrated network and chemoinformatics tool for systematic prediction of drug-target interactions and drug repositioning. Brief Bioinform. 2017;18(2):333–47. https://doi.org/10.1093/bib/bbw012.

Wu ZR, Li WH, Liu GX, Tang Y. Network-based methods for prediction of drug-target interactions. Front Pharmacol. 2018a;9:1134. https://doi.org/10.3389/fphar.2018.01134.

Wu ZR, Lu WQ, Yu WW, Wang TDY, Li WH, Liu GX, Zhang HK, Pang XF, Huang J, Liu MY, Cheng FX, Tang Y. Quantitative and systems pharmacology 2. in silico polypharmacology of G protein-coupled receptor ligands via network-based approaches. Pharmacol Res. 2018b;129:400–13. https://doi.org/10.1016/j.phrs.2017.11.005.

Wu ZR, Peng YY, Yu ZH, Li WH, Liu GX, Tang Y. NetInfer: a web server for prediction of targets and therapeutic and adverse effects via network-based inference methods. J Chem Inf Model. 2020;60(8):3687–91. https://doi.org/10.1021/acs.jcim.0c00291.

Xiang J, Chun E, Liu C, Jing L, Al-Sahouri Z, Zhu L, Liu W. Successful strategies to determine high-resolution structures of GPCRs. Trends Pharmacol Sci. 2016;37(12):1055–69. https://doi.org/10.1016/j.tips.2016.09.009.

Yamanishi Y, Araki M, Gutteridge A, Honda W, Kanehisa M. Prediction of drug-target interaction networks from the integration of chemical and genomic spaces. Bioinformatics. 2008;24(13):I232–40. https://doi.org/10.1093/bioinformatics/btn162.

Yang SY. Pharmacophore modeling and applications in drug discovery: challenges and recent advances. Drug Discov Today. 2010;15(11-12):444–50. https://doi.org/10.1016/j.drudis.2010.03.013.

Yang HB, Lou CF, Sun LX, Li J, Cai YC, Wang Z, Li WH, Liu GX, Tang Y. admetSAR 2.0: web-service for prediction and optimization of chemical ADMET properties. Bioinformatics. 2019;35(6):1067–9. https://doi.org/10.1093/bioinformatics/bty707.

Yao ZJ, Dong J, Che YJ, Zhu MF, Wen M, Wang NN, Wang S, Lu AP, Cao DS. TargetNet: a web service for predicting potential drug-target interaction profiling via multi-target SAR models. J Comput Aid Mol Des. 2016;30(5):413–24. https://doi.org/10.1007/s10822-016-9915-2.

Yildirim MA, Goh KI, Cusick ME, Barabasi AL, Vidal M. Drug-target network. Nat Biotechnol. 2007;25(10):1119–26. https://doi.org/10.1038/nbt1338.

Yu H, Chen JX, Xu X, Li Y, Zhao HH, Fang YP, Li XX, Zhou W, Wang W, Wang YH. A systematic prediction of multiple drug-target interactions from chemical, genomic, and pharmacological data. PLoS One. 2012;7(5):e37608. https://doi.org/10.1371/journal.pone.0037608.

Zhou T, Ren J, Medo M, Zhang YC. Bipartite network projection and personal recommendation. Phys Rev E. 2007;76(4):046115. https://doi.org/10.1103/PhysRevE.76.046115.

Zhou T, Kuscsik Z, Liu JG, Medo M, Wakeling JR, Zhang YC. Solving the apparent diversity-accuracy dilemma of recommender systems. Proc Natl Acad Sci U S A. 2010;107(10):4511–5. https://doi.org/10.1073/pnas.1000488107.

The Current State of Precision Medicine and Targeted-Cancer Therapies: Where Are We?

Attila A. Seyhan

Contents

A. A. Seyhan (✉)
Laboratory of Translational Oncology and Experimental Cancer Therapeutics, Warren Alpert Medical School, Brown University, Providence, RI, USA

The Joint Program in Cancer Biology, Brown University and Lifespan Health System, Providence, RI, USA

Legorreta Cancer Center at Brown University, Warren Alpert Medical School, Brown University, Providence, RI, USA

Department of Pathology and Laboratory Medicine, Warren Alpert Medical School, Brown University, Providence, RI, USA
e-mail: Attila_Seyhan@brown.edu

M. T. Scotti, C. L. Bellera (eds.), *Drug Target Selection and Validation*, Computer–Aided Drug Discovery and Design,
https://doi.org/10.1007/978-3-030-95895-4_7

Abstract

Precision or personalized medicine is a strategy that allows the selection of specific treatments based on the patient's genetic or genomic changes that may occur in an individual's tumor. The goal of precision medicine is simply to deliver the right treatment to the right patient at the right dose and the right time.

Precision medicine involves the collection and validation of population-wide data, including multi-omics molecular profiling data such as epigenetic, genetic, genomic, proteomic, and metabolomic alterations to identify targetable/actionable molecular targets. Personalized medicine on the other hand involves the collection and analysis of an individual patient's tumor or liquid biopsy molecular profiling data and clinical data for their treatment with targeted therapies developed to be specific to those genetic or genomic abnormalities. The process of precision medicine may also involve the use of machine learning algorithms and artificial intelligence (AI) to both design a drug combination based on a patient's data and follow specific drug dosing regimens.

However, there are some challenges. For example, considerations for molecular profiling should include the type of tissues to utilize, timing of profiling in the disease course, comprehensiveness of the molecular panel as well as the interpretation of molecular data, which remains a challenge. Because resistance to targeted therapies is also a concern, resampling and retesting of tumors, including a dynamic sampling of liquid biopsy specimens after clinical progression, may help to make better treatment decisions. In addition, analysis, and interpretation of large molecular profiling datasets to identify clinically relevant markers for disease susceptibility and treatment efficacy and pairing these molecular signatures to specific targeted therapies remains a challenge. Furthermore, successful deployment of both precision and personalized medicine requires their effective integration, with genome-guided drug pairing (driven by population data) followed by AI-guided dynamic therapeutic intervention (driven by individual data).

Collectively, the successful deployment of precision medicine requires significant resources ranging from large-scale data generation to analysis capabilities, and coordination and effective collaboration between all relevant stakeholders such as scientists, clinicians, patients, biopharmaceutical companies, insurers, and regulatory agencies.

Keywords

Precision medicine · Personalized medicine · Biomarkers · Omics technologies · Targeted therapies

1 Introduction

Cancer causes 9.6 million deaths and 18.1 million new cases each year worldwide (Bray et al. 2018). The characteristics of cancer comprise sustaining proliferative signaling, evading growth suppressors, resisting cell death, enabling replicative immortality, inducing angiogenesis, and activating invasion and metastasis triggered by genome instability, chronic inflammation, evasion of the patient immune system, and others (Hanahan and Weinberg 2011).

Cancer is a complex heterogeneous multi-layered disease with multiple complex networks of interactions located at different levels such as cellular, tissue, and organ. At the cellular level, the networks consist of epigenetics and genes, miRNAs, signaling molecules (proteins, lipids, ions), and metabolic intermediates, whereas at the tissue level, the networks consist of interactions between different cell types and between the cells and the supporting stromal tissue, and at the organismic level, the networks consist of interactions between different body systems (endocrine, nervous, immune, etc.) (Paul 2020). In addition, cancer pathogenesis is a dynamic process that involves a sequence of epigenetic–genetic–environmental interactions in a progressive process where malfunction in any of the interconnected systems, including DNA repair, apoptotic and immune functions, and others influence gene expression, and tumorigenesis (Knox 2010). There are over 100 types of cancers, located in different organs and sub tissues, and cell types (Krzyszczyk et al. 2018). Based on their molecular subtypes, some cancers (e.g., colon, breast, and non-Hodgkin's lymphoma) have even more subclasses (Alizadeh et al. 2000; Guinney et al. 2015; Perou et al. 2000; Davies et al. 2002). Furthermore, the expression of biomarkers within the same tumor can vary based on their location or stage of cancer. Despite this, cancer patients who have the same type and stage of disease all receive the same types of standard cancer treatments, including surgery, radiation therapy, chemotherapy, and immunotherapy (Krzyszczyk et al. 2018).

It is now well recognized that due to disease, patient, and tumoral heterogeneity as well as some other risk factors that affect the individual's susceptibilities and risks for a specific disease, cancer treatments are effective only in a subset of the patient population.

Hitherto, cancer patients who have the same type and stage of disease all receive the same treatment although individual patients with the same disease respond differently to the same treatment.

For example, tumors can have different underlying genetic causes and may express different proteins in one patient versus another or within the same patient where the primary tumor can have different underlying genetic causes versus tumors that have metastasized to other parts of the body.

In addition, each tumor is dynamic and undergoes epigenetic, genetic, genomic, physiologic, anatomic, and other biological functions especially in metastasized tumors, and these changes may be specific to that individual patient resulting in variable responses to treatment.

Furthermore, the efficacy of treatments can be influenced by several other factors, including the type, stage, and location of cancer as well as the individual patient's age and overall health and environmental risk factors.

Likewise, physiological, environmental, and lifestyle factors also influence an individual's susceptibility to certain diseases and disease progression as well as response to treatments (Nicholson 2006; Hartmanshenn et al. 2016). The impact of physiology on the absorption, distribution, metabolism, and elimination (ADME) of drugs in the body has been well established (Kesisoglou et al. 2016; Rowland et al. 2011).

In addition, physiological differences due to age, sex, ethnicity, and stage of the disease also affect the individual's pharmacokinetic response to drugs and response to treatment (Hartmanshenn et al. 2016). For instance, drug clearance appears to be lower in geriatric patients compared to young adults, thus leading to different pharmacokinetic responses (Reeve et al. 2015).

Likewise, healthy diets and moderate exercise are associated with lower risks for disease, whereas unhealthy lifestyle choices, including smoking and alcohol consumption, have been associated with cancer and other diseases (Mashberg et al. 1993).

It is also worth mentioning that individual's socioeconomic status and their access to healthcare, healthcare inequality (inequity implies unfairness and injustice), cancer health disparities (disparity implies a difference of some kind) (Meghani and Gallagher 2008; Singh and Jemal 2017), and other risk factors also affect the individual's susceptibilities and risks for a specific disease and treatment outcomes.

Because of this, the traditional "one size fits all" approach for cancer therapy is only effective in a subset of patients and ineffective in many other patients resulting in serious side effects.

This inherent variability of cancer imparts itself to the emerging field of *precision medicine* also known as personalized medicine, which is a strategy tailored for each patient's tumor genetic or genomic makeup. Thus, precision medicine could be described in terms of the genetically or genomically altered targets involved in targeted therapy (Bashraheel et al. 2020).

Therefore, a more effective approach would be the targeted and thus more precise treatment approach for each specific subtype of cancer, based on patient's epigenetic, genetic, genomic, and other omics (e.g., transcriptomics, metabolomics, proteomics, etc.) data (Krzyszczyk et al. 2018) as well as patient's socioeconomic status, physiological, environmental, lifestyle, and other risk factors.

2 Precision (and Personalized) Medicine

Precision medicine, also known as personalized medicine, is an approach that is applied at the individual patient level using the patient's own molecular and clinical data. It is a strategy that allows the selection of treatment according to a patient's genetic or genomic characteristics allowing the treatment to be customized according to the genetic or genomic changes that may occur in an individual's tumor. Thus, precision medicine could be described in terms of the genetically or genomically altered targets involved in targeted therapy. Personalized medicine on the other hand is a medical treatment that is customized to the individual characteristics of each patient. This approach has emerged only after the development of advanced genomics, diagnostics, digital devices, and imaging as well as bioinformatics, artificial intelligence, and other analytics tools.

However, the National Research Council Committee states in its report that the term "personalized medicine" is sometimes misinterpreted as implying that unique treatments can be designed for each patient. Because of this, the Committee, therefore, preferred the term "precision medicine" over "personalized medicine."

Regardless of the semantics, the precision medicine (or personalized medicine) approach has changed the traditional clinical practice and how we treat patients from a "one size fits all approach" to "the right drug, for the right patient, at the right time approach."

For example, in precision oncology, precision oncology pairs one driver mutation with one drug, which may be further optimized by treating molecularly complex and heterogeneous cancers with combinations of targeted drugs (Sicklick et al. 2019). The literature to date supports the therapeutic benefits of a precision medicine approach to cancer therapy (Bashraheel et al. 2020).

Besides genomics and other "omics" based molecular profiling technology platforms, precision medicine also incorporates other disciplines such as basic biological, physical, computational, bioinformatics, statistics, clinical, and health science (Hawgood et al. 2015). More recently machine learning and artificial intelligence (AI) have been gaining wider acceptance that allows analysis and interpretation of massive data generated from these molecular profiling efforts to develop a mechanism-based and data-driven approach to map out individual patient's responses to a specific disease state and treatment.

Although the conventional therapeutic approach for cancer has focused on a single type of intervention for all patients, in recent years, the therapeutic benefits of a precision medicine approach to therapy have been gaining recognition (Bashraheel et al. 2020; Sicklick et al. 2019).

The field has gained wide recognition and acceptance across basic and clinical research, the pharmaceutical industry, biopharmaceutical and commercial testing laboratories, regulatory agencies, and others. For example, there are increasing efforts including the commercial resources to collect individual tumor molecular profiling and patient outcome data and other relevant data including medication and current treatment information, and others to characterize molecular differences between tumors. Some precision medicine products and resources are already available (e.g., Tempus, Foundation Medicine, Caris Life Sciences, and many others) to link these differences to an effective drug such as precision medicine for cancer treatments, which could benefit patient outcomes (Schwartzberg et al. 2017). However, the clinical interpretation of molecular changes from these efforts is at the core of providing the value of precision medicine.

3 Precision Medicine: How It All Began

Precision medicine is an emerging field whereby it aims to deliver the right treatment to the right patient at the right dose and the right time. In precision medicine, patients with tumor profiles that share the same genetic or genomic alterations are treated with drugs that target those alterations regardless of the tumor type, which is now becoming a tumor-agnostic targeted therapy strategy.

Precision medicine began ca. in 1998 with the discovery of the *BCR-ABL* rearrangement in chronic myeloid leukemia that was successfully targeted by the drug imatinib. This resulted in significant clinical remissions leading to the approval by the U.S. Food and Drug Administration in 2001 (Schwartzberg et al. 2017).

Incidentally, for the first time in history, the first draft of the human genome sequence was delivered in the same year (Venter et al. 2001). This was subsequently followed by the first cancer genome (Sjoblom et al. 2006).

The discovery of several nonoverlapping driver mutations and tyrosine kinase (TK) inhibitors in non-small cell lung cancer and melanoma led to assays of alterations conducted by polymerase chain reaction (PCR) enabling these biomarkers to be used for treatment decisions in solid tumors, which signified the potential of molecular profiling.

Likewise, the advent of next-generation sequencing (NGS) on formalin-fixed, paraffin-embedded (FFPE) tissue enabled massive parallel ultra-high-throughput sequencing with scalability and speed for the identification of genetic alterations in a large number of genes rapidly and cost-effectively.

These efforts led to the rapid identification of many actionable somatic mutations such as the driver mutations forming the foundation of precision oncology (Stratton et al. 2009).

Driver mutations are those where mutations in oncogenes make them constitutively active leading to the progression of the disease: whereas tumor suppressor genes keep uncontrolled cellular proliferation and hence tumor formation under control. When inactivated through mutation or allele loss, tumor suppressor genes often cause uncontrolled cellular proliferation, tumor pathogenesis, and cancer progression.

Multiple processes are involved in aberrant genetic, epigenetic, and genomic function, leading to altered expression of the protein coded for by the gene. To capture potential alterations at the genetic, epigenetic, and genomic levels, it is believed that a multi-omics approach is the best strategy to capture these alterations.

Ideally, a genomic test with clinical utility should be predictive of treatment response from a targeted agent. An early example in solid tumor oncology was testing the tumor specimens of patients for *HER2* gene amplification (by fluorescent in situ hybridization) or protein overexpression (by immunohistochemistry). In this case, *HER2*-positive tumors predicted the patient's response to HER2-targeted trastuzumab therapies; conversely, *HER2*-negative tumors did not benefit from this therapy.

It is also critical to understand the difference between germline alterations and somatic alterations. The recently incorporated BRCA germline testing for all patients with pancreatic cancer is a good example of this. Germline testing involves an extensive coverage of BRCA, whereas current somatic testing covers only certain regions of the BRCA gene (El-Deiry et al. 2019).

The field of molecular profiling now has advanced into the field of single-cell RNA sequencing for gaining insights into those alterations at the single-cell level for deciphering tumor heterogeneity, the development of biomarkers, pathway or gene networks, and signatures for application to specific patient groups. In addition, the field of molecular profiling also adapted the multiplex gene testing for other biologic species, including mRNA, microRNA (miRNA), and proteins to predict response to a paired agent.

Despite the advancements, precision medicine is still evolving and the results obtained from these approaches to treatment have been mixed (Ravoori et al. 2015; What 2015).

As reported previously (Le Tourneau et al. 2015), a prospective phase II clinical trial of molecular profiling to link matched therapy did not show superior outcomes for the matched group but was undermined from methodologic study design problems.

Previous large retrospective studies have demonstrated that a significant proportion of patients (80–90%) who have undergone molecular profiling tests would have potentially actionable genomic alterations, though the definition of actionable genomic alterations may vary (Sholl et al. 2016; Meric-Bernstam et al. 2015; Johnson et al. 2014; Schwaederle et al. 2015).Nevertheless, evidence suggests that only a few patients to date actually received genomically targeted therapy in a clinical trial.

4 The Precision Medicine Process and Integration into Cancer Treatment

As illustrated in Fig. 1, precision medicine aims at the development of specialized treatments for each specific subtype of cancer, based on the molecular profiling of key patient genetic and omics data (epigenomics, genomics, transcriptomics, metabolomics, proteomics, etc.) (Krzyszczyk et al. 2018).

For example, identification of a mutation in the anaplastic lymphoma kinase (ALK) that leads to tumor formation in ca. 5% of non-small-cell lung cancers (Soda et al. 2007) led to the development of ALK blockers such as crizotinib and ceritinib; subsequently, leading to the FDA approval of drugs for patients who test positive for the ALK mutation. Another example is the development of the poly ADP ribose polymerase inhibitor olaparib in the treatment of BRCA-mutant ovarian cancer (Ledermann et al. 2014).

In addition to the development of targeted therapies based on these molecular alterations, companion diagnostics (CDx)—which are molecular assays that detect levels of proteins, genes or gene products, or specific mutations to help guide the selection of a specific targeted therapy according to individual patient's molecular profile and the type of cancer they have—have been increasingly used in recent years (Verma 2012).

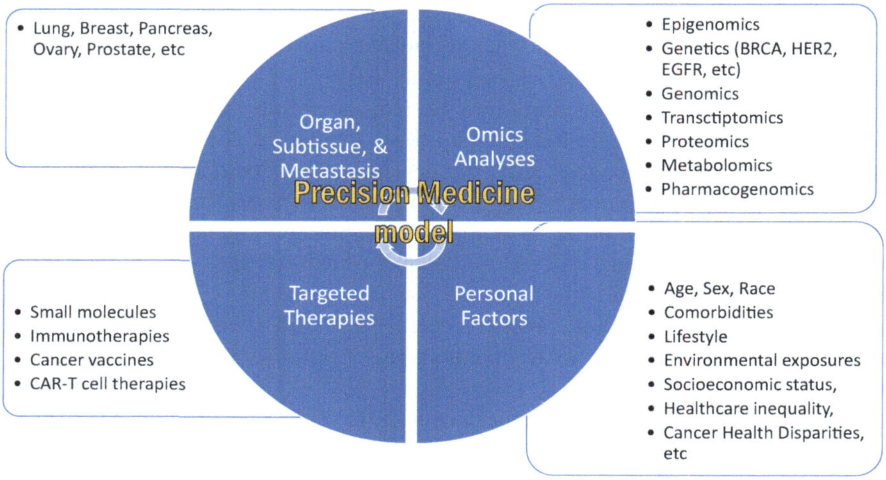

Fig. 1 The precision medicine approach. The precision medicine approach is characterized by individualized treatments customized to specific tissues, gene mutations or genomic, or other molecular alterations and personal factors of the patient relevant to each unique case of cancer. Companion diagnostics help to identify targeted therapies that may be most effective for a specific patient's tumor. Adapted from (Krzyszczyk et al. 2018)

A CDx device can be an in vitro diagnostic device or an imaging tool that provides information that is required for the safe and effective use of a corresponding therapeutic product. The use of a CDx device is specified in the instructions for use in the labeling of the diagnostic device, as well as in the labeling of the therapeutic product. For example, Table 1 exemplifies lists of cleared or approved CDx devices.

For instance, HERCEPTEST (Dako Denmark A/S) and HER2 FISH (PharmDx Kit) have been used to test HER2 protein and gene overexpression in breast, metastatic gastric, or gastroesophageal junction adenocarcinoma fixed tissues, BRACAnalysis CDx (Myriad Genetic Labs) detects and classifies DNA variants in the protein-coding region of the BRCA1/2 genes using patient's whole blood samples. These companion diagnostics allow for the selection of a specific treatment that is more likely to be effective for each patient based on the specific characteristics of the patient's tumor or other liquid biopsy tissue molecular profiles.

As a result, the FDA has supported the precision medicine approach with their approval of these and other technologies since 1998, when trastuzumab was approved for the treatment of HER2 receptor-positive breast cancer.

Furthermore, the ratification of the Precision Medicine Initiative in 2015 has also contributed additional support for precision medicine by requiring the FDA to develop new platforms to evaluate precision medicine diagnostics and therapies.

Consequently, the integration of the precision medicine approach into cancer research and treatment has already gained wide acceptance among cancer researchers, patient care providers, commercial test providers, biopharmaceutical industry, and regulatory agencies such as FDA and CMS oversight, leading to major improvements in patient outcomes.

Table 1 Examples of list of cleared or approved companion diagnostic devices (in vitro and imaging tools) (Adapted from: https://www.fda.gov/medical-devices/in-vitro-diagnostics/list-cleared-or-approved-companion-diagnostic-devices-in-vitro-and-imaging-tools)

Diagnostic name	PMA/510(k)/ HDE	Diagnostic manufacturer	Indication(s) Trade name (Generic)—NDA/BLA	Device indication for a specific group of oncology therapeutic products and trade name (Generic)—NDA/BLA
BRACAnalysis CDx	P140020/S016 P140020/S019 P140020/S020	Myriad Genetic Laboratories, Inc.	Breast cancer • Lynparza (olaparib)—NDA 208558 • Talzenna (talazoparib)—NDA 211651 Ovarian cancer • Lynparza (olaparib)—NDA 208558 • Rubraca (rucaparib)—NDA 209115 Pancreatic cancer • Lynparza (olaparib)—NDA 208558 Metastatic castrate resistant prostate cancer (mCRPC) • Lynparza (olaparib)—NDA 208558	
therascreen EGFR RGQ PCR Kit	P120022/S018	Qiagen Manchester, Ltd.	Non-small cell lung cancer • Iressa (gefitinib)—NDA 206995 • Gilotrif (afatinib)- NDA 201292 • Vizimpro (dacomitinib)- NDA 211288	

(continued)

Table 1 (continued)

Diagnostic name	PMA/510(k)/ HDE	Diagnostic manufacturer	Indication(s) Trade name (Generic)—NDA/BLA	Device indication for a specific group of oncology therapeutic products and trade name (Generic)—NDA/BLA
cobas EGFR Mutation Test v2	P120019 P120019/S007 P120019/S016 P120019/S018 P120019/ S019P120019/ S031	Roche Molecular Systems, Inc.	Non-small cell lung cancer • EGFR exon 19 deletions • EGFR exon 21(L858R) Tissue and plasma • See next column	Non-small cell lung cancer (tissue) "identifying patients with NSCLC whose tumors have EGFR exon 19 deletions or exon 21 (L858R) substitution mutations and are suitable for treatment with a tyrosine kinase inhibitor approved by FDA for that indication" List of tyrosine kinase inhibitors approved by FDA for this indication: • Tarceva (erlotinib)—NDA 021743 • Tagrisso (osimertinib)—NDA 208065 • Iressa (gefitinib)—NDA 206995 • Gilotrif (afatinib)—NDA 201292 • Vizimpro (dacomitinib)— NDA 211288 Non-small cell lung cancer (plasma) "identifying patients with NSCLC whose tumors have EGFR exon 19 deletions or exon 21 (L858R) substitution mutations and are suitable for treatment with a tyrosine kinase inhibitor approved by FDA for that indication" List of tyrosine kinase inhibitors approved by FDA for this indication: • Tarceva (erlotinib)—NDA 021743 • Tagrisso (osimertinib)—NDA 208065 • Iressa (gefitinib)—NDA 206995

Diagnostic name	PMA/510(k)/ HDE	Diagnostic manufacturer	Indication(s) Trade name (Generic)—NDA/BLA	Device indication for a specific group of oncology therapeutic products and trade name (Generic)—NDA/BLA
PD-L1 IHC 22C3 pharmDx	P150013 P150013/S006 P150013/S009 P150013/S011 P150013/S014 P150013/S016 P150013/S020 P150013/S021	Dako North America, Inc.	EGFR T790M (Tissue and plasma) • Tagrisso (osimertinib)—NDA 208065Non-small cell lung cancer (NSCLC), gastric or gastroesophageal junction adenocarcinoma, cervical cancer, urothelial carcinoma, head and neck squamous cell carcinoma (HNSCC), and esophageal squamous cell carcinoma (ESCC) and triple-negative breast cancer (TNBC) • KEYTRUDA (pembrolizumab)—BLA 125514 • Libtayo (cemiplimab-rwlc)—BLA 761097	
Abbott Real*Time* IDH1	P170041	Abbott Molecular, Inc.	Acute myeloid leukemia • Tibsovo (ivosidenib)—NDA 211192	
MRDx BCR-ABL Test	K173492	MolecularMD Corporation	Chronic myeloid leukemia • Tasigna (nilotinib)—NDA 022068/S026	
FoundationOne CDx	P170019 P170019/S004 P170019/S006 P170019/S008 P170019/S011 P170019/S013 P170019/S015 P170019/S016 P170019/S017	Foundation Medicine, Inc.	Non-small cell lung cancer • Gilotrif (afatinib)—NDA 201292 • Iressa (gefitinib)—NDA 206995 • Tarceva (erlotinib)—NDA 021743 • Tagrisso (osimertinib) NDA 208065 • Alecensa (alectinib)—NDA 208434 • Xalkori (crizotinib)—NDA 202570 • Zykadia (ceritinib)—NDA 205755 • Tafinlar (dabrafenib)—NDA 202806 in combination with Mekinist (trametinib)—NDA 204114 • Tabrecta (capmatinib)—NDA 213591	

(continued)

Table 1 (continued)

Diagnostic name	PMA/510(k)/ HDE	Diagnostic manufacturer	Indication(s) Trade name (Generic)—NDA/BLA	Device indication for a specific group of oncology therapeutic products and trade name (Generic)—NDA/BLA
			Melanoma • Tafinlar (dabrafenib)—NDA 202806 • Zelboraf (vemurafenib)—NDA 202429 • Mekinist (trametinib)—NDA 204114 or Cotellic (cobimetinib)—NDA 206192 in combination with Zelboraf (vemurafenib)—NDA 202429 Breast cancer • Herceptin (trastuzumab)—BLA 103792 • Perjeta (pertuzumab)—BLA 125409 • Kadcyla (ado-trastuzumab emtansine)—BLA 125427 • Piqray (alpelisib)—NDA 212526 Colorectal cancer • Erbitux (cetuximab)—BLA 125084 • Vectibix (panitumumab)—BLA 125147 Ovarian cancer • Rubraca (rucaparib)—NDA 209115 • Lynparza (olaparib)—NDA 208558 Cholangiocarcinoma • Pemazyre (pemigatinib)—NDA 213736 Metastatic castrate resistant prostate cancer (mCRPC) • Lynparza (olaparib)—NDA 208558 Solid tumors (TMB ≥ 10 mutations per megabase) • Keytruda (pembrolizumab)—BLA 125514 Solid tumors (NTRK1/2/3 fusions) • Vitrakvi (larotrectinib)—NDA 210861, 211710	

Diagnostic name	PMA/510(k)/ HDE	Diagnostic manufacturer	Indication(s) Trade name (Generic)—NDA/BLA	Device indication for a specific group of oncology therapeutic products and trade name (Generic)—NDA/BLA
VENTANA ALK (D5F3) CDx Assay	P140025/S006 P140025/S014	Ventana Medical Systems, Inc.	Non-small cell lung cancer • Zykadia (ceritinib)—NDA 205755 • Xalkori (crizotinib)—NDA 202570 • Alecensa (alectinib)—NDA 208434 • Lorbrena (lorlatinib)—NDA 210868	
Abbott RealTi*me* IDH2	P170005	Abbott Molecular, Inc.	Acute myeloid leukemia • Idhifa (enasidenib)—NDA 209606	
Praxis Extended RAS Panel	P160038	Illumina, Inc.	Colorectal cancer • Vectibix (panitumumab)—NDA 125147	
Oncomine Dx Target Test	P160045 P160045/S019	Life Technologies Corporation	Non-small cell lung cancer • Tafinlar (dabrafenib) in combination with Mekinist (trametinib)—NDA 202806 and NDA 204114 • Xalkori (crizotinib)—NDA 202570 • Iressa (gefitinib)—NDA 206995 • Gavreto (pralsetinib)—NDA 213721	
LeukoStrat CDx FLT3 Mutation Assay	P160040	Invivoscribe Technologies, Inc.	Acute myelogenous leukemia • Rydapt (midostaurin)—NDA 207997 • Xospata (gilterinib)—NDA 211349	
FoundationFocus CDx*BRCA* Assay	P160018	Foundation Medicine, Inc.	Ovarian cancer • Rubraca (rucaparib)—NDA 209115	
Vysis CLL FISH Probe Kit	P150041	Abbott Molecular, Inc.	B-cell chronic lymphocytic leukemia • Venclexta (venetoclax)—NDA 208573	
KIT D816V Mutation Detection by PCR for Gleevec Eligibility in Aggressive Systemic Mastocytosis (ASM)	H140006	ARUP Laboratories, Inc.	Aggressive systemic mastocytosis • Gleevec (imatinib mesylate)—NDA 021335	

(continued)

Table 1 (continued)

Diagnostic name	PMA/510(k)/ HDE	Diagnostic manufacturer	Indication(s) Trade name (Generic)—NDA/BLA	Device indication for a specific group of oncology therapeutic products and trade name (Generic)—NDA/BLA
PDGFRB FISH for Gleevec Eligibility in Myelodysplastic Syndrome / Myeloproliferative Disease (MDS/MPD)	H140005	ARUP Laboratories, Inc.	Myelodysplastic syndrome/myeloproliferative disease • Gleevec (imatinib mesylate)—NDA 021335	
cobas KRAS Mutation Test	P140023	Roche Molecular Systems, Inc.	Colorectal cancer • Erbitux (cetuximab)—BLA 125084 • Vectibix (panitumumab)—BLA 125147	
therascreen KRAS RGQ PCR Kit	P110030 P110027	Qiagen Manchester, Ltd.	Colorectal cancer • Erbitux (cetuximab)—BLA 125084 • Vectibix (panitumumab)—BLA 125147	
Dako EGFR pharmDx Kit	P030044/S002	Dako North America, Inc.	Colorectal cancer • Erbitux (cetuximab)—BLA 125084 • Vectibix (panitumumab)—BLA 125147	
FerriScan	DEN130012/ K124065	Resonance Health Analysis Services Pty Ltd	Non-transfusion-dependent thalassemia • Exjade (deferasirox)—NDA 021882	
Dako c-KIT pharmDx	P040011	Dako North America, Inc.	Gastrointestinal stromal tumors • Gleevec (imatinib mesylate)—NDA 021335 • Glivec (imatinib mesylate)—NDA 021588	
INFORM HER-2/neu	P940004	Ventana Medical Systems, Inc.	Breast cancer • Herceptin (trastuzumab)—BLA 103792	
PathVysion HER-2 DNA Probe Kit	P980024	Abbott Molecular Inc.	Breast cancer • Herceptin (trastuzumab)—BLA 103792	

Diagnostic name	PMA/510(k)/ HDE	Diagnostic manufacturer	Indication(s) Trade name (Generic)—NDA/BLA	Device indication for a specific group of oncology therapeutic products and trade name (Generic)—NDA/BLA
PATHWAY anti-Her2/neu (4B5) Rabbit Monoclonal Primary Antibody	P990081/ S001-S028 P990081/S039	Ventana Medical Systems, Inc.	Breast cancer • Herceptin (trastuzumab)—BLA 103792 • Kadcyla (ado-trastuzumab emtansine)—BLA 125427	
InSite Her-2/neu KIT	P040030	Biogenex Laboratories, Inc.	Breast cancer • Herceptin (trastuzumab)—BLA 103792	
SPOT-LIGHT HER2 CISH Kit	P050040/ S001-S003	Life Technologies Corporation	Breast cancer • Herceptin (trastuzumab)—BLA 103792	
Bond Oracle HER2 IHC System	P090015	Leica Biosystems	Breast cancer • Herceptin (trastuzumab)—BLA 103792	
HER2 CISH pharmDx Kit	P100024	Dako Denmark A/S	Breast cancer • Herceptin (trastuzumab)—BLA 103792	
INFORM HER2 Dual ISH DNA Probe Cocktail	P100027 P100027/S030	Ventana Medical Systems, Inc.	Breast cancer • Herceptin (trastuzumab)—BLA 103792 • Kadcyla (ado-trastuzumab emtansine)—BLA 125427	
HercepTest	P980018/S018	Dako Denmark A/S	Breast cancer • Herceptin (trastuzumab)—BLA 103792 • Perjeta (pertuzumab)—BLA 125409 • Kadcyla (ado-trastuzumab emtansine)—BLA 125427 Gastric and gastroesophogeal cancer • Herceptin (trastuzumab)—BLA 103792	
HER2 FISH pharmDx Kit	P040005 P040005/S005 P040005/S006 P040005/S009	Dako Denmark A/S	Breast cancer • Herceptin (trastuzumab)—BLA 103792 • Perjeta (pertuzumab)—BLA 125409 • Kadcyla (ado-trastuzumab emtansine)—BLA 125427 Gastric and gastroesophogeal cancer • Herceptin (trastuzumab)—BLA 103792	

(continued)

Table 1 (continued)

Diagnostic name	PMA/510(k)/ HDE	Diagnostic manufacturer	Indication(s) Trade name (Generic)—NDA/BLA	Device indication for a specific group of oncology therapeutic products and trade name (Generic)—NDA/BLA
THXID BRAF Kit	P120014	bioMérieux Inc.	Melanoma • Braftovi (encorafenib) in combination with Mektovi (binimetinib)—NDA 210496 and NDA 210498 • Mekinist (tramatenib)—NDA 204114 • Tafinlar (dabrafenib)—NDA 202806	
Vysis ALK Break Apart FISH Probe Kit	P110012 P110012/S020	Abbott Molecular Inc.	Non-small cell lung cancer (NSCLC) • Xalkori (crizotinib)—NDA 202570 • Alunbrig (brigatinib)—NDA 208772	
cobas 4800 BRAF V600 Mutation Test	P110020/S016	Roche Molecular Systems, Inc.	Melanoma • Zelboraf (vemurafenib)—NDA 202429 • Cotellic (cobimetinib)—NDA 206192 in combination with Zelboraf (vemurafenib)—NDA 202429	
VENTANA PD-L1(SP142) Assay	P160002/S006 P160002/S009 P160002/S012	Ventana Medical Systems, Inc.	Urothelial carcinoma, Triple-Negative Breast Carcinoma (TNBC) and Non small cell lung cancer (NSCLC) • TECENTRIQ (atezolizumab)—BLA 761034	
therascreen FGFR RGQ RT-PCR Kit	P180043	QIAGEN Manchester Ltd.	Urothelial cancer • BALVERSA (erdafitinib)—NDA 212018	
therascreen PIK3CA RGQ PCR Kit	P190001 P190004	QIAGEN GmbH	Breast cancer (tissue and plasma) • PIQRAY (alpelisib)—NDA 212526	
Myriad myChoice® CDx	P190014 P190014/S003	Myriad Genetic Laboratories, Inc.	Ovarian cancer • Zejula® (niraparib)—NDA 208447 • Lynparza (olaparib)—NDA 208558	
therascreen BRAF V600E RGQ PCR Kit	P190026	QIAGEN GmbH	Colorectal cancer • BRAFTOVI (encorafenib)—NDA 210496 in combination with Erbitux (cetuximab)—BLA 125084	

Diagnostic name	PMA/510(k)/ HDE	Diagnostic manufacturer	Indication(s) Trade name (Generic)—NDA/BLA	Device indication for a specific group of oncology therapeutic products and trade name (Generic)—NDA/BLA
PD-L1 IHC 28–8 pharmDx	P150025/S013	Dako North America, Inc.	Non-small cell lung cancer (NSCLC) • OPDIVO (nivolumab) (BLA 125554) in combination with YERVOY (ipilimumab) (BLA 125377)	
cobas EZH2 Mutation Test	P200014	Roche Molecular Systems, Inc.	Follicular lymphoma tumor • Tazverik (tazemetostat)—NDA 213400	
VENTANA HER2 Dual ISH DNA Probe Cocktail	P190031	Ventana Medical Systems, Inc.	Breast cancer • Herceptin (trastuzumab)—BLA 103792	
Guardant360® CDx	P200010 P200010/S001	Guardant Health, Inc.	Non-small cell lung cancer (plasma) • Tagrisso (osimertinib)—NDA 208065 • Rybrevant (amivantamab-vmjw)—BLA 761210	
FoundationOne® Liquid CDx	P190032 P200006 P200016	Foundation Medicine, Inc.	Non-small cell lung cancer (plasma) • Iressa (gefitinib)—NDA 206995 • Tagrisso (osimertinib)—NDA 208065 • Tarceva (erlotinib)—NDA 021743 • Alecensa (alectinib)—NDA 208434 Metastatic castrate resistant prostate cancer (mCRPC) (plasma) • Rubraca (rucaparib)—NDA 209115 • Lynparza (olaparib)—NDA 206162, 208558 Ovarian cancer (plasma) • Rubraca (rucaparib)—NDA 209115 Breast cancer (plasma) • Piqray (alpelisib)—NDA 212526	
VENTANA MMR RxDx Panel	P200019	Ventana Medical Systems, Inc.	Endometrial Carcinoma (EC) • Jemperli (dostarlimab-gxly)—BLA 761174	

However, there are some challenges, including regulatory challenges, economic concerns (cost savings for pharmaceutical companies, cost of the tests, and insurance coverage), feasibility, associated ethics (socioeconomic advantage, data privacy concerns, informed patient consent), and others, that must be addressed before precision medicine can become fully integrated into standardized care.

Nevertheless, precision medicine is rapidly evolving and becoming more efficient and sophisticated; however, many of the questions raised above must be addressed before precision medicine is fully integrated into standard of cancer care.

5 How Precise Is Precision Medicine?

By definition, precision medicine refers to the tailoring of disease prevention and treatment based on the characteristics of each patient. However, a broader definition of precision medicine is a more integrated multidisciplinary approach that comprises not just the genetic, genomic, transcriptomic, or other "omics" makeup of an individual, but also other factors such as environmental exposures, lifestyle, tobacco use, alcohol use, unhealthy diet, obesity, sun exposure, and physical inactivity as well as individuals socioeconomic status, healthcare inequality (inequity implies unfairness and injustice), cancer health disparities (disparity implies a difference of some kind) (Meghani and Gallagher 2008; Singh and Jemal 2017), and other risk factors that affect the individual's susceptibilities and risks for a specific disease and treatment outcomes.

Thus, a more comprehensive understanding of the risk factors that dictate a person's life and disease might enable us to better incorporate these factors to deliver therapies that are both more precise and personalized.

In literature, the definition of precision medicine has evolved and the term "precision oncology" has been used to define diverse strategies in oncology ranging from the use of targeted therapies generally to using data from next-generation sequencing to select therapy for a person independent of cancer type (Prasad and Gale 2017). For example, in early iterations of precision medicine, a broad definition of precision medicine that included blood-typing, targeted therapies, and even immune therapy was used (Collins and Varmus 2015), whereas a narrower definition of precision medicine was used by others, which includes the use of next-generation sequencing of tissues to guide targeted therapy.

To understand what precision oncology means in current contexts, Prasad and Gale (2017) analyzed the use of "precision oncology" in the literature by searching Google Scholar for "precision oncology" over three intervals and by classifying 50 articles. In their analysis, they found that in the earliest versions, "precision oncology" predominantly described targeted therapies (e.g., VEGF or BCR/ABL1 targeted therapies, including bevacizumab and imatinib). Thereafter, precision oncology was used to define the selection of therapies based on data from analyses of biomarkers. These include the use of crizotinib in lung cancer patients with EML4-ALK rearrangements or adjuvant chemotherapy based on the data from genomic testing (e.g., the Oncotype DX panel) in breast cancer. Since 2016, the term precision oncology has been used to describe the use of data from next-generation sequencing to guide therapies.

More recently, precision oncology has been used to describe a tumor agnostic targeted therapy independent of cancer type as currently defined (based on anatomy and histology) and instead by mutation. For example, patients with BRAF V600E mutations would be treated with a BRAF inhibitor irrespective of if they have breast cancer, melanoma, colorectal cancer, acute myeloid leukemia, multiple myeloma, and other types of cancers (Smiech et al. 2020).

The idea of precision medicine is that treatments can be customized to the genetic alterations in each patient's cancer leading the way to the treatments a patient's tumor is most likely to respond to, and not treating those patients that are not likely to benefit from this treatment. As such, information about these genetic alterations in a tumor can help clinicians to decide which targeted therapy will work best for a patient.

Currently, several clinical studies are investigating treating patients with targeted therapies that target the cancer-causing epigenetic, genetic, or genomic alterations in their tumors, regardless of the organ site of origin where the tumor develops in the body.

6 Advent of Targeted Cancer Therapy

As discussed in detail by Dupont et al. (2021), therapeutic target discovery has been focused on the druggable genome, which was described as genes and gene products known or predicted to interact with orally bioavailable compounds (Hopkins and Groom 2002).

In addition to the presence of a protein structure that can be bound by small molecules, other potential druggable proteins are those that play important biological functions, and thus targeting those proteins might provide therapeutic benefit for the patient. For example, several protein families that play key biological functions have been the focal point of cancer drug development, such as G protein-coupled receptor (GPCR), protein kinases, and nuclear/hormone receptors (Kinch 2014).

With the development of novel therapeutic agents, such as monoclonal antibodies and small-molecule kinase inhibitors, a targeted therapy strategy has benefited many cancer patients.

As illustrated in Table 2, many targeted therapies have been approved by the FDA in the past several decades for the treatment of a variety of cancer types with specific genetic, genomic, or epigenetic alterations paired to specific targeted therapies. These targeted therapies include a variety of drug categories, such as antibody-drug conjugates, cytotoxic drugs, hormone therapy, immunotherapy, monoclonal antibody, protein kinase inhibitors, radioactive therapeutic agents, and small-molecule inhibitors (Dupont et al. 2021).

For example, the anti-CD20 monoclonal antibody rituximab was approved for the treatment of low-grade B cell lymphoma in 1997, and thereafter, the anti-HER2

Table 2 List of targeted therapies approved by the FDA

Drug	Primary targets	Drug category	Classification	Cancer type (FDA-approved)	Indications	Year FDA-approved
Ado-trastuzumab emtansine	ERBB2	Antibody-drug conjugate	ERBB2-directed antibody and microtubule inhibitor conjugate	Breast cancer	HER2+ breast cancer	2013
Belantamab mafodotin-blmf	BCMA	Antibody-drug conjugate	BCMA-directed antibody and microtubule inhibitor conjugate	Multiple myeloma	Multiple myeloma	2020
Brentuximab vedotin	TNFRSF8	Antibody-drug conjugate	CD30-directed antibody and antimitotic agent MMAE conjugate	Lymphoma	cHL; sALCL; PTCL	2011
Enfortumab vedotin-ejfv	Nectin-4	Antibody-drug conjugate	Nectin-4-directed antibody and microtubule inhibitor conjugate	Urothelial cancer	Urothelial cancer	2019
Fam-trastuzumab deruxtecan-nxki	ERBB2	Antibody-drug conjugate	HER2-directed antibody and topoisomerase inhibitor conjugate	Breast cancer	HER2+ breast cancer	2019
Gemtuzumab ozogamicin	CD33; DNA	Antibody-drug conjugate	CD33-directed antibody and cytotoxic drug conjugate	Leukemia	CD33-positive AML	2000
Ibritumomab tiuxetan	CD20	Antibody-drug conjugate	CD20-directed radiotherapeutic antibody	Lymphoma	NHL	2002
Inotuzumab ozogamicin	CD22	Antibody-drug conjugate	CD22-directed antibody-drug conjugate	Leukemia	ALL	2017
Moxetumomab pasudotox-tdfk	CD22	Antibody-drug conjugate	CD22-directed antibody-drug conjugate	Leukemia	Hairy cell leukemia	2018
Polatuzumab vedotin-piiq	CD79b	Antibody-drug conjugate	CD79b-directed antibody–drug conjugate	Lymphoma	DLBCL	2019
Sacituzumab Govitecan-HZIY	TOP1	Antibody-drug conjugate	Trop-2-directed antibody and topoisomerase inhibitor conjugate	Breast cancer	mTNBC	2020

Drug	Primary targets	Drug category	Classification	Cancer type (FDA-approved)	Indications	Year FDA-approved
Altretamine	DNA	Cytotoxic drug	Alkylating agent	Ovarian cancer	Recurrent ovarian cancer	1990
Arsenic trioxide	Unknown	Cytotoxic drug	Arsenic derivative	Leukemia	APL characterized by the presence of the t(15;17) translocation or PML/RAR-alpha gene expression	2000
Asparaginase E.coli	L-asparagine	Cytotoxic drug	Antimetabolite	Leukemia	ALL	1978
Asparaginase Erwinia chrysanthemi	L-asparagine	Cytotoxic drug	Antimetabolite	Leukemia	ALL	2011
Azacitidine	DNMT1; RNA; DNA	Cytotoxic drug	Nucleoside analogue	Leukemia; Melodysplasia	CMML; AML; MDS	2004
Bendamustine	DNA	Cytotoxic drug	Alkylating agent	Leukemia; Lymphoma	CLL; NHL; DLBCL	2008
Bleomycin sulfate	DNA; LIG1; LIG3	Cytotoxic drug	Cytotoxic glycopeptide antibiotics	Head and neck cancer; Lymphoma; Testicular cancer	HNSCC; Lymphoma; Testicular carcinoma	1973
Busulfan	DNA	Cytotoxic drug	Alkylating agent	Leukemia	CML	1954
Cabazitaxel	TUBA4A; TUBB1	Cytotoxic drug	Cytoskeletal disruptor	Prostate cancer	Metastatic castration-resistant prostate cancer	2010
Calaspargase pegol-mknl	L-asparagine	Cytotoxic drug	Antimetabolite	Leukemia	ALL	2018
Capecitabine	DNA synthesis; RNA synthesis; Protein synthesis; TYMS	Cytotoxic drug	Nucleoside precursor	Breast cancer; Colorectal cancer	Metastatic breast cancer; Metastatic CRC; Dukes'C colorectal cancer	1998

(continued)

Table 2 (continued)

Drug	Primary targets	Drug category	Classification	Cancer type (FDA-approved)	Indications	Year FDA-approved
Carboplatin	DNA	Cytotoxic drug	Platinum-based drug	Ovarian cancer; Lung cancer	Advanced ovarian carcinoma; Lung cancer	1989
Carmustine	DNA; RNA	Cytotoxic drug	Alkylating agent	Brain cancer; Lymphoma; Multiple myeloma	Glioblastoma; Brainstem glioma; Medulloblastoma; Astrocytoma; Ependymoma; Metastatic brain tumors; Lymphoma; Multiple Myeloma	1977
Chlorambucil	DNA	Cytotoxic drug	Alkylating agent	Leukemia; Lymphoma	CLL; Malignant lymphoma	1957
Cisplatin	DNA	Cytotoxic drug	Platinum-based drug	Testicular cancer; Ovarian cancer; Bladder cancer	Testicular cancer; Ovarian cancer; Bladder cancer	1978
Cladribine	DNA; POL; RRM1; RRM2; RRM2B; POLA1; POLE; POLE2; POLE3; POLE4; PNP	Cytotoxic drug	Nucleoside precursor	Leukemia	Hairy cell leukemia	1993
Clofarabine	DNA; POLA1; RRM1	Cytotoxic drug	Nucleoside precursor	Leukemia	ALL	2004

Drug	Primary targets	Drug category	Classification	Cancer type (FDA-approved)	Indications	Year FDA-approved
Cyclophosphamide	DNA	Cytotoxic drug	Alkylating agent	Lymphoma; Multiple myeloma; Leukemia; Neuroblastoma; Ovarian cancer; Retinoblastoma; Breast cancer	Malignant lymphoma; Multiple myeloma; Leukemias; Neuroblastoma; Ovarian adenocarcinoma; Retinoblastoma; Breast carcinoma	1959
Cytarabine	DNA; POLB	Cytotoxic drug	Nucleoside analogue	Leukemia	AML; ALL; CML; Meningeal leukemia	1969
Dacarbazine	DNA; POLA2	Cytotoxic drug	Alkylating agent	Melanoma; Lymphoma; Soft tissue sarcoma	Melanoma; Hodgkin's lymphoma; Soft tissue sarcoma	1975
Dactinomycin	DNA	Cytotoxic drug	Cytotoxic antibiotic	Soft tissue sarcoma; Gestational trophoblastic neoplasia; Testicular cancer; Kidney cancer	Ewing sarcoma; Rhabdomyosarcoma; Recurrent solid tumors; Testicular cancer; Wilms tumor	1964
Daunorubicin	DNA; TOP2A; TOP2B	Cytotoxic drug	Topoisomerase inhibitor	Leukemia	AML; ALL	1998
Decitabine	DNMT1; DNA	Cytotoxic drug	Nucleoside analogue	Leukemia; Myelodysplastic syndrome	AML; CML; MDS	2006
Docetaxel	TUBA4A; TUBB1	Cytotoxic drug	Cytoskeletal disruptor	Breast cancer; Lung cancer; Prostate cancer; Gastric cancer; Head and neck cancer	Breast cancer; NSCLC; Castration-resistant prostate cancer; Gastric adenocarcinoma; HNSCC	1996

(continued)

Table 2 (continued)

Drug	Primary targets	Drug category	Classification	Cancer type (FDA-approved)	Indications	Year FDA-approved
Doxorubicin hydrochloride	DNA; TOP2A; NOLC1	Cytotoxic drug	Topoisomerase inhibitor	Ovarian cancer; Breast cancer; Multiple myeloma; Leukemia; Lymphoma; Soft tissue sarcoma; Gastric cancer; Lung cancer; Thyroid cancer; Bladder cancer; Kidney cancer; Brain cancer; Bone cancer	Ovarian cancer; Breast cancer; Multiple myeloma; ALL; AML; Lymphoma; STS; AIDS-related Kaposi's sarcoma; Gastric carcinoma; Bronchogenic carcinoma; Thyroid carcinoma; Transitional cell bladder carcinoma; Wilms tumor; Neuroblastoma; Bone sarcoma	1974
Epirubicin hydrochloride	DNA; TOP2A; CHD1	Cytotoxic drug	Topoisomerase inhibitor	Breast cancer; Gastric cancer; Esophageal cancer	Metastatic breast cancer; Gastric cancer; Esophageal cancer	1999
Eribulin mesylate	TUBA4A; TUBB1	Cytotoxic drug	Cytoskeletal disruptor	Breast cancer; Liposarcoma	Breast cancer; Liposarcoma	2010
Estramustine	MAP1A; MAP2; ESR1; ESR2	Cytotoxic drug	Derivative of estradiol with a nitrogen mustard moiety	Prostate cancer	Prostate cancer	1981
Etoposide	TOP2A; TOP2B	Cytotoxic drug	Topoisomerase inhibitor	Testicular cancer; Lung cancer	Testicular cancer; SCLC	1983
Floxuridine	TYMS	Cytotoxic drug	Nucleoside analogue	Gastric cancer	Gastrointestinal adenocarcinoma	1970
Fludarabine phosphate	DNA; POLA1; RRM1	Cytotoxic drug	Nucleoside precursor	Leukemia	CLL	1991

Drug	Primary targets	Drug category	Classification	Cancer type (FDA-approved)	Indications	Year FDA-approved
Fluorouracil	DNA; RNA; TYMS	Cytotoxic drug	Nucleoside analogue	Breast cancer; Colorectal cancer; Gastric cancer; Pancreatic cancer	Breast adenocarcinoma; Adenocarcinoma of the colon and rectum; Gastric adenocarcinoma; Pancreatic adenocarcinoma	1970
Gemcitabine	DNA; RRM1; TYMS; CMPK1	Cytotoxic drug	Nucleoside analogue	Ovarian cancer; Breast cancer; Lung cancer; Pancreatic cancer	Ovarian cancer; Metastatic breast cancer; NSCLC; Pancreatic cancer	1996
Hydroxyurea	RRM1	Cytotoxic drug	Antimetabolite	Leukemia; Head and neck cancer	CML; HNSCC	1967
Idarubicin hydrochloride	DNA; TOP2A	Cytotoxic drug	Topoisomerase inhibitor	Leukemia	AML	1997
Ifosfamide	DNA	Cytotoxic drug	Alkylating agent	Testicular cancer; Lymphoma; Sarcoma	Germ cell testicular cancer; Lymphoma; Sarcoma	1988
Irinotecan	DNA; TOP1; TOP1MT	Cytotoxic drug	Topoisomerase inhibitor	Colorectal cancer; Pancreatic cancer	CRC; Pancreatic adenocarcinoma	1996
Ixabepilone	TUBB3	Cytotoxic drug	Cytoskeletal disruptor	Breast cancer	Metastatic or locally advanced breast cancer	2007
Lomustine	DNA	Cytotoxic drug	Alkylating agent	Brain cancer; Lymphoma	Glioma; Meduloblastoma; Hodgkin lymphoma	1976

(continued)

Table 2 (continued)

Drug	Primary targets	Drug category	Classification	Cancer type (FDA-approved)	Indications	Year FDA-approved
Lurbinectedin	DNA	Cytotoxic drug	Alkylating agent	Lung cancer	SCLC	2020
Mechlorethamine hydroxychloride	DNA	Cytotoxic drug	Alkylating agent	Leukemia; Lymphoma; Myeloproliferative disorder; Lung cancer	CML; Hodgkin lymphoma; Mycosis fungoides; Polycythemia vera; Bronchogenic carcinoma	1949
Melphalan	DNA	Cytotoxic drug	Alkylating agent	Multiple myeloma; Ovarian cancer; Liver cancer	Multiple myeloma; Non-resectable epithelial carcinoma of the ovary; Cholangiocarcinoma	1964
Mercaptopurine	HPRT1; PPAT; IMPDH!; IMPDH2	Cytotoxic drug	Nucleoside analogue	Leukemia	ALL	1953
Methotrexate	DHFR	Cytotoxic drug	Antimetabolite	Leukemia; Lymphoma; Breast cancer; Head and neck cancer; Lung cancer; Bone cancer; Gestational trophoblastic disease	ALL; Mycosis fungoides; NHL; Breast cancer; Epidermoid cancers of the head and neck; Lung cancer; Non metastatic osteosarcoma; Gestational choriocarcinoma; Chorioadenoma destruens	1953

Drug	Primary targets	Drug category	Classification	Cancer type (FDA-approved)	Indications	Year FDA-approved
Mitomycin	DNA	Cytotoxic drug	Alkylating agent	Gastric cancer; Pancreatic cancer; Urothelial cancer	Gastric or pancreatic adenocarcinoma; Low-grade upper tract urothelial cancer	2002
Mitotane	CYP11B1	Cytotoxic drug	Cytotoxic antibiotic	Adrenocortical carcinoma	Adrenocortical carcinoma	1970
Mitoxantrone hydrochloride	DNA; TOP2A	Cytotoxic drug	Cytotoxic antibiotic	Leukemia	AML	1987
Nelarabine	DNA; POLA1; RRM1	Cytotoxic drug	Nucleoside precursor	Lymphoma; Leukemia	T-cell lymphoblastic lymphoma; T-cell acute lymphoblastic leukemia	2005
Omacetaxine mepesuccinate	RPL3	Cytotoxic drug	Protein synthesis inhibitor	Leukemia	CML	2012
Oxaliplatin	DNA	Cytotoxic drug	Platinum-based drug	Colorectal cancer	CRC	2002
Paclitaxel	TUBA4A; TUBB1	Cytotoxic drug	Cytoskeletal disruptor	Ovarian cancer; Breast cancer; Lung cancer; Kaposi's sarcoma; Pancreatic cancer; Gastric cancer	Ovarian cancer; Breast cancer; Lung cancer; Kaposi's sarcoma; Pancreas adenocarcinoma; Gastric or gastroesophageal junction adenocarcinoma	1992
Pegaspargase	L-asparagine	Cytotoxic drug	Antimetabolite	Leukemia	ALL	1994
Pemetrexed	DHFR; GART; TYMS	Cytotoxic drug	Antimetabolite	Lung cancer; Mesothelioma	NSCLC; Malignant mesothelioma	2004
Pentostatin	ADA	Cytotoxic drug	Nucleoside analogue	Leukemia	Hairy cell leukemia	1991
Pralatrexate	DHFR; TYMS	Cytotoxic drug	Antimetabolite	Lymphoma	PTCL	2009

(continued)

Table 2 (continued)

Drug	Primary targets	Drug category	Classification	Cancer type (FDA-approved)	Indications	Year FDA-approved
Procarbazine hydrochloride	DNA; MAOA; MAOB	Cytotoxic drug	Alkylating agent	Lymphoma	Hodgkin lymphoma	1969
Streptozocin	DNA	Cytotoxic drug	Alkylating agent	Pancreatic cancer	Metastatic islet cell cancer of pancreas	1982
Temozolomide	DNA	Cytotoxic drug	Alkylating agent	Brain cancer	Anaplastic astrocytoma; Glioblastoma	1999
Teniposide	TOP2A	Cytotoxic drug	Topoisomerase inhibitor	Leukemia	Refractory childhood ALL	1992
Thioguanine	DNA	Cytotoxic drug	Nucleoside analogue	Leukemia	AML	1966
Thiotepa	DNA	Cytotoxic drug	Alkylating agent	Breast cancer; Ovarian cancer; Bladder cancer	Adenocarcinoma of the breast or ovary; Bladder superficial papillary carcinoma	1959
Topotecan hydrochloride	DNA; TOP1; TOP1MT	Cytotoxic drug	Topoisomerase inhibitor	Lung cancer; Cervical cancer; Ovarian cancer	SCLC; Cervical cancer; Metastatic ovarian cancer	1996
Trabectedin	DNA	Cytotoxic drug	Alkylating agent	Soft tissue sarcoma	Unresectable or metastatic liposarcoma or leiomyosarcoma	2015
Trifluridine/tipiracil	DNA; TYMS; TYMP	Cytotoxic drug	Nucleoside analogue combined with thymidine phosphorylase inhibitor	Colorectal cancer; Gastric cancer	Colorectal cancer; Gastric or gastroesophageal junction adenocarcinoma	2015
Valrubicin	DNA; TOP2A	Cytotoxic drug	Topoisomerase inhibitor	Bladder cancer	BCG-refractory bladder carcinoma in situ	1998

Drug	Primary targets	Drug category	Classification	Cancer type (FDA-approved)	Indications	Year FDA-approved
Vinblastine sulfate	TUBA1A; TUBB; TUBD1; TUBE1; TUBG1	Cytotoxic drug	Cytoskeletal disruptor	Lymphoma; Testicular cancer; Kaposi's sarcoma; Choriocarcinoma; Breast cancer	Hodgkin lymphoma; Lymphocytic lymphoma; Histiocytic lymphoma; Mycosis fungoides; Testicular cancer; Kaposi's sarcoma; Choriocarcinoma; Breast cancer; Letterer-Siwe disease	1965
Vincristine sulfate	TUBA4A; TUBB	Cytotoxic drug	Cytoskeletal disruptor	Acute leukemia; Lymphoma; Soft tissue cancer; Neuroblastoma; Kidney cancer	Acute leukemia; Lymphoma; Rhabdomyosarcoma; Neuroblastoma; Wilms tumor	1963
Vinorelbine tartrate	TUBB	Cytotoxic drug	Cytoskeletal disruptor	Lung cancer	NSCLC	1994
Denileukin diftitox	IL2RA; IL2RB; IL2RG; protein synthesis	Cytotoxin fusion protein	IL2 combined with diphtheria toxin	Lymphoma	T-cell lymphoma expressing CD25	1999
Tagraxofusp-erzs	CD123	Cytotoxin fusion protein	IL-3 conjugated diphtheria toxin	Leukemia	Blastic plasmacytoid dendritic cell neoplasm	2018
Abarelix	GNRHR	Hormone therapy	GnRH receptor inhibitor	Prostate cancer	Advanced symptomatic prostate cancer	2003

(continued)

Table 2 (continued)

Drug	Primary targets	Drug category	Classification	Cancer type (FDA-approved)	Indications	Year FDA-approved
Abiraterone acetate	CYP17A1	Hormone therapy	Anti-androgen	Prostate cancer	Metastatic castration-resistant prostate cancer; metastatic high-risk castration-sensitive prostate cancer	2011
Anastrozole	CYP19A1	Hormone therapy	Aromatase inhibitor	Breast cancer	HR+ HER2- breast cancer	1995
Apalutamide	AR	Hormone therapy	Anti-androgen	Prostate cancer	Metastatic castration-sensitive prostate cancer; Non-metastatic castration-resistant prostate cancer	2018
Bicalutamide	AR	Hormone therapy	Anti-androgen	Prostate cancer	Metastatic prostate carcinoma	1995
Darolutamide	AR	Hormone therapy	Anti-androgen	Prostate cancer	Non-metastatic castration-resistant prostate cancer	2019
Degarelix	GNRHR	Hormone therapy	GnRH receptor inhibitor	Prostate cancer	Advanced prostate cancer	2008
Enzalutamide	AR	Hormone therapy	Anti-androgen	Prostate cancer	Castration-resistant prostate cancer; Metastatic castration-sensitive prostate cancer	2012
Exemestane	CYP19A1	Hormone therapy	Aromatase inhibitor	Breast cancer	ER+ Breast cancer	1999

Drug	Primary targets	Drug category	Classification	Cancer type (FDA-approved)	Indications	Year FDA-approved
Fluoxymesterone	AR; ESR1; NR3C1; PRLR	Hormone therapy	Androgen	Breast cancer	Breast cancer	1956
Flutamide	AR	Hormone therapy	Anti-androgen	Prostate cancer	Metastatic prostate carcinoma	1989
Fulvestrant	ESR1	Hormone therapy	Anti-estrogen	Breast cancer	HR+ HER2- breast cancer	2002
Goserelin acetate	GNRHR; LHCGR	Hormone therapy	GnRH analogue	Prostate cancer; Breast cancer	Prostate cancer; Breast cancer	1989
Lanreotide	SSTR2; SSTR5	Hormone therapy	Somatostatin analogue	Gastroenteropancreatic neuroendocrine tumor	GEP-NET	2007
Letrozole	CYP19A1	Hormone therapy	Aromatase inhibitor	Breast cancer	HR+ breast cancer	1997
Leuprolide acetate	GNRHR	Hormone therapy	GnRH analogue	Prostate cancer	Prostate cancer	2002
Methyltestosterone	AR	Hormone therapy	Androgen	Breast cancer	Advanced inoperable metastatic breast cancer	1973
Nilutamide	AR	Hormone therapy	Anti-androgen	Prostate cancer	Metastatic prostate cancer	1996
Tamoxifen citrate	ESR1; ESR2	Hormone therapy	Anti-estrogen	Breast cancer	ER+ breast cancer	1977

(continued)

Table 2 (continued)

Drug	Primary targets	Drug category	Classification	Cancer type (FDA-approved)	Indications	Year FDA-approved
Toremifene citrate	ESR1	Hormone therapy	Anti-estrogen	Breast cancer	Metastatic breast cancer, ER+ or unknown	1997
Triptorelin pamoate	GNRH1	Hormone therapy	GnRH analogue	Prostate cancer	Advanced prostate cancer	2000
Aldesleukin	IL2RA; IL2RB; IL2RG	Immunotherapy	Immunostimulant	Melanoma; Kidney cancer	Melanoma; Renal cell carcinoma	1992
Axicabtagene ciloleucel	CD18	Immunotherapy	CAR-T cell therapy	Lymphoma	Diffuse large B-cell lymphoma	2017
Brexucabtagene autoleucel	CD19	Immunotherapy	CAR-T cell therapy	Lymphoma	MCL	2020
Interferon Alfa-2	IFNAR1; IFNAR2	Immunotherapy	Immunostimulant	Leukemia; Melanoma; Lymphoma; Kaposi's sarcoma	Hairy cell leukemia; Melanoma; Follicular NHL; AIDS-related Kaposi's sarcoma	1986
Sipuleucel-T	ACPP	Immunotherapy	Autologous cell immunotherapy	Prostate cancer	Castrate resistant (hormone refractory) prostate cancer	2010
Talimogene laherparepvec		Immunotherapy	Oncolytic virus with antitumor and immune-stimulating activities	Melanoma	Melanoma	2015
Tisagenlecleucel	CD19	Immunotherapy	CAR-T cell therapy	Leukemia	ALL	2017
Alemtuzumab	CD52	Monoclonal antibody	CD52-directed cytolytic antibody	Leukemia	CLL	2001

Drug	Primary targets	Drug category	Classification	Cancer type (FDA-approved)	Indications	Year FDA-approved
Atezolizumab	PD-L1	Monoclonal antibody	PD-L1 blocking antibody	Urothelial carcinoma; Lung cancer; Breast cancer; Hepatocellular carcinoma; Melanoma	Urothelial carcinoma; NSCLC; SCLC; TNBC; HCC; Melanoma	2016
Avelumab	PD-L1	Monoclonal antibody	PD-L1 blocking antibody	Skin cancer; Urothelial carcinoma; Kidney cancer	MCC; Urothelial carcinoma; RCC	2017
Bevacizumab	VEGFA	Monoclonal antibody	VEGF inhibitor	Colorectal cancer; Lung cancer; Brain cancer; Kidney cancer; Cervical cancer; Ovarian cancer; Hepatocellular carcinoma	CRC; Non-squamous NSCLC; Glioblastoma; RCC; Cervical cancer; Ovarian cancer; HCC	2004
Blinatumomab	CD19; CD3	Monoclonal antibody	Bi-specific T-cell Engager	Leukemia	ALL	2014
Cemiplimab-rwlc	PD-1	Monoclonal antibody	PD-1 blocking antibody	Skin cancer	Cutaneous squamous cell carcinoma (cSCC)	2018
Cetuximab	EGFR	Monoclonal antibody	EGFR antagonist	Head and neck cancer; Colorectal cancer	Squamous cell carcinoma of the head and neck; K-RAS wild-type EGFR expressing CRC	2004
Daratumumab	CD38	Monoclonal antibody	CD38-directed cytolytic antibody	Multiple myeloma	Multiple myeloma	2015
Denosumab	TNFSF11	Monoclonal antibody	RANKL inhibitor	Bone cancer	Giant cell tumor of bone	2010
Dinutuximab	GD2	Monoclonal antibody	GD2-binding monoclonal antibody	Neuroblastoma	Pediatric high-risk neuroblastoma	2015

(continued)

Table 2 (continued)

Drug	Primary targets	Drug category	Classification	Cancer type (FDA-approved)	Indications	Year FDA-approved
Durvalumab	PD-L1	Monoclonal antibody	PD-L1 blocking antibody	Urothelial carcinoma; Lung cancer	Urothelial carcinoma; NSCLC; ES-SCLC	2017
Elotuzumab	SLAMF7	Monoclonal antibody	SLAMF7-directed immunostimulatory antibody	Multiple myeloma	Multiple myeloma	2015
Ipilimumab	CTLA4	Monoclonal antibody	CTLA-4-blocking antibody	Melanoma; Kidney cancer; Colorectal cancer; Hepatocellular carcinoma; Lung cancer	Melanoma; RCC; CRC; HCC; NSCLC with no EGFR or ALK genomic tumor alterration; Malignant pleural mesothelioma	2011
Isatuximab-IRFC	CD38	Monoclonal antibody	CD38-directed cytolytic antibody	Multiple myeloma	Multiple myeloma	2020
Mogamulizumab-kpkc	CCR4	Monoclonal antibody	CCR4-directed monoclonal antibody	Cutaneous T-cell lymphoma	Mycosis fungoides	2018
Necitumumab	EGFR	Monoclonal antibody	EGFR antagonist	Lung cancer	NSCLC	2015

Drug	Primary targets	Drug category	Classification	Cancer type (FDA-approved)	Indications	Year FDA-approved
Nivolumab	PD-1	Monoclonal antibody	PD-1 blocking antibody	Melanoma; Lung cancer; Kidney cancer; Lymphoma; Colorectal cancer; Urothelial carcinoma; Hepatocellular carcinoma; Esophageal cancer; Head and Neck cancer	Melanoma; NSCLC; SCLC; Malignant pleural mesothelioma; RCC; cHL; HNSCC; CRC; HCC; Urothelial carcinoma; Esophageal squamous cell carcinoma	2014
Obinutuzumab	CD20	Monoclonal antibody	CD20-directed cytolytic antibody	Leukemia; Lymphoma	CLL; Follicular lymphoma	2013
Ofatumumab	CD20	Monoclonal antibody	CD20-directed cytolytic antibody	Leukemia	CLL	2009
Olaratumab	PDGF-Rα	Monoclonal antibody	PDGF-Rα blocking antibody	Soft tissue sarcoma	STS	2016
Panitumumab	EGFR	Monoclonal antibody	EGFR antagonist	Colorectal cancer	Wild-type RAS metastatic CRC	2006

(continued)

Table 2 (continued)

Drug	Primary targets	Drug category	Classification	Cancer type (FDA-approved)	Indications	Year FDA-approved
Pembrolizumab	PD-1	Monoclonal antibody	PD-1 blocking antibody	Melanoma; Lung cancer; Head and neck cancer; Lymphoma; Urothelial carcinoma; Gastric cancer; Esophageal cancer; Cervical cancer; Hepatocallular carcinoma; Skin cancer Kidney cancer; Endometrial cancer	Melanoma; NSCLC; SCLC; HNSCC; cHL; PMBCL; Urothelial carcinoma; PD-L1 expressing gastric cancer, esophageal cancer, or cervical cancer; HCC; MCC; cSCC; RCC; Endometrial carcinoma; TMB-H cancer; MSI-H or dMMR solid cancer	2014
Pertuzumab	ERBB2	Monoclonal antibody	HER2 receptor antagonist	Breast cancer	HER2+ breast cancer	2012

Drug	Primary targets	Drug category	Classification	Cancer type (FDA-approved)	Indications	Year FDA-approved
Ramucirumab	VEGFR2	Monoclonal antibody	VEGFR2 antagonist	Gastric cancer; Lung cancer; Colorectal cancer; Hepatocellular carcinoma	Gastric or gastroesophageal junction adenocarcinoma; NSCLC; CRC; Hepatocellular carcinoma	2014
Rituximab	CD20	Monoclonal antibody	CD20-directed cytolytic antibody	Lymphoma; Leukemia	CD20-positive NHL; CD20-positive CLL	1997
Tafasitamab-cxix	CD19	Monoclonal antibody	CD19-directed cytolytic antibody	Lymphoma	Diffuse large B-cell lymphoma	2020
Tositumomab/Iodine I 131 Tositumomab	CD20	Monoclonal antibody	CD20-directed radiotherapeutic antibody	Lymphoma	CD20-positive NHL	2003
Trastuzumab	ERBB2	Monoclonal antibody	HER2 receptor antagonist	Breast cancer; Gastric cancer	HER2+ breast cancer; HER2+ metastatic gastric or gastroesophageal junction adenocarcinoma	1998
Abemaciclib	CDK4/6	Protein kinase inhibitor	Serine/Threonine kinase inhibitor	Breast cancer	HR+ HER2- breast cancer	2017
Acalabrutinib	BTK	Protein kinase inhibitor	Tyrosine kinase inhibitor	Lymphoma	MCL	2017
Afatinib dimaleate	EGFR; ERBB2; ERBB4	Protein kinase inhibitor	Tyrosine kinase inhibitor	Lung cancer	NSCLC	2013

(continued)

Table 2 (continued)

Drug	Primary targets	Drug category	Classification	Cancer type (FDA-approved)	Indications	Year FDA-approved
Alectinib	ALK; RET	Protein kinase inhibitor	Tyrosine kinase inhibitor	Lung cancer	ALK-positive NSCLC	2015
Avapritinib	PDGFR; KIT	Protein kinase inhibitor	Tyrosine kinase inhibitor	Gastrointestinal cancer	GIST with PDGFRA exon 18 mutation	2020
Axitinib	VEGFR	Protein kinase inhibitor	Tyrosine kinase inhibitor	Kidney cancer	Advanced RCC	2012
Binimetinib	MEK1/2	Protein kinase inhibitor	Serine/Tyrosine/Threonine kinase inhibitor	Melanoma	Melanoma with BRAF V600E/K mutation	2018
Bosutinib	BCR-ABL	Protein kinase inhibitor	Tyrosine kinase inhibitor	Leukemia	Ph + CML	2012
Brigatinib	ALK; EGFR	Protein kinase inhibitor	Tyrosine kinase inhibitor	Lung cancer	ALK-positive NSCLC	2017
Cabozantinib	VEGFR2; MET; RET	Protein kinase inhibitor	Tyrosine kinase inhibitor	Thyroid cancer; Kidney cancer; Hepatocellular carcinoma	Medullary thyroid cancer; Advanced RCC; Hepatocellular carcinoma	2012
Capmatinib hydrochloride	MET	Protein kinase inhibitor	Tyrosine kinase inhibitor	Lung cancer	NSCLC with a mutation that leads to MET exon 14 skipping	2020
Ceritinib	ALK	Protein kinase inhibitor	Tyrosine kinase inhibitor	Lung cancer	ALK-positive NSCLC	2014
Cobimetinib fumarate	MEK1/2	Protein kinase inhibitor	Serine/Tyrosine/Threonine kinase inhibitor	Melanoma	Melanoma with BRAF V600E/K mutation	2015
Crizotinib	ALK; MET	Protein kinase inhibitor	Tyrosine kinase inhibitor	Lung cancer	ALK- or ROS1-positive NSCLC	2011

Drug	Primary targets	Drug category	Classification	Cancer type (FDA-approved)	Indications	Year FDA-approved
Dabrafenib mesylate	BRAF	Protein kinase inhibitor	Serine/Threonine kinase inhibitor	Melanoma; Lung cancer; Thyroid cancer	Melanoma with BRAF V600E/K; NSCLC with BRAF V600E; ATC with BRAF V600E	2013
Dacomitinib	EGFR	Protein kinase inhibitor	Tyrosine kinase inhibitor	Lung cancer	NSCLC with EGFR exon 19 deletion or L858R mutation	2018
Dasatinib	BCR-ABL	Protein kinase inhibitor	Tyrosine kinase inhibitor	Leukemia	Ph + CML; Ph + ALL	2006
Encorafenib	BRAF	Protein kinase inhibitor	Serine/Threonine kinase inhibitor	Melanoma; Colorectal cancer	Melanoma with BRAF V600E/K mutation; CRC with BRAF V600E mutation	2018
Entrectinib	TRKA/B/C; ROS1	Protein kinase inhibitor	Tyrosine kinase inhibitor	Lung cancer; Solid tumors	ROS1-positive NSCLC; Solid tumors with NTRK gene fusion	2019
Erdafitinib	FGFR	Protein kinase inhibitor	Tyrosine kinase inhibitor	Urothelial carcinoma	Urothelial carcinoma with susceptible FGFR2/3 alteration	2019
Erlotinib hydrochloride	EGFR	Protein kinase inhibitor	Tyrosine kinase inhibitor	Pancreatic cancer; Lung cancer	Pancreatic cancer; NSCLC with EGFR exon 19 deletion or L858R mutation	2004

(continued)

Table 2 (continued)

Drug	Primary targets	Drug category	Classification	Cancer type (FDA-approved)	Indications	Year FDA-approved
Everolimus	**mTOR**	Protein kinase inhibitor	Serine/Threonine kinase inhibitor	Breast cancer; Kidney cancer; Neuroendocrine tumors	HR+ HER2- breast cancer; RCC; NET of gastrointestinal, pancreatic or lung origin	2009
Fedratinib	JAK2	Protein kinase inhibitor	Tyrosine kinase inhibitor	Myelofibrosis	Myelofibrosis	2019
Gefitinib	EGFR	Protein kinase inhibitor	Tyrosine kinase inhibitor	Lung cancer	NSCLC	2003
Gilteritinib	FLT3	Protein kinase inhibitor	Tyrosine kinase inhibitor	Leukemia	FLT3-mutated AML	2018
Ibrutinib	BTK	Protein kinase inhibitor	Tyrosine kinase inhibitor	Lymphoma	MCL; CLL; SLL; WM; MZL	2013
Imatinib mesylate	BCR-ABL; KIT; PDGFR	Protein kinase inhibitor	Tyrosine kinase inhibitor	Leukemia; Gastrointestinal cancer	Ph + CML; Ph + ALL; MDS/MPD with PDGFR gene re-arrangements; ASM; CEL; DFSP; KIT+ GIST	2001
Lapatinib ditosylate	EGFR; ERBB2	Protein kinase inhibitor	Tyrosine kinase inhibitor	Breast cancer	HER2+ breast cancer	2007
Larotrectinib	TRKA/B/C	Protein kinase inhibitor	Tyrosine kinase inhibitor	Solid tumors	Solid tumors with NTRK gene fusion	2018

Drug	Primary targets	Drug category	Classification	Cancer type (FDA-approved)	Indications	Year FDA-approved
Lenvatinib	VEGFR; FGFR; PDGFRA; RET; KIT	Protein kinase inhibitor	Tyrosine kinase inhibitor	Thyroid cancer; Kidney cancer; Hepatocellular carcinoma; Endometrial carcinoma	DTC; RCC; HCC; Endometrial carcinoma	2015
Lorlatinib	ALK	Protein kinase inhibitor	Tyrosine kinase inhibitor	Lung cancer	ALK-positive NSCLC	2018
Midostaurin	FLT3	Protein kinase inhibitor	Tyrosine kinase inhibitor	Leukemia	FLT3-mutated AML; ASM; Mast cell leukemia; SM-AHN	2017
Neratinib	ERBB2	Protein kinase inhibitor	Tyrosine kinase inhibitor	Breast cancer	HER2+ breast cancer	2017
Nilotinib	BCR-ABL	Protein kinase inhibitor	Tyrosine kinase inhibitor	Leukemia	Ph + CML	2007
Osimertinib	EGFR T790M	Protein kinase inhibitor	Tyrosine kinase inhibitor	Lung cancer	Lung cancer	2015
Palbociclib	CDK4/6	Protein kinase inhibitor	Serine/Threonine kinase inhibitor	Breast cancer	HR+ HER2- breast cancer	2015
Pazopanib hydrochloride	VEGFR; PDGFRA/B; KIT	Protein kinase inhibitor	Tyrosine kinase inhibitor	Kidney cancer; Soft tissue sarcoma	RCC; STS	2009
Pemigatinib	FGFR	Protein kinase inhibitor	Tyrosine kinase inhibitor	Cholangiocarcinoma	Cholangiocarcinoma with FGFR2 fusion or other rearrangement	2020
Pexidartinib	CSF1R; KIT; FLT3	Protein kinase inhibitor	Tyrosine kinase inhibitor	Tenosynovial giant cell tumor	Tenosynovial giant cell tumor	2019

(continued)

Table 2 (continued)

Drug	Primary targets	Drug category	Classification	Cancer type (FDA-approved)	Indications	Year FDA-approved
Ponatinib hydrochloride	BCR-ABL	Protein kinase inhibitor	Tyrosine kinase inhibitor	Leukemia	Ph + CML or Ph + ALL, including T315I-positive	2012
Pralsetinib	RET	Protein kinase inhibitor	Tyrosine kinase inhibitor	Lung cancer	RET fusion-positive NSCLC	2020
Regorafenib	VEGFR; RET; KIT; PDGFRA/B; FGFR1/2; RAF1; BRAF	Protein kinase inhibitor	Serine/Tyrosine/Threonine kinase inhibitor	Colorectal cancer; Gastric cancer; Hepatocellular carcinoma	Colorectal cancer; GIST; Hepatocellular carcinoma	2012
Ribociclib succinate	CDK4/6	Protein kinase inhibitor	Serine/Threonine kinase inhibitor	Breast cancer	HR+ HER2- breast cancer	2017
Ripretinib	KIT; PDGFRA	Protein kinase inhibitor	Tyrosine kinase inhibitor	Gastrointestinal cancer	GIST	2020
Ruxolitinib phosphate	JAK1/2	Protein kinase inhibitor	Tyrosine kinase inhibitor	Myeloproliferative disorder	Myelofibrosis; Polycythemia vera	2011
Selpercatinib	RET	Protein kinase inhibitor	Tyrosine kinase inhibitor	Lung cancer; Thyroid cancer	RET fusion-positive NSCLC; RET-mutant MTC; RET fusion-positive thyroid cancer	2020
Selumetinib sulfate	MEK1/2	Protein kinase inhibitor	Serine/Tyrosine/Threonine kinase inhibitor	Neurofibroma	Neurofibroma	2020
Sorafenib tosylate	BRAF; RAF1; VEGFR; PDGFRB; KIT; FLT3; RET	Protein kinase inhibitor	Serine/Tyrosine/Threonine kinase inhibitor	Kidney cancer; Hepatocellular carcinoma; Thyroid cancer	RCC; Hepatocellular carcinoma; DTC	2005

Drug	Primary targets	Drug category	Classification	Cancer type (FDA-approved)	Indications	Year FDA-approved
Sunitinib malate	VEGFR; FLT3; RET; KIT; PDGFRA/B; CSF1R	Protein kinase inhibitor	Tyrosine kinase inhibitor	Gastrointestinal cancer; Kidney cancer; Pancreatic cancer	GIST; RCC; Pancreatic NET	2006
Temsirolimus	mTOR	Protein kinase inhibitor	Serine/Threonine kinase inhibitor	Kidney cancer	RCC	2007
Trametinib	MEK1/2	Protein kinase inhibitor	Serine/Tyrosine/Threonine kinase inhibitor	Melanoma; Lung cancer; Thyroid cancer	Melanoma with BRAF V600E/K; NSCLC with BRAF V600E; ATC with BRAF V600E	2013
Tucatinib	ERBB2	Protein kinase inhibitor	Tyrosine kinase inhibitor	Breast cancer	HER2+ breast cancer	2020
Vandetanib	EGFR; VEGFR2	Protein kinase inhibitor	Tyrosine kinase inhibitor	Thyroid cancer	MTC	2011
Vemurafenib	BRAF	Protein kinase inhibitor	Serine/Threonine kinase inhibitor	Melanoma; Histiocytic sarcoma	Melanoma with BRAF V600E mutation; Erdheim-Chester disease with BRAF V600E mutation	2011
Zanubrutinib	BTK	Protein kinase inhibitor	Tyrosine kinase inhibitor	Lymphoma	MCL	2019

(continued)

Table 2 (continued)

Drug	Primary targets	Drug category	Classification	Cancer type (FDA-approved)	Indications	Year FDA-approved
Alpelisib	PIK3CA	Kinase inhibitor	PI3K inhibitor	Breast cancer	PIK3CA-mutated HR+ HER2- breast cancer	2019
Copanlisib	PIK3CA; PIK3CD	Kinase inhibitor	PI3K inhibitor	Lymphoma	Follicular lymphoma	2017
Duvelisib	PIK3CD; PI3KCG	Kinase inhibitor	PI3K inhibitor	Leukemia; Lymphoma	CLL; SLL; FL	2018
Idelalisib	PIK3CD	Kinase inhibitor	PI3K inhibitor	Leukemia; Lymphoma	CLL; SLL; FL	2014
Iobenguane sulfate I-131	Sodium-dependent noradrenaline transporter	Radioactive therapeutic agent	Norepinephrine transporter-directed radioactive molecule	Neuroendocrine tumor	Pheochromocytoma; Paraganglioma	2018
Lutetium Lu 177 dotatate	SSTR1; SSTR2; SSTR3; SSTR4; SSTR5	Radioactive therapeutic agent	Radiolabeled somatostatin analogue	Gastroenteropancreatic neuroendocrine tumor	Somatostatin receptor-positive GEP-NET	2018
Radium 223 dichloride	Unknown	Radioactive therapeutic agent		Prostate cancer	Castration-resistant prostate cancer; Symptomatic bone metastases	2013
Alitretinoin	RARA; RARB; RARG; RXRA; RXRB; RXRG	Small-molecule	Retinoid	Kaposi's sarcoma; Leukemia	AIDS-related Kaposi's sarcoma; APL	1999

Drug	Primary targets	Drug category	Classification	Cancer type (FDA-approved)	Indications	Year FDA-approved
Bexarotene	RXRA; RXRB; RXRG	Small-molecule	Retinoid	Lymphoma	Cutaneous T-cell lymphoma	1999
Dexamethasone	NR3C1	Small-molecule	Corticosteroid	Leukemia; Lymphoma; Multiple myeloma	Leukemia; Lymphoma; Multiple myeloma	1958
Imiquimod	TLR7	Small-molecule	TLR7 agonist	Skin cancer	Superficial basal cell carcinoma	1997
Lenalidomide	CRBN	Small-molecule	Immunomodulatory drug (IMiD)	Multiple myeloma; Lymphoma; Myelodisplasia	Multiple myeloma; MCL; FL; MZL; DLBCL; MDS	2005
Pomalidomide	CRBN	Small-molecule	Immunomodulatory drug (IMiD)	Multiple myeloma; Kaposi's sarcoma	Multiple myeloma; AIDS-related Kaposi's sarcoma	2013
Porfimer	FCGR1A; LDLR	Small-molecule	Photoactivated radical generator	Esophageal cancer; Lung cancer	Esophageal cancer; Endobronchial NSCLC	1995
Thalidomide	CRBN	Small-molecule	Immunomodulatory drug (IMiD)	Multiple myeloma	Multiple myeloma	1998
Belinostat	HDAC1; HDAC2; HDAC3; HDAC6	Small-molecule inhibitor	Histone deacetylase inhibitor	Lymphoma	PTCL	2014
Bortezomib	PSMB1; PSMB5	Small-molecule inhibitor	Proteasome inhibitor	Multiple myeloma; Lymphoma	Multiple myeloma; MCL	2003

(continued)

Table 2 (continued)

Drug	Primary targets	Drug category	Classification	Cancer type (FDA-approved)	Indications	Year FDA-approved
Carfilzomib	PSMB1; PSMB2; PSMB5; PSMB8; PSMB9; PSMB10	Small-molecule inhibitor	Proteasome inhibitor	Multiple myeloma	Multiple myeloma	2012
Enasidenib	IDH2	Small-molecule inhibitor	IDH2 inhibitor	Leukemia	AML with IDH2 mutation	2017
Glasdegib	SMO	Small-molecule inhibitor	Hedgehog pathway inhibitor	Leukemia	AML	2018
Ivosidenib	IDH1	Small-molecule inhibitor	IDH1 inhibitor	Leukemia	AML with a susceptible IDH1 mutation	2018
Ixazomib	PSMB1; PSMB2; PSMB5; PSMD1; PSMD2	Small-molecule inhibitor	Proteasome inhibitor	Multiple myeloma	Multiple myeloma	2015
Niraparib	PARP	Small-molecule inhibitor	PARP inhibitor	Ovarian cancer; Prostate cancer	Ovarian, fallopian tube, or primary peritoneal cancer; BRCA-mutated castration-resistant prostate cancer	2017

Drug	Primary targets	Drug category	Classification	Cancer type (FDA-approved)	Indications	Year FDA-approved
Olaparib	PARP1; PARP2; PARP3	Small-molecule inhibitor	PARP inhibitor	Ovarian cancer; Breast cancer; Pancreatic cancer; Prostate cancer	BRCA-mutated ovarian cancer, HER2- breast cancer or pancreatic adenocarcinoma; HRR-mutated prostate cancer	2014
Panobinostat	HDAC	Small-molecule inhibitor	Histone deacetylase inhibitor	Multiple myeloma	Multiple myeloma	2015
Romidepsin	HDAC1; HDAC2; HDAC4; HDAC6	Small-molecule inhibitor	Histone deacetylase inhibitor	Lymphoma	Cutaneous or peripheral T-cell lymphoma	2009
Rucaparib camsylate	PARP	Small-molecule inhibitor	PARP inhibitor	Ovarian cancer; BRCA-mutated prostate cancer	Ovarian cancer; Prostate cancer	2016
Selinexor	XPO1	Small-molecule inhibitor	Inhibitor of nuclear transport	Multiple myeloma; Lymphoma	Multiple myeloma; DLBCL	2019
Sonidegib	SMO	Small-molecule inhibitor	Hedgehog pathway inhibitor	Skin cancer	Basal cell carcinoma	2015
Talazoparib tosylate	PARP	Small-molecule inhibitor	PARP inhibitor	Breast cancer	BRCA-mutated HER2- breast cancer	2018
Tazemetostat hydrobromide	EZH2	Small-molecule inhibitor	Methyltransferase inhibitor	Soft tissue sarcoma; Lymphoma	Epitheloid sarcoma; EZH2-mutated follicular lymphoma	2020
Venetoclax	BCL-2	Small-molecule inhibitor	BCL-2 inhibitor	Leukemia; Lymphoma	CLL; AML; SLL	2016
Vismodegib	SMO	Small-molecule inhibitor	Hedgehog pathway inhibitor	Skin cancer	Basal cell carcinoma	2012

(continued)

Table 2 (continued)

Drug	Primary targets	Drug category	Classification	Cancer type (FDA-approved)	Indications	Year FDA-approved
Vorinostat	HDAC1; HDAC2; HDAC3; HDAC6	Small-molecule inhibitor	Histone deacetylase inhibitor	Lymphoma	Cutaneous T-cell lymphoma	2006
Ziv-aflibercept	PGF; VEGFA; VEGFB	Small-molecule inhibitor	VEGF inhibitor	Colorectal cancer	Colorectal cancer	2012

ALL acute lymphoblastic leukemia; AML acute myeloid leukemia; APL acute promyelocytic leukemia; ASM aggressive systemic mastocytosis; ATC anaplastic thyroid cancer; cHL classical Hodgkin lymphoma; CLL chronic lymphocytic leukemia; CML chronic myelogenous leukemia; CMML chronic myelomonocytic leukemia; CRC colorectal cancer; cSCC cutaneous squamous cell carcinoma; DLBCL diffuse large B-cell lymphoma; dMMR mismatch repair deficient; DTC differentiated thyroid cancer; ES-SCLC extensive-stage small cell lung cancer; FL follicular lymphoma; GEP-NET gastroenteropancreatic neuroendocrine tumor; GIST gastrointestinal stromal tumor; GM-CSF Granulocyte-macrophage colony-stimulating factor; HCC hepatocellular carcinoma; HNSCC head and neck squamous cell cancer; MCC merkel cell carcinoma; MCL mantle cell lymphoma; MDS myelodysplastic syndrom; MPD myeloproliferative disorder; MPM malignant pleural mesothelioma; MSI-H microsatellite instability-high; MTC medullary thyroid cancer; mTNBC metastatic triple negative breast cancer;MZL marginal zone lymphoma; NET neuroendocrine tumor; NHL non-Hodgkin lymphoma; NSCLC non-small cell lung cancer; Ph + Philadelphia translocation-positive; PMBCL primary mediastinal large B-cell lymphoma; PTCL peripheral T-cell lymphoma; RCC renal cell cancer; sALCL systemic anaplastic large cell lymphoma; SLL small lymphocytic lymphoma; SM-AHN systemic mastocytosis with associated hematological neoplasm; STS soft tissue sarcoma; TMB-H tumor mutational burden-high; TNBC triple negative breast cancer; WM Waldenström's macroglobulinemia

(ERB-B2 receptor tyrosine kinase) humanized monoclonal antibody trastuzumab was approved in 1998 for the treatment of HER2 overexpression in breast cancer.

Monoclonal PD-1 blocking antibody pembrolizumab was approved in 2014 for the treatment of melanoma, lung cancer, head and neck cancer, lymphoma, urothelial carcinoma, gastric cancer, esophageal cancer, cervical cancer, hepatocellular carcinoma, skin cancer, kidney cancer, endometrial cancer.

More recently, the anti-CD19 cytolytic monoclonal antibody tafasitamab-cxix was approved for the treatment of diffuse large B-cell lymphoma in 2020. As of 2020, ca 28 monoclonal antibodies have been approved by the US Food and Drug Agency (FDA) for the treatment of various cancers (Drugs@FDA, data in October 2020, Table 2) (Dupont et al. 2021).

Other examples of other antibody-based therapies include antibody-drug conjugates that are engineered to deliver cytotoxic drugs to specific tumor cells enhancing specificity and efficacy while reducing cytotoxicity. However, antibody-drug conjugates are large due to the size of antibodies, thus limiting epitope availability to membrane-associated targets and extracellular targets.

Similarly, inhibiting protein activity with small-molecule inhibitors is another strategy for targeted therapy. Examples of these types of targeted therapy include protein kinase inhibitors because protein kinases are key components of intracellular signaling pathways involved in various cellular functions such as cell proliferation, growth, differentiation, cell survival, angiogenesis, and apoptosis, and these pathways are often dysregulated in cancer.

A small-molecule protein kinase inhibitor imatinib was developed using rational drug design to target the fusion protein BCR-ABL driving chronic myeloid leukemia (CML) and was approved in 2001, which led to the development of numerous small-molecule kinase inhibitors (Table 3) (Bhullar et al. 2018; Kannaiyan and Mahadevan 2018; Pottier et al. 2020).

Because most kinase inhibitors interact with the kinase domain such as those of the first-generation kinase inhibitors, which are ATP-competitive inhibitors, these kinase inhibitors need extremely high binding affinity with the target due to the competition with high intracellular ATP concentrations.

Recently, the next generation non-ATP-competitive allosteric kinase inhibitors, such as mTOR inhibitors temsirolimus and everolimus, followed by MEK inhibitors trametinib, cobimetinib, binimetinib, and selumetinib that bind kinases outside of the ATP pocket, were developed and approved for cancer therapy in 2007 and 2009 and from 2013 to 2020.

These non-ATP-competitive allosteric kinase inhibitors are more specific than ATP-competitive inhibitors, as they target protein domains bearing less homology within the kinases than the ATP pocket, which results in reduced off-target effects and associated toxicity profiles (Panicker et al. 2019).

Small-molecule covalent kinase inhibitors have also been developed and approved in 2013 by the FDA such as the BTK inhibitor ibrutinib; and afatinib, which targets EGFR, HER2, HER4, and some EGFR mutants resistant to the first-generation EGFR inhibitors.

Table 3 Pharmacogenomic biomarkers cited in the oncology drug labeling (Adapted from: https://www.fda.gov/drugs/science-and-research-drugs/table-pharmacogenomic-biomarkers-drug-labeling)

1	Drug	Biomarker	Therapeutic area	Labeling sections
2	Abemaciclib (1)	ESR (Hormone Receptor)	Oncology	Indications and Usage, Adverse Reactions, Clinical Studies
3	Abemaciclib (2)	ERBB2 (HER2)	Oncology	Indications and Usage, Adverse Reactions, Clinical Studies
4	Ado-Trastuzumab Emtansine	ERBB2 (HER2)	Oncology	Indications and Usage, Dosage and Administration, Adverse Reactions, Clinical Pharmacology, Clinical Studies
5	Afatinib	EGFR	Oncology	Indications and Usage, Dosage and Administration, Adverse Reactions, Clinical Studies
6	Alectinib	ALK	Oncology	Indications and Usage, Dosage and Administration, Adverse Reactions, Clinical Pharmacology, Clinical Studies
7	Alpelisib (1)	ERBB2 (HER2)	Oncology	Indication and Usage, Dosage and Administration, Adverse Reactions, Clinical Studies
8	Alpelisib (2)	ESR (Hormone Receptor)	Oncology	Indication and Usage, Dosage and Administration, Adverse Reactions, Clinical Studies
9	Alpelisib (3)	PIK3CA	Oncology	Indication and Usage, Dosage and Administration, Adverse Reactions, Clinical Studies
10	Anastrozole	ESR, PGR (Hormone Receptor)	Oncology	Indications and Usage, Adverse Reactions, Drug Interactions, Clinical Studies
11	Arsenic Trioxide	PML-RARA	Oncology	Indications and Usage, Clinical Studies
12	Atezolizumab (1)	CD274 (PD-L1)	Oncology	Indications and Usage, Dosage and Administration, Warnings and Precautions, Adverse Reactions, Clinical Pharmacology, Clinical Studies
13	Atezolizumab (2)	Gene Signature (T-effector)	Oncology	Clinical Studies
14	Atezolizumab (3)	EGFR	Oncology	Indications and Usage, Adverse Reactions, Clinical Studies
15	Atezolizumab (4)	ALK	Oncology	Indications and Usage, Adverse Reactions, Clinical Studies

(continued)

Table 3 (continued)

1	Drug	Biomarker	Therapeutic area	Labeling sections
16	Atezolizumab (5)	BRAF	Oncology	Indications and Usage, Dosage and Administration, Adverse Reactions, Clinical Studies
17	Avapritinib	PDGFRA	Oncology	Indications and Usage, Dosage and Administration, Clinical Studies
18	Avelumab	CD274 (PD-L1)	Oncology	Clinical Studies
19	Belinostat	UGT1A1	Oncology	Dosage and Administration, Clinical Pharmacology
20	Binimetinib (1)	BRAF	Oncology	Indications and Usage, Dosage and Administration, Warnings and Precautions, Adverse Reactions, Use in Specific Populations, Clinical Studies
21	Binimetinib (2)	UGT1A1	Oncology	Clinical Pharmacology
22	Blinatumomab	BCR-ABL1 (Philadelphia chromosome)	Oncology	Adverse Reactions, Clinical Studies
23	Bosutinib	BCR-ABL1 (Philadelphia chromosome)	Oncology	Indications and Usage, Dosage and Administration, Warnings and Precautions, Adverse Reactions, Use in Specific Populations, Clinical Studies
24	Brentuximab Vedotin (1)	ALK	Oncology	Clinical Studies
25	Brentuximab Vedotin (2)	TNFRSF8 (CD30)	Oncology	Indications and Usage, Dosage and Administration, Adverse Reactions, Use in Specific Populations, Clinical Studies
26	Brigatinib	ALK	Oncology	Indications and Usage, Dosage and Administration, Adverse Reactions, Clinical Studies
27	Busulfan	BCR-ABL1 (Philadelphia chromosome)	Oncology	Clinical Studies
28	Cabozantinib	RET	Oncology	Clinical Studies
29	Capmatinib	MET	Oncology	Indications and Usage, Dosage and Administration, Clinical Studies
30	Capecitabine	DPYD	Oncology	Warnings and Precautions, Patient Counseling Information
31	Ceritinib	ALK	Oncology	Indications and Usage, Dosage and Administration, Warning and Precautions, Adverse Reactions, Clinical Studies
32	Cetuximab (1)	EGFR	Oncology	Indications and Usage, Dosage and Administration, Adverse Reactions, Clinical Studies

(continued)

Table 3 (continued)

1	Drug	Biomarker	Therapeutic area	Labeling sections
33	Cetuximab (2)	RAS	Oncology	Indications and Usage, Dosage and Administration, Warnings and Precautions, Adverse Reactions, Clinical Studies
34	Cisplatin	TPMT	Oncology	Adverse Reactions
35	Cobimetinib	BRAF	Oncology	Indications and Usage, Dosage and Administration, Adverse Reactions, Clinical Studies
36	Crizotinib (1)	ALK	Oncology	Indications and Usage, Dosage and Administration, Adverse Reactions, Use in Specific Populations, Clinical Pharmacology, Clinical Studies
37	Crizotinib (2)	ROS1	Oncology	Indications and Usage, Dosage and Administration, Adverse Reactions, Use in Specific Populations, Clinical Studies
38	Dabrafenib (1)	BRAF	Oncology	Indications and Usage, Dosage and Administration, Warnings and Precautions, Adverse Reactions, Clinical Pharmacology, Clinical Studies, Patient Counseling Information
39	Dabrafenib (2)	G6PD	Oncology	Warnings and Precautions, Adverse Reactions, Patient Counseling Information
40	Dabrafenib (3)	RAS	Oncology	Dosage and Administration, Warnings and Precautions
41	Dacomitinib	EGFR	Oncology	Indications and Usage, Dosage and Administration, Adverse Reactions, Use in Specific Populations, Clinical Studies
42	Dasatinib	BCR-ABL1 (Philadelphia chromosome)	Oncology	Indications and Usage, Dosage and Administration, Warnings and Precautions, Adverse Reactions, Use in Specific Populations, Clinical Studies
43	Denileukin Diftitox	IL2RA (CD25 antigen)	Oncology	Indications and Usage, Clinical Studies
44	Dinutuximab	MYCN	Oncology	Clinical Studies
45	Docetaxel	ESR, PGR (Hormone Receptor)	Oncology	Clinical Studies
46	Durvalumab	CD274 (PD-L1)	Oncology	Clinical Pharmacology, Clinical Studies
47	Duvelisib	Chromosome 17p	Oncology	Clinical Studies

(continued)

Table 3 (continued)

1	Drug	Biomarker	Therapeutic area	Labeling sections
48	Enasidenib	IDH2	Oncology	Indications and Usage, Dosage and Administration, Clinical Pharmacology, Clinical Studies
49	Encorafenib (1)	BRAF	Oncology	Indications and Usage, Dosage and Administration, Warnings and Precautions, Adverse Reactions, Use in Specific Populations, Clinical Pharmacology, Clinical Studies
50	Encorafenib (2)	RAS	Oncology	Dosage and Administration, Warnings and Precautions, Clinical Studies
51	Enfortumab Vedotin-ejfv	NECTIN4	Oncology	Clinical Studies
52	Entrectinib (1)	ROS1	Oncology	Indications and Usage, Dosage and Administration, Adverse Reactions, Use in Specific Populations, Clinical Pharmacology, Clinical Studies
53	Entrectinib (2)	NTRK	Oncology	Indications and Usage, Dosage and Administration, Adverse Reactions, Use in Specific Populations, Clinical Pharmacology, Clinical Studies
54	Erdafitinib (1)	FGFR	Oncology	Indications and Usage, Dosage and Administration, Adverse Reactions, Clinical Studies, Patient Counseling Information
55	Erdafitinib (2)	CYP2C9	Oncology	Use in Specific Populations, Clinical Pharmacology
56	Eribulin (1)	ERBB2 (HER2)	Oncology	Clinical Studies
57	Eribulin (2)	ESR, PGR (Hormone Receptor)	Oncology	Clinical Studies
58	Erlotinib	EGFR	Oncology	Indications and Usage, Dosage and Administration, Adverse Reactions, Clinical Studies
59	Everolimus (1)	ERBB2 (HER2)	Oncology	Indications and Usage, Dosage and Administration, Warnings and Precautions, Adverse Reactions, Use in Specific Populations, Clinical Pharmacology, Clinical Studies

(continued)

Table 3 (continued)

1	Drug	Biomarker	Therapeutic area	Labeling sections
60	Everolimus (2)	ESR (Hormone Receptor)	Oncology	Indications and Usage, Dosage and Administration, Warnings and Precautions, Adverse Reactions, Use in Specific Populations, Clinical Pharmacology, Clinical Studies
61	Exemestane	ESR, PGR (Hormone Receptor)	Oncology	Indications and Usage, Dosage and Administration, Clinical Studies
62	Fam-Trastuzumab Deruxtecan-nxki	ERBB2 (HER2)	Oncology	Indications and Usage, Warnings and Precautions, Adverse Reactions, Use in Specific Populations, Clinical Pharmacology, Clinical Studies
63	Fluorouracil (2)	DPYD	Oncology	Warnings and Precautions, Patient Counseling Information
64	Flutamide	G6PD	Oncology	Warnings
65	Fulvestrant (1)	ERBB2 (HER2)	Oncology	Indications and Usage, Adverse Reactions, Clinical Studies
66	Fulvestrant (2)	ESR, PGR (Hormone Receptor)	Oncology	Indications and Usage, Adverse Reactions, Clinical Pharmacology, Clinical Studies
67	Gefitinib (1)	EGFR	Oncology	Indications and Usage, Dosage and Administration, Clinical Studies
68	Gefitinib (2)	CYP2D6	Oncology	Clinical Pharmacology
69	Gilteritinib	FLT3	Oncology	Indications and Usage, Dosage and Administration, Clinical Studies
70	Goserelin	ESR, PGR (Hormone Receptor)	Oncology	Indications and Usage, Clinical Studies
71	Ibrutinib (1)	Chromosome 17p	Oncology	Indications and Usage, Clinical Studies
72	Ibrutinib (2)	Chromosome 11q	Oncology	Clinical Studies
73	Ibrutinib (3)	MYD88	Oncology	Clinical Studies
74	Imatinib (1)	KIT	Oncology	Indications and Usage, Dosage and Administration, Clinical Studies

(continued)

Table 3 (continued)

1	Drug	Biomarker	Therapeutic area	Labeling sections
75	Imatinib (2)	BCR-ABL1 (Philadelphia chromosome)	Oncology	Indications and Usage, Dosage and Administration, Warnings and Precautions, Adverse Reactions, Use in Specific Populations, Clinical Pharmacology, Clinical Studies
76	Imatinib (3)	PDGFRB	Oncology	Indications and Usage, Dosage and Administration, Clinical Studies
77	Imatinib (4)	FIP1L1-PDGFRA	Oncology	Indications and Usage, Dosage and Administration, Clinical Studies
78	Inotuzumab Ozogamicin	BCR-ABL1 (Philadelphia chromosome)	Oncology	Clinical Studies
79	Ipilimumab (1)	HLA-A	Oncology	Clinical Studies
80	Ipilimumab (2)	Microsatellite Instability, Mismatch Repair	Oncology	Indications and Usage, Dosage and Administration, Warnings and Precautions, Adverse Reactions, Use in Specific Populations, Clinical Studies
81	Ipilimumab (3)	CD274 (PD-L1)	Oncology	Indications and Usage, Dosage and Administration, Use in Specific Populations, Clinical Studies
82	Ipilimumab (4)	ALK	Oncology	Indications and Usage, Adverse Reactions, Clinical Studies
83	Ipilimumab (5)	EGFR	Oncology	Indications and Usage, Adverse Reactions, Clinical Studies
84	Irinotecan	UGT1A1	Oncology	Dosage and Administration, Warnings and Precautions, Clinical Pharmacology
85	Isatuximab- irfc (1)	Chromosome 17p	Oncology	Clinical Studies
86	Isatuximab- irfc (2)	Chromosome 4p;14q	Oncology	Clinical Studies
87	Isatuximab- irfc (3)	Chromosome 14q;16q	Oncology	Clinical Studies
88	Ivosidenib	IDH1	Oncology	Indications and Usage, Dosage and Administration, Clinical Pharmacology, Clinical Studies
89	Ixabepilone (1)	ERBB2 (HER2)	Oncology	Clinical Studies
90	Ixabepilone (2)	ESR, PGR (Hormone Receptor)	Oncology	Clinical Studies

(continued)

Table 3 (continued)

1	Drug	Biomarker	Therapeutic area	Labeling sections
91	Lapatinib (1)	ERBB2 (HER2)	Oncology	Indications and Usage, Dosage and Administration, Adverse Reactions, Use in Specific Populations, Clinical Studies
92	Lapatinib (2)	ESR, PGR (Hormone Receptor)	Oncology	Indications and Usage, Dosage and Administration, Adverse Reactions, Use in Specific Populations, Clinical Studies
93	Lapatinib (3)	HLA-DQA1	Oncology	Clinical Pharmacology
94	Lapatinib (4)	HLA-DRB1	Oncology	Clinical Pharmacology
95	Larotrectinib	NTRK	Oncology	Indications and Usage, Dosage and Administration, Adverse Reactions, Clinical Studies
96	Lenvatinib	Microsatellite Instability, Mismatch Repair	Oncology	Indications and Usage, Adverse Reactions, Clinical Studies
97	Letrozole	ESR, PGR (Hormone Receptor)	Oncology	Indications and Usage, Adverse Reactions, Clinical Studies
98	Lorlatinib (1)	ALK	Oncology	Indications and Usage, Adverse Reactions, Clinical Studies
99	Lorlatinib (2)	ROS1	Oncology	Adverse Reactions
100	Lutetium Dotatate Lu-177	SSTR	Oncology	Indications and Usage, Adverse Reactions, Clinical Pharmacology, Clinical Studies
101	Margetuximab-cmkb (1)	ERBB2 (HER2)	Oncology	Indications and Usage, Adverse Reactions, Clinical Pharmacology, Clinical Studies
102	Margetuximab-cmkb (2)	FCGR2A (CD32A)	Oncology	Clinical Pharmacology
103	Margetuximab-cmkb (3)	FCGR2B (CD32B)	Oncology	Clinical Pharmacology
104	Margetuximab-cmkb (4)	FCGR3A (CD16A)	Oncology	Clinical Pharmacology
105	Mercaptopurine (1)	TPMT	Oncology	Dosage and Administration, Warnings and Precautions, Adverse Reactions, Clinical Pharmacology
106	Mercaptopurine (2)	NUDT15	Oncology	Dosage and Administration, Warnings and Precautions, Clinical Pharmacology
107	Midostaurin (1)	FLT3	Oncology	Indications and Usage, Dosage and Administration, Adverse Reactions, Clinical Studies
108	Midostaurin (2)	NPM1	Oncology	Clinical Studies
109	Midostaurin (3)	KIT	Oncology	Clinical Studies

(continued)

Table 3 (continued)

1	Drug	Biomarker	Therapeutic area	Labeling sections
110	Neratinib (1)	ERBB2 (HER2)	Oncology	Indications and Usage, Adverse Reactions, Clinical Studies
111	Neratinib (2)	ESR, PGR (Hormone Receptor)	Oncology	Clinical Studies
112	Nilotinib (1)	BCR-ABL1 (Philadelphia chromosome)	Oncology	Indications and Usage, Dosage and Administration, Warnings and Precautions, Adverse Reactions, Use in Specific Populations, Clinical Pharmacology, Clinical Studies
113	Nilotinib (2)	UGT1A1	Oncology	Clinical Pharmacology
114	Niraparib	BRCA, Genomic Instability (Homologous Recombination Deficiency)	Oncology	Indication and Usage, Dosage and Administration, Clinical Studies
115	Nivolumab (1)	BRAF	Oncology	Adverse Reactions, Clinical Studies
116	Nivolumab (2)	CD274 (PD-L1)	Oncology	Indications and Usage, Dosage and Administration, Use in Specific Populations, Clinical Pharmacology, Clinical Studies
117	Nivolumab (3)	Microsatellite Instability, Mismatch Repair	Oncology	Indications and Usage, Clinical Studies
118	Nivolumab (4)	EGFR	Oncology	Indications and Usage, Adverse Reactions, Clinical Studies
119	Nivolumab (5)	ALK	Oncology	Indications and Usage, Adverse Reactions, Clinical Studies
120	Obinutuzumab	MS4A1 (CD20 antigen)	Oncology	Clinical Studies
121	Olaparib (1)	BRCA	Oncology	Indications and Usage, Dosage and Administration, Warnings and Precautions, Adverse Reactions, Clinical Studies
122	Olaparib (2)	ERBB2 (HER2)	Oncology	Indications and Usage, Dosage and Administration, Adverse Reactions, Clinical Studies
123	Olaparib (3)	ESR, PGR (Hormone Receptor)	Oncology	Indications and Usage, Clinical Studies

(continued)

Table 3 (continued)

1	Drug	Biomarker	Therapeutic area	Labeling sections
124	Olaparib (4)	BRCA, Genomic Instability (Homologous Recombination Deficiency)	Oncology	Indications and Usage, Dosage and Administration, Adverse Reactions, Clinical Studies
125	Olaparib (5)	Homologous Recombination Repair	Oncology	Indications and Usage, Dosage and Administration, Adverse Reactions, Clinical Studies
126	Olaparib (6)	PPP2R2A	Oncology	Clinical Studies
127	Olaratumab	PDGFRA	Oncology	Clinical Studies
128	Omacetaxine	BCR-ABL1 (Philadelphia chromosome)	Oncology	Clinical Studies
129	Osimertinib	EGFR	Oncology	Indications and Usage, Dosage and Administration, Adverse Reactions, Clinical Studies
130	Palbociclib (1)	ESR (Hormone Receptor)	Oncology	Indications and Usage, Adverse Reactions, Clinical Studies
131	Palbociclib (2)	ERBB2 (HER2)	Oncology	Indications and Usage, Adverse Reactions, Clinical Studies
132	Panitumumab (1)	EGFR	Oncology	Adverse Reactions, Clinical Pharmacology, Clinical Studies
133	Panitumumab (2)	RAS	Oncology	Indications and Usage, Dosage and Administration, Warnings and Precautions, Adverse Reactions, Clinical Studies
134	Pazopanib (1)	UGT1A1	Oncology	Clinical Pharmacology
135	Pazopanib (2)	HLA-B	Oncology	Clinical Pharmacology
136	Pembrolizumab (1)	BRAF	Oncology	Adverse Reactions, Clinical Studies
137	Pembrolizumab (2)	CD274 (PD-L1)	Oncology	Indications and Usage, Dosage and Administration, Adverse Reactions, Use in Specific Populations, Clinical Studies
138	Pembrolizumab (3)	Microsatellite Instability, Mismatch Repair	Oncology	Indications and Usage, Dosage and Administration, Adverse Reactions, Use in Specific Populations, Clinical Studies
139	Pembrolizumab (4)	EGFR	Oncology	Indications and Usage, Adverse Reactions, Clinical Studies
140	Pembrolizumab (5)	ALK	Oncology	Indications and Usage, Adverse Reactions, Clinical Studies

(continued)

Table 3 (continued)

1	Drug	Biomarker	Therapeutic area	Labeling sections
141	Pembrolizumab (6)	Tumor Mutational Burden	Oncology	Indications and Usage, Dosage and Administration, Clinical Studies
142	Pemigatinib	FGFR2	Oncology	Indication and Usage, Dosage and Administration, Clinical Studies
143	Pertuzumab (1)	ERBB2 (HER2)	Oncology	Indications and Usage, Dosage and Administration, Warnings and Precautions, Adverse Reactions, Clinical Pharmacology, Clinical Studies
144	Pertuzumab (2)	ESR, PGR (Hormone Receptor)	Oncology	Clinical Studies
145	Ponatinib	BCR-ABL1 (Philadelphia chromosome)	Oncology	Indications and Usage, Warnings and Precautions, Adverse Reactions, Use in Specific Populations, Clinical Studies
146	Pralsetinib	CCDC6-RET, KIF5B-RET, RET	Oncology	Indications and Usage, Dosage and Administration, Adverse Reactions, Use in Specific Populations, Clinical Pharmacology, Clinical Studies
147	Raloxifene	ESR (Hormone Receptor)	Oncology	Clinical Studies
148	Ramucirumab (1)	EGFR	Oncology	Indications and Usage, Dosage and Administration, Adverse Reactions, Clinical Studies
149	Ramucirumab (2)	RAS	Oncology	Clinical Studies
150	Rasburicase (1)	G6PD	Oncology	Boxed Warning, Contraindications, Warnings and Precautions
151	Rasburicase (2)	CYB5R	Oncology	Boxed Warning, Contraindications, Warnings and Precautions
152	Regorafenib	RAS	Oncology	Indications and Usage, Clinical Studies
153	Ribociclib (1)	ESR, PGR (Hormone Receptor)	Oncology	Indications and Usage, Adverse Reactions, Clinical Studies
154	Ribociclib (2)	ERBB2 (HER2)	Oncology	Indications and Usage, Adverse Reactions, Clinical Studies
155	Rituximab	MS4A1 (CD20 antigen)	Oncology	Indications and Usage, Dosage and Administration, Adverse Reactions, Use in Specific Populations, Clinical Studies

(continued)

Table 3 (continued)

1	Drug	Biomarker	Therapeutic area	Labeling sections
156	Rucaparib (1)	BRCA	Oncology	Indications and Usage, Dosage and Administration, Adverse Reactions, Clinical Studies
157	Rucaparib (2)	CYP2D6	Oncology	Clinical Pharmacology
158	Rucaparib (3)	CYP1A2	Oncology	Clinical Pharmacology
159	Rucaparib (4)	BRCA, Loss of Heterozygosity (Homologous Recombination Deficiency)	Oncology	Warnings and Precautions, Adverse Reactions, Clinical Studies
160	Sacituzumab Govitecan-hziy	UGT1A1	Oncology	Warnings and Precautions, Clinical Pharmacology
161	Selpercatinib	RET	Oncology	Indications and Usage, Dosage and Administration, Adverse Reactions, Use in Specific Populations, Clinical Studies
162	Talazoparib (1)	BRCA	Oncology	Indications and Usage, Dosage and Administration, Adverse Reactions, Clinical Studies
163	Talazoparib (2)	ERBB2 (HER2)	Oncology	Indications and Usage, Adverse Reactions, Clinical Studies
164	Tamoxifen (1)	ESR, PGR (Hormone Receptor)	Oncology	Indications and Usage, Adverse Reactions, Clinical Pharmacology, Clinical Studies
165	Tamoxifen (2)	F5 (Factor V Leiden)	Oncology	Warnings and Precautions
166	Tamoxifen (3)	F2 (Prothrombin)	Oncology	Warnings and Precautions
167	Tamoxifen (4)	CYP2D6	Oncology	Clinical Pharmacology
168	Thioguanine (1)	TPMT	Oncology	Dosage and Administration, Warnings, Precautions, Clinical Pharmacology
169	Thioguanine (2)	NUDT15	Oncology	Dosage and Administration, Warnings, Precautions, Clinical Pharmacology
170	Tipiracil and Trifluridine (1)	ERBB2 (HER2)	Oncology	Indications and Usage, Adverse Reactions, Clinical Studies
171	Tipiracil and Trifluridine (2)	RAS	Oncology	Indications and Usage, Clinical Studies
172	Toremifene	ESR (Hormone Receptor)	Oncology	Indications and Usage, Clinical Studies
173	Trametinib (1)	BRAF	Oncology	Indications and Usage, Dosage and Administration, Adverse Reactions, Clinical Pharmacology, Clinical Studies
174	Trametinib (2)	G6PD	Oncology	Adverse Reactions

(continued)

Table 3 (continued)

1	Drug	Biomarker	Therapeutic area	Labeling sections
175	Trametinib (3)	RAS	Oncology	Warnings and Precautions
176	Trastuzumab (1)	ERBB2 (HER2)	Oncology	Indications and Usage, Dosage and Administration, Clinical Pharmacology, Clinical Studies
177	Trastuzumab (2)	ESR, PGR (Hormone Receptor)	Oncology	Clinical Studies
178	Tretinoin	PML-RARA	Oncology	Indications and Usage, Warnings, Clinical Pharmacology
179	Tucatinib	ERBB2 (HER2)	Oncology	Indications and Usage, Adverse Reactions, Clinical Studies
180	Vemurafenib (1)	BRAF	Oncology	Indications and Usage, Dosage and Administration, Warnings and Precautions, Adverse Reactions, Use in Specific Populations, Clinical Pharmacology, Clinical Studies, Patient Counseling Information
181	Vemurafenib (2)	RAS	Oncology	Warnings and Precautions, Adverse Reactions
182	Venetoclax (1)	Chromosome 17p	Oncology	Clinical Studies
183	Venetoclax (2)	Chromosome 11q	Oncology	Clinical Studies
184	Venetoclax (3)	TP53	Oncology	Clinical Studies
185	Venetoclax (4)	IDH1	Oncology	Clinical Studies
186	Venetoclax (5)	IDH2	Oncology	Clinical Studies
187	Venetoclax (6)	IGH	Oncology	Clinical Studies
188	Venetoclax (7)	NPM1	Oncology	Clinical Studies
189	Venetoclax (8)	FLT3	Oncology	Clinical Studies
190	Vincristine	BCR-ABL1 (Philadelphia chromosome)	Oncology	Indications and Usage, Adverse Reactions, Clinical Studies
191	Gemtuzumab Ozogamicin	CD33	Oncology	Indications and Usage, Dosage and Administration, Adverse Reactions, Clinical Studies

These small-molecule covalent kinase inhibitors bind to their target with covalent bonds, which leads to higher binding affinity and specificity, potency, prolonged pharmacodynamics, and reduced off-target effects, and associated toxicity profiles (Liu et al. 2013). The driving forces for the development of these small-molecule covalent kinase inhibitors were to achieve greater target specificity, reduced off-target effects, and to a certain extent overcome drug resistance.

7 Increasing the Precision of Cancer Therapeutics with Targeted Therapies

As part of precision medicine, the emerging field of targeted cancer therapy is an approach in which drugs are developed to target specific genetic or genomic aberrations directly or indirectly in each cancer or across other cancer types. Furthermore, these genetic or genomic aberrations might also function as prognostic biomarkers to monitor disease progress and patient outcome or predictive biomarkers that can predict the response to therapeutic agents (Bashraheel et al. 2020; Collins and Varmus 2015; Smith and Prasad 2021).

The earlier generation of cancer drugs included cytotoxic antineoplastic agents that were mostly non-specific and developed primarily based on these drugs' ability to kill rapidly dividing cells regardless of their potential mechanism of action. Because these cytotoxic drugs tend to act on fast-growing cells, they also kill non-cancerous healthy cells such as rapidly dividing blood cells, cells of the intestinal tract, and hair follicles. This leads to severe toxicity to healthy tissues and associated side effects.

Since then, cancer treatments have evolved from these non-specific, antineoplastic cytotoxic agents, to more specific genome- and immune-targeted therapies, which have altered the outcomes of many cancer patients (Bashraheel et al. 2020; Dupont et al. 2021; Kato et al. 2017). Because of this, a new strategy has emerged to discover and develop more targeted and cancer cell-specific drugs, and thus use fewer toxic drugs.

Consequently, there have been significant developments regarding "targeted" therapeutics that target key signaling molecules or genetic aberrations.

This new drug discovery strategy has resulted in the development of drugs that specifically target specific genetic or genomic alterations present in patient's cancer cells and thus resulted in the development of a novel drug class referred to as "targeted therapies" (Table 2) (Dupont et al. 2021).

Targeted therapy exploits mutation and tumor-specific aberrations, key signaling molecules, or vulnerabilities for each of the cancer types.

As discussed in the literature, targeted cancer therapies often involve small-molecule chemotherapeutic agents, monoclonal antibodies, antibody-drug conjugates, and immunotherapy that attack, directly or indirectly, a specific genetic biomarker found in a given cancer (Dupont et al. 2021; Smith and Prasad 2021).

For example, the monoclonal antibodies trastuzumab and pertuzumab target HER2 and are used when treating HER2-positive breast cancer. Although targeted oncology has improved survival by years for some incurable cancers such as metastatic breast and lung cancer, as few as 8% of patients with advanced cancer are eligible for targeted cancer therapies, and even fewer patients benefit from them.

The development of the targeted therapies was further accelerated by the availability of molecular profiling of patient's tumors using the next generation genome and RNA sequencing and other molecular characterization techniques.

For example, recent large scale massively parallel molecular profiling efforts such as whole genome pan-tumor multiplexed next-generation DNA and RNA sequencing of patient's tumors have revealed that particular oncogene mutations or altered expression of a particular gene frequently occur in more than one tumor type leading to new approaches for tumor-specific, or tumor-agnostic drug-development strategies.

Likewise, the development of machine learning algorithms and artificial intelligence that can analyze massively parallel datasets generated from these efforts are providing additional opportunities toward personalized and precision medicine.

In addition to the development of genome-driven targeted therapies, the emergence of immune oncology-based targeted agents such as immune checkpoint blocking agents is now providing new opportunities toward more specific, effective, and durable targeted therapies that can be adapted to patient's genetic and genomic abnormalities and vulnerabilities.

Therefore, targeted therapies that aim at the "Achilles' heel" of the disease alongside adjuvant therapies combined with immune checkpoint blocking agents to stimulate the patient's immune response have resulted in significantly improved overall survival benefits to patients with a certain type of cancer who previously had limited therapeutic options.

For example, Table 2 illustrates a list of FDA-approved molecular entities and therapeutic biological products for cancer treatment (as of October 2020) (Dupont et al. 2021).

For example, Fig. 2 illustrates how some of the cancer therapies developed in the past decade exemplify the extent to which newer therapies have become increasingly more precise, both at the disease level (i.e., targeted only cancer cells while sparing the normal cells to a great extent, and at the molecular level (i.e., targeted toward a specific molecule or a cell signaling pathway). Note that there are some therapies more precise than others that include several types of immunotherapies (Ravoori et al. 2015; What 2015).

For example, immune checkpoint inhibitors are engineered therapeutic monoclonal antibodies that blockade checkpoint ligands on immune cells and are not targeted toward a genetic component of the tumor.

On the other hand, CAR-T cell therapies are more precise because they are genetically modified immune cells obtained from patients so that the cells can recognize and attack specific cancer cells through antigen recognition.

Hitherto, therapeutic antibodies and cell signaling inhibitors are among the most specific and thus precise targeted therapies in cancer. Examples of therapeutic antibodies include those that target EGFR and block it (cetuximab), and those that can recognize a target on a cancer cell and mark it, to which a toxic drug can be delivered precisely (antibody-drug conjugates such as TDM-1). In addition, there are those that target specific target alterations in cell signaling pathways, such as imatinib (c-KIT, PDGFRB, FIP1L1-PDGFRABCR-ABL1), and ceritinib (ALK Fusions).

Cancer therapeutics that are not necessarily designed based on tumor genetics can also be more effective in some patients due to the patients' genetic predispositions.

Studies have demonstrated that immunotherapies, such as anti-CTLA4 and anti-PD1/PD-L1 checkpoint blocking monoclonal antibodies, which are not strictly

Type of cancer therapy		Examples
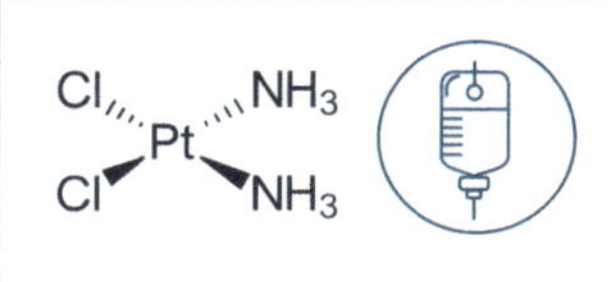	Chemotherapy	5-Flurouracil Carboplatin Cisplatin
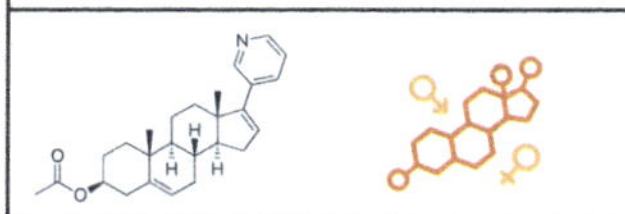	Hormone therapy	Abiraterone acetate Fulvestrant
	Epigenetic therapy	Azacitidine Decitabine
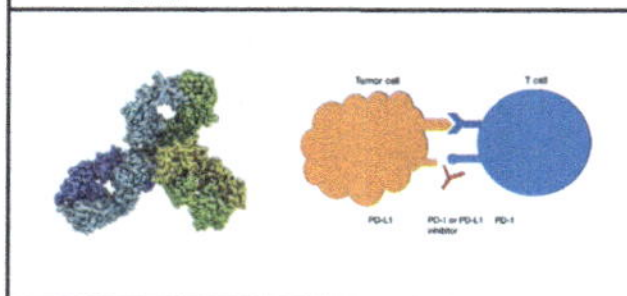	Immune stimulators Checkpoint inhibitors	Aldesleukin Pembrolizumab
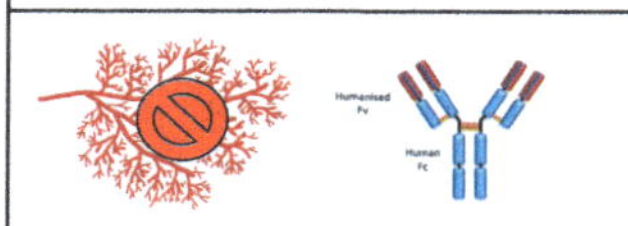	Angiogenesis inhibitors	Bevacizumab Regorafenib
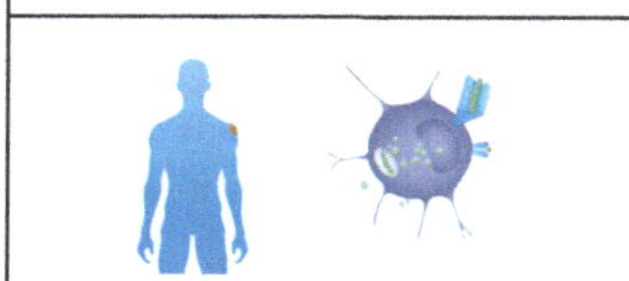	Vaccines	Sipuleucel-T DCVax-L
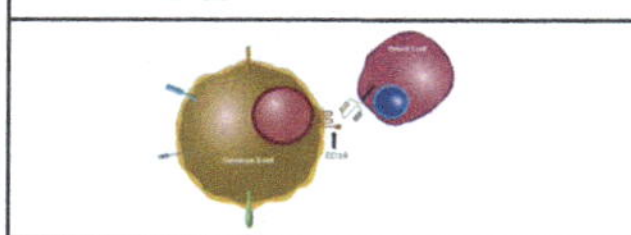	Adoptive Immunotherapy	Anti-CD19 CAR-T cell therapy CART-Meso
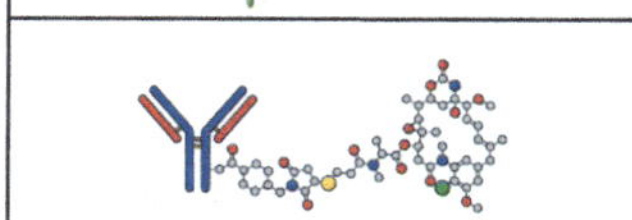	Therapeutic antibodies	Cetuximab TDM-1 (Trastuzumab emtansine)
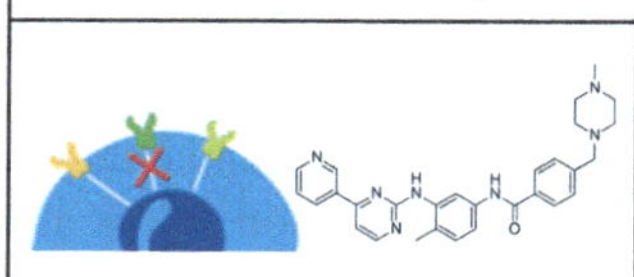	Cell signaling inhibitors	Ibrutinib Imatinib Ceritinib

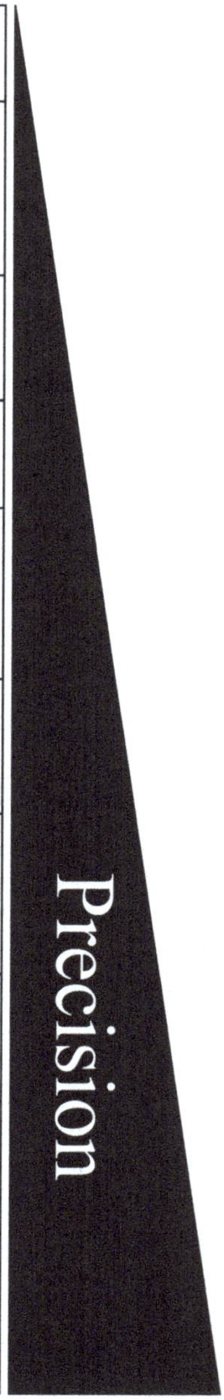

Fig. 2 Increasing the precision of cancer therapeutics through the use of more targeted therapies. Adapted from (Ravoori et al. 2015; What 2015)

precision medicine, are more effective in patients with advanced melanoma whose tumors have certain genetic alterations.

This makes molecular targeting of immunotherapy feasible. Such studies also underscore the importance of companion diagnostics, tests that can identify the right patients for a specific treatment, as an integral part of precision medicine.

Together, immunotherapies, targeted therapies, companion diagnostics, and combinations are defining the current and future cancer treatment approaches.

Despite these advances, successful application of precision medicine needs an integrated approach that incorporates not just the genetic or genomic molecular profiling data of a person but also information regarding environmental exposures and lifestyle, as well as person's socioeconomic status and access to healthcare, healthcare inequality (inequity implies unfairness and injustice), cancer health disparities (disparity implies a difference of some kind) (Meghani and Gallagher 2008; Singh and Jemal 2017), and other risk factors that influence the individual's susceptibilities and risks for a specific disease and treatment outcomes.

Therefore, a better understanding of the factors that dictate an individual's susceptibilities and risks for a specific disease will lead to better integration of these characteristics to develop and provide therapies that are both precise and personalized to the individual.

8 Tumor-Agnostic-Targeted Therapies

Conventionally, the tissue and organ origin of the tumor and anatomic staging have been used to select patient populations for individual cancer clinical trials. However, it has become more evident that cancer is a disease that manifests itself at the cellular and molecular level, with many cancer subtypes being described as a function of biomarkers or tumor genetic mutations. Since then, there have been new approaches to the way clinicians design and perform clinical trials. Increasingly common are trials tailored to detect enhanced efficacy in a patient subpopulation, e.g., patients with a known biomarker signature or whose tumors harbor a specific genetic mutation (Renfro et al. 2017).

Therapies designed to target cancers that harbor specific molecular signatures have provided new opportunities for targeted and more precise drug development efforts.

For example, the advances in the genomics profiling and the availability of The Cancer Genome Atlas, the Genomics Evidence Neoplasia Information Exchange, and other largescale pan-tumor sequencing efforts, and the broad application of multiplexed next-generation sequencing efforts led to the recognition that particular oncogene mutations are frequently present in more than one tumor type albeit at differing frequencies. This finding has given new strategies for novel, tumor-agnostic, drug-development strategies.

Because many of these molecular signatures are also present in many other tumor types irrespective of their organ or tissue of origin with varying frequency, tumor-agnostic genomically enriched trials, such as tumor-agnostic "basket trial," has been explored to assess the effect of these targeted treatments. These genomically enriched trial strategies and their impact on drug development and approval has recently gained some grounds. In addition, the tumor agnostic trials also provide more patient access to these trials, in particular for patients whose tumors have these molecular signatures infrequently (Offin et al. 2018).

As discussed in the literature, agnostic-histology approval of new drugs in oncology has been facilitated by the recent advances in various multi-omics high-throughput technologies. These technologies have enabled a more comprehensive understanding of the molecular alterations associated with a variety of cancer types. These efforts have led to the identification of novel actionable targets, biomarkers, and subsequently the development of novel drugs specific to their targets (Hierro et al. 2017, 2019).

In addition, applying biomarkers in molecularly selected populations brought about yet another benefit to patients that can be stratified for a specific targeted therapy and sparing those who may not benefit from these therapies and associated toxicities; although they have also brought another layer of complexity (Hunter 2016).

As a result of these developments, we are witnessing the emergence of the recognition of tumors as genetic diseases, which is accelerating the development of new drugs that are designed to target those molecular aberrations (Hierro et al. 2019).

For example, several classifications have already been developed to guide clinicians to prioritize molecular biomarkers to stratify patients for the most appropriate targeted therapies, e.g., European Society for Medical Oncology (ESMO) Scale for Clinical Actionability of molecular Targets (ESCAT) (Mateo et al. 2018).

As discussed in literature, a tumor-agnostic approach for targeted therapies in oncology is gaining some acceptance (Ravoori et al. 2015; What 2015). In a tumor-agnostic approach, treatments are chosen based on specific mutations in a tumor rather than the organ of origin (Offin et al. 2018).

In addition, some of the targeted therapies are being developed as tumor-agnostic drugs (such as anti-PD1, anti-PDL-1, and anti-CTLA4 therapies) (Table 4), which target a specific mutation, or genetic or genomic alteration found across different types of tumors and could be used in treating cancers from multiple organ types. As such, histology-agnostic trials might represent the ideal scenario for gaining a better understanding of the functional relevance of a certain molecular aberration across multiple tumor types (Hierro et al. 2019; Lacombe et al. 2014). For example, Pembrolizumab is approved for all cancers that have mismatch repair deficiency, a biomarker theorized to confer sensitivity to immune unmasking.

Other examples include studies that supported the agnostic-histology approval of pembrolizumab for MSI-H/dMMR solid tumors, which was granted accelerated approval by the FDA on May 23, 2017 (Hierro et al. 2019).

A strong preclinical and clinical rationale that ensured the homogeneity of effects of this alteration among different tumors and tissues made this trial successful. For example, MSI tumors were found across multiple histologies at varying frequencies (40) (Hause et al. 2016), sharing common features that are predictive of response to immunotherapy, such as PD-L1 expression, high tumor mutational burden (TMB), and lymphocyte T infiltration (Howitt et al. 2015).

Table 4 Selected biomarkers used in clinical trials are specific to the mechanisms targeted by targeted drugs and immunotherapies

Indication	Select biomarkers (2000–2018)			Select biomarkers (2016–2018[b])		
Breast cancer	HER2	ER[a]	PR[a]	Ki67	PD-L1	TILs
Leukemia	MRD	BCR-ABL1	FLT3	CD19	BTK	TP53
Lung	EGFR	ALK	PD-L1	ROS1	KRAS	PD1
Lymphoma	CD20	CD4	MRD	CD19	MYC	CD30
Prostate cancer	PSA	AR	CTC	ATM	BRCA2	BRCA1
Melanoma	BRAF	CD8	TIL	PD-L1	PD1	CD4
Head & neck cancer	HPV	EGFR	p16	PD-L1	CD8	TILs
Colorectal cancer	KRAS	EGFR	BRAF	MSI	NRAS	CD8
Renal cancer	MTOR	VEGF	EGFR	PD-L1	CD8	PD1
Brain/glioma/CNS	EGFR	MTOR	IDH1/2	CD4	BRAF	CD8
Myeloma	MRD	HLA-A	PSA	LDH	IL-6	EGFR
Stomach cancer	HER2	KIT	EGFR	PD-L1	TILs	PD1
Ovarian cancer	BRCA1	BRCA2	CD8	PD-L1	TILs	CD8
MDS[a]	HLA-A	MRD	FLT3	CD4	CD8	CD3
Pancreatic cancer	EGFR	KRAS	PSA	CD8	PD-L1	IFN
Pan-tumor	MSI[c]	NTRK[c]	TMB	FGFR		

Top biomarkers used by indication and years
[a]ER, Estrogen receptor; PR, progesterone receptor; MDS, myelodysplastic syndromes
[b]Includes biomarkers not already included in top 3 from 2000 to 2018
[c]Biomarkers with therapies approved for pan-tumor use. Source: Aggregate Analysis of ClinicalTrials.gov (AACT) database (https://aact.ctti-clinicaltrials.org/)

Likewise, it was also reported that MSI tumors undergo an immune-editing process. This process leads to genetic advantages for immune escape, most likely due to a common resistance feature across MSI tumors (Grasso et al. 2018). In addition, assessment of the efficacy with a reliable endpoint further improved the interpretation of data across histologies within and between trials.

There are many examples of several studies that supported the agnostic-histology approval of pembrolizumab for MSI-H/dMMR solid tumors by the FDA, including KEYNOTE-016 (NCT01876511), KEYNOTE-158 (NCT02628067), KEYNOTE-164 (NCT02460198), KEYNOTE-012 (NCT01848834), KEYNOTE-028 (NCT02054806) (Hierro et al. 2019).

Furthermore, with the arrival of these best-in-class molecular targeted agents, including novel tyrosine kinase inhibitors such as Debio347 FGFR1/2/3), TAS-120 (FGFR1/2/3/4, Larotrectin ib. (TRK), Entrectinib (RXDX-101) (TRK/ALK/ROS1), Merestinib (LY2801653) (TRK/MET), Repotrectinib (TPX-0005) (TRK/ALK/ROS1), Loxo-195 (TRK), Loxo-292 (RET), Agerafenib (RXDX-105) (RET/BRAF), and BLU-667 (RET), they have been increasingly used under agnostic-histology clinical development (Hierro et al. 2019).

Moreover, entrectinib (Rozlytrek) and larotrectinib (Vit- rakvi) (the neurotrophin receptor kinase inhibitors) are approved for all solid tumors that contain neurotrophin receptor kinase fusions.

However, such a tumor-agnostic strategy has not always been so successful. For example, vemurafenib (Zelboraf) works well for BRAF-mutated melanoma and colorectal cancer but seems to be not very effective in BRAF-mutated myeloma.

9 The Limitations and Challenges Associated with Targeted Therapies

However, targeted therapies have some limitations. Tumors are heterogeneous, and some specific cells in tumors survive or escape treatments resulting in drug resistance. For example, cancer cells can develop resistance to targeted therapies either through mutation of the target itself so that the targeted therapy can no longer interact well with the target, and/or the tumor finds new ways to keep its growth that is no longer dependent on the target.

In addition to the genetics of the tumor, other factors that affect treatment response and outcomes include tumor burden and growth kinetics, tumor heterogeneity, physical barriers, the immune system and the tumor microenvironment, undruggable cancer drivers and the many consequences, cancer cell plasticity, phenotype switching, therapeutic pressure, lifestyle and environmental factors, among many others (Vasan et al. 2019).

In these instances, several strategies have been used to address resistance to targeted therapy. For example, a combination targeted therapy may be a good option to tackle therapy resistance (Eroglu and Ribas 2016; Tolcher et al. 2018). Recent data suggest that using two targeted therapies that target different parts of the cell signaling pathway that is altered in melanoma by the BRAF V600E mutation was more effective against the emergence of resistance and disease progression than using only the one targeted therapy (Flaherty et al. 2012).

Another strategy is to use a targeted therapy in combination with one or more chemotherapy drugs. In this example, the targeted therapy trastuzumab (HER2/neu receptor inhibitor) has been used in combination with a traditional chemotherapy drug, docetaxel, to treat patients with metastatic breast cancer with HER2/neu overexpression.

In addition, drugs for some identified targets such as mutated Ras are difficult to develop because of the target's structure and/or the way its function is regulated in the cell. Ras, a signaling protein, is mutated in almost one-quarter of all cancers and the majority of certain cancer types, including pancreatic cancer.

For example, mutations that lead to RAS gene activation play important roles in the oncogenic transformation of almost one-quarter of all human cancer (Teng et al. 2021; Simanshu et al. 2017; Cox et al. 2014). Previous studies have shown that directly targeting RAS mutants by genetic knockdown or silencing approaches is effective in inhibiting RAS-driven cancers (Acunzo et al. 2017; Sunaga et al. 2011). However, approaches used in these studies to suppress the expression of KRAS mutant genes do not fully phenocopy pharmacological inhibition of KRAS proteins. Several recent clinical trials of G12C allele-specific covalent inhibitors exhibited high dose tolerance and effective antitumor effect in certain patients, indicating that selective inhibition of KRAS (G12C) mutant cancers is a potential therapeutic strategy (Moore et al. 2020; Hallin et al. 2020; Canon et al. 2019). Likewise, a monobody, called 12VC1, that recognizes the active state of both KRAS (G12V) and KRAS(G12C) up to 400-times more tightly than wild-type KRAS has been reported (Teng et al. 2021).

Despite these efforts, there has not been a successful development of Ras inhibitors of Ras signaling with the current drug development approaches and recent evidence indicating that even a successful inhibition of Ras in preclinical models might lead to drug resistance. For example, a recent report demonstrated that loss of KRAS, albeit a genetic KO model, reduces but does not abolish the tumorigenic capacity of PDAC cells and BRAF and MYC partially rescue tumorigenesis of KRAS KO cells in immunocompetent mice (Ischenko et al. 2021).

In addition, some targeted immunotherapies such as immune checkpoint inhibitor therapy may lead to adverse effects, including hypothyroidism, and more major adverse events such as colitis and pneumonitis, which can be fatal and require immediate intervention.

There could also be drug interactions with common medications, including antacids and warfarin. Furthermore, targeted therapies are costly resulting in "financial toxicities," which is a problem for patients with cancer.

Nevertheless, the precision medicine strategy for cancer treatment is still evolving. There are several new treatments designed to target a specific genetic or genomic alteration linked to their corresponding biomarkers that are being explored in many precision medicine clinical trials.

10 Precision Medicine and the Future of Clinical Trials

As discussed by Ho et al. (2020), precision medicine has already revealed important information on the mechanisms of disease pathology and relevant biological pathways and targets that can directly be targeted by new drugs engineered specifically for those targets to intervene with disease progression as well as biomarkers that reflect treatment response. Such understanding has collectively mediated substantial advances in the precision medicine and targeted therapy fields toward improving patient treatment outcomes (van der Meel et al. 2019; Cheng et al. 2017; Bednar et al. 2017).

With the emergence of additional data, genomic testing, novel biomarkers, and relevant targeted therapies will further improve precision medicine in such ways that it will enable the broader application of using patient-specific genomics or other-omics based biomarker-driven targeted and personalized therapy that may predict and/or induce improved response rates of cancer patients and patients with other malignancies (Yan et al. 2017; Pannone et al. 2017; Das et al. 2018; Mohanty et al. 2016; Smyth et al. 2017; Weinshilboum and Wang 2017; Chia et al. 2017).

Alongside the development of genome-driven targeted therapies, many biomarkers that are linked to those genetic or genomic alterations have been developed alongside the targeted therapies (Ho et al. 2020).

With the guidance of these biomarkers, some clinical trials are accepting patients with specific types and stages of cancer, while others accept patients with a variety of cancer types and stages. To be eligible for precision medicine trials, a patient's tumor must have a genetic alteration that can be targeted by a treatment that is specific to that genetic alteration.

To find out what type of genetic alterations are in a patient's cancer, a tumor or liquid biopsy is obtained at baseline after the diagnosis, which then is profiled to determine the genetic changes that may be causing the tumor growth.

If there is a targeted drug or other treatment approved for a specific type of cancer, the patients' tumors are tested to find out whether the genetic change targeted by the treatment is present in the patient's tumor. For instance, patients with lung, breast, colorectal cancers, and melanoma, and some leukemias usually are tested for specific genetic alterations after their diagnosis. Because additional genetic changes that can drive cancer could occur throughout the treatment, patients are tested if cancer relapses or gets worse.

In some instances where there is not an approved targeted therapy for the type of cancer, patients' tumors may be analyzed to see if they can participate in a precision medicine clinical trial testing the new generation targeted therapies.

Several clinical trials are being conducted to bring biomarker-driven targeted therapy and precision medicine to the patient. For example, in a study (NCT02597738), lung cancer patients were paired to genomic analysis of murine and human specimens and further coupled with imaging analysis in a co-clinical trial to identify genomic signatures to improve liquid biopsies (NCT02597738) (Ho et al. 2020).

Likewise, in a precision medicine-guided study for treating cancer pain, patients were screened for the cytochrome p450 2D6 (CYP2D6) and μ-opioid receptor (OPRM1) genotypes to monitor their response to opioid treatment (NCT02664350) (Ho et al. 2020). Furthermore, in another study, the precision medicine approach was used by combining four predictive markers (HER2; TP53, CHEK2, and RB1) for breast cancer with massive parallel genetic screening to generate datasets that can be employed to personalize treatment regimens (NCT02624973) (Ho et al. 2020).

As described by Ho et al. (2020), these developments have been actively explored in many other clinical trials (NCT02795156, NCT03903835).

11 Omics Technologies for Precision Medicine

Recent scientific and technological breakthroughs, including those involved in various multi-omics profiling, single-cell sequencing, and computational tools, and recently emerging machine learning and artificial intelligence platform technologies have significantly improved our understanding of disease pathogenesis and changed the way clinicians diagnose and treat disease (Ho et al. 2020). These advances have led to more precise, predictable, and powerful precision medicine and patient-centric targeted therapies and improved patient outcomes.

As discussed in the literature (Schwartzberg et al. 2017), genetic, genomic, and epigenetic alterations, as well as lifestyle and environmental factors all contribute to the health of an individual and different pathological states of disease. As such, more comprehensive clinical phenotyping, combined with advanced molecular profiling of patients' tumors and other biospecimens is now allowing scientists to construct causal network models and molecular signatures for better disease progress prognosis and predict patient's response to specific therapies.

Despite all these developments, the use of "omics" technologies and single-cell sequencing efforts are generating massive amounts of data sets such that analysis and interpretation of these data sets have become a major challenge requiring advanced computational and statistical approaches, including machine learning and artificial intelligence tools to analyze and help interpret these large data sets.

It has been reported that less than a quarter of people with the most common cancers benefit from precision medicine today (Boehm et al. 2021; Zehir et al. 2017; Stockley et al. 2016). It is still too early for clinicians to predict drug activity and mechanism exclusively based on a mutation or another molecular alteration in an individual tumor alone.

The challenge is now to turn the large multi-omics data sets of disease into actionable information by more precisely defining a disease, thus allowing earlier diagnosis and predicting drug activity and mechanism for more precise treatment options as well as for the development of more specific targeted therapies.

This requires significant resources ranging from well-annotated biobank specimens, including patient tumor and adjacent healthy tissue, blood, and other specimens to comprehensive molecular profiling and clinical data, systematic and methodical laboratory experiments to identify tumor vulnerabilities, novel targets, novel molecular signatures, and biomarkers by way of manipulating a given target in a given tumor type that might affect tumor growth, as well as the development of novel drugs for those targets.

Furthermore, all this requires a team of interdisciplinary experts, large-scale data generation and analysis capabilities, and coordination.

12 Molecular Profiling and Biomarkers in Precision Medicine

Molecular profiling is the assessment of nucleic acids, including DNA, mRNA, micro RNA (miRNA), long noncoding RNAs (lncRNA), and other RNA species, as well as proteins, metabolites, and other biological macromolecules within a patient's specimens, including cells obtained from a tumor biopsy or through the capture of tumor cells circulating in the bloodstream. Other macromolecules include cell-free DNA (cfDNA), circulating tumor cells, or circulating microvesicles and exosomes for the analysis of DNA, mRNAs, miRNAs and proteins released by tumor cells into the circulation to create a detailed molecular network to help guide better treatment decisions.

Because analysis of DNA-level alterations does not necessarily provide information about the biological alterations, this necessitates analysis of factors, including transcriptome and proteome, to obtain a detailed molecular network to help guide better treatment decisions (El-Deiry et al. 2019).

The molecular profiling efforts have come a long way from a few small, predictive, disease-specific, evidence-based panels of biomarkers to more a comprehensive so-called composite (i.e., a panel of several and different types of biomarkers) biomarker panel testing that measures levels of or alterations in a variety of genes or gene products that can be used as biomarkers of both response prediction (a patient outcome/response to a specific therapy) and a patient's prognosis (patient survival regardless of treatment received) (El-Deiry et al. 2019).

These newly developed biomarkers have been employed for the development of next-generation targeted therapies, which have been explored therapeutically in several clinical trials. Consequently, evidence-based biomarker analysis combined with targeted therapy of various cancers has led to the development of a commercially available panel of genomic markers (e.g. Foundation Medicine's FoundationOne CDx) that were recently approved by the FDA and concurrently accepted by the Centers for Medicare & Medicaid Services (CMS) (El-Deiry et al. 2019).

13 More Than Mutations: Cancer Research Needs a Better Map

As suggested by Boehm et al. (2021), there is a need to move beyond tumor sequencing data to identify vulnerabilities in cancers. Cancer cells harbor numerous genetic and epigenetic aberrations that trigger tumor growth when altered in various ways, including mutations, duplications, insertions, deletions, copied, or overexpressed that become essential for the cells' continued proliferation. However, this also could become the weakness or "the Achilles' heel" of these oncogenes because their proliferation and survival can often be impaired by the inactivation of this single genetic or epigenetic abnormality that functions as an oncogene. In oncology, this is termed "dependency" (or vulnerability), which could be a gene, protein, or other molecular feature that a tumor depends on for its growth and survival.

This dependency phenomenon, termed "oncogene addiction" (Weinstein and Joe 2006, 2008), forms the basis of the rationale for molecular targeted therapy. This strategy often relies on the output of various omics methods, including integrative genomics as well as systems biology to determine the state of oncogene addiction in a specific cancer. Upon determining the factors involved in oncogene addiction, the combination therapy strategy is often followed to avoid the development of resistance to a specific therapy such that cancers in theory can no longer escape from a given state of oncogene addiction.

For example, there are targeted therapies that can suppress the function of such gene products directly or in various other ways (e.g., vemurafenib, which blocks the cell-signaling enzyme B-RAF to treat melanoma and other cancers, osimertinib, which blocks the cell receptor EGFR to treat certain lung cancers, and more recently, sotorasib, which blocks a mutant form of the K-Ras cell-signaling protein) (Hong et al. 2020). The addictions involving mutant genes are the type of vulnerability most readily revealed by sequencing of the tumor.

Another type of vulnerability is referred to as synthetic lethality in which a gene becomes indispensable for a tumor's growth because of alterations in other genes (Luo et al. 2009; Seyhan et al. 2011, 2012; Sharma et al. 2017). For instance, a subset of breast and ovarian cancer patients have mutations in the BRCA1 gene, and drugs used in BRCA1 mutant patients are designed to target not the BRCA1 protein itself, but a protein involved in the DNA-repair mechanism that these cancers depend on for their survival. Thus, more comprehensive understanding of cancer vulnerabilities could transform cancer treatment.

For this reason, the Cancer Dependency Map (DepMap) portal developed at the Broad Institute of MIT and Harvard has provided a platform to empower the researchers to discover novel cancer vulnerabilities by providing open access to key cancer dependencies analytical and visualization tools aimed to assess the effects of known drugs across hundreds of cancer lines (Corsello et al. 2020; Garnett et al. 2012; Iorio et al. 2016; Barretina et al. 2012; Haverty et al. 2016).

The Cancer Dependency Map integrates the work of multiple experimental and computational research projects at the Sanger Institute with the shared aim of identifying dependencies in cancer cells that could be exploited to develop new therapies.

The other model aimed to functionally disrupt nearly every gene in hundreds of cancer models (Tsherniak et al. 2017; Behan et al. 2019; Dempster et al. 2019). It is argued that this pilot project represents just a small proportion, probably less than 10%, of a "complete" Cancer Dependency Map (Boehm et al. 2021).

14 The Role of Companion Diagnostics in Precision and Personalized Medicine

As discussed in the literature (Keeling et al. 2020; Seyhan and Carini 2019) and elsewhere, companion diagnostics has become an integral part of precision and personalized medicine.

It has been well established that many drugs are ineffective in a large proportion of the patient populations they intend to treat. Studies have shown that the percentages of patients for whom drugs are ineffective in target patient populations are up to or more than 40% in conditions such as cancer, diabetes, asthma, arthritis, Alzheimer's disease, and others (https://www.appliedclinicaltrialsonline.com/view/personalized-medicine-and-companion-diagnostics).

Therefore, new tools are required to achieve more precise, targeted, and personalized medicine. In this context, companion diagnostic (CDx) has become a new toolset to identify the right patients, for whom the benefit-to-risk profile may be favorably influenced with a specific treatment.

As a result of these developments, to accurately identify patients for the right treatment, CDx has been developed and implemented alongside the treatments that target a specific genetic or biological target present in only some subset of patient populations.

For example, many genetic or genomic biomarkers linked to drugs with pharmacogenomic information found in the drug labeling have been reported in the literature (Seyhan and Carini 2019) and on the FDA website (Table 3) (https://www.fda.gov/drugs/science-and-research-drugs/table-pharmacogenomic-biomarkers-drug-labeling). The list of biomarkers includes germline or somatic gene variants (i.e., polymorphisms, mutations), functional deficiencies with a genetic etiology, altered gene expression signatures, chromosomal abnormalities, and selected protein biomarkers that are used to select treatments for patients (https://www.fda.gov/drugs/science-and-research-drugs/table-pharmacogenomic-biomarkers-drug-labeling).

The FDA has begun issuing statements that a CDx might be required when a drug works on a specific genetic or biological target present in only some patients. The first example of a CDx was for Herceptin and the accompanying HER2 diagnostic test to identify ca. one-third of breast cancer patients with high HER2 expression as the best candidates for treatment (Table 1). Toward this, a guideline was issued by the FDA in August 2014 on companion diagnostics, which provides key insights to pharmaceutical companies developing targeted therapeutics along with their associated companion diagnostic devices. The FDA considers CDx for targeted therapeutics to be essential for the safe and effective use of these targeted therapeutics as well as identify specific patients most likely to benefit from a specific targeted therapy, identify patients who may not draw benefit from a specific therapy or are at increased risk of adverse events, and/or permit monitoring of patient response to inform treatment or dosage adjustments.

(http://www.fda.gov/downloads/MedicalDevices/DeviceRegulationandGuidance/GuidanceDocuments/UCM262327.pdf).

However, the success of biomarker-driven precision and personalized treatments depends on the accurate identification of patients who have the required biomarker signature. To accurately identify those patients, CDx tests have been developed sometimes alongside the targeted therapies. In this context, a biomarker identified in a patient using the CDx test could then be used to guide therapeutic decisions (Table 1).

Pharmacogenomics (the study of how genes affect a person's response to particular drugs), which is a part of precision medicine, combines pharmacology (the science of drugs) and genomics (the study of genes and their functions) to develop safe and effective medications, and doses that are customized to a person's genetic or genomic variations. (https://medlineplus.gov/genetics/understanding/precision-medicine/precisionvspersonalized/).

As of June 2021, there were approximately 191 pharmacogenomic biomarkers cited in the drug labeling that are oncology relevant (Table 3) (https://www.fda.gov/drugs/science-and-research-drugs/table-pharmacogenomic-biomarkers-drug-labeling). These biomarkers include germline or somatic gene variants (i.e., polymorphisms, mutations), functional deficiencies with a genetic etiology, altered gene expression signatures, chromosomal abnormalities, and selected protein biomarkers that are used to select treatments for patients (Seyhan and Carini 2019).

The FDA's Center for Drug Evaluation and Research's (CDER) annual report, indicated that by the end of 2019, ca. 200 FDA-approved therapies with a biomarker indicated in the drug label were recommended for prescribing (Keeling et al. 2020) and in 2018, 2019, and 2020 many of those approved therapies involved expanding previously approved therapies with new precision medicine indications. In 2018 and 2019, new Precision Medicine indications were approved by the FDA for ca. 20 drugs each year (https://www.fda.gov/drugs/new-drugs-fda-cders-new-molecular-entities-and-new-therapeutic-biological-products/new-drug-therapy-approvals-2019) and ca. 25 drugs in 2020 for various cancers, including cancers of lung, thyroid, breast, urothelial, colorectal, prostate, bladder, and rare cancers such as multiple myeloma, neuroblastoma, myelodysplastic syndromes and chronic myelomonocytic leukemia, gastrointestinal stromal tumor, epithelioid sarcoma, metastatic cholangiocarcinoma (bile duct cancer), neurofibromatosis type 1, and mesothelioma.

For example, a recent report indicated increased use of biomarkers in oncology clinical trials from 15% of trials in 2000 to 55% of trials in 2018 (Vadas et al. 2020).

In recent years, the importance of biomarkers has been widely recognized and they have been utilized in all stages of drug development including enrichment, patient stratification, safety, and efficacy.

For example, since 2000, there has been significant growth in the use of biomarkers in oncology clinical trials across most cancers, ranging from 25% to 60% of trials from 2000 to 2018, although use is highest in breast, leukemia, lung, lymphoma, prostate, melanoma, and head and neck cancers with lung notably outpacing others.

In a comprehensive review published recently (El-Deiry et al. 2019), the authors present the currently accepted genes or gene products that act as predictive and risk assessment biomarkers for each specific cancer type, more specifically, solid tumors. This comprehensive review also addresses the details on when these biomarkers during the disease course or the levels of these biomarkers should be determined. Based on the compelling evidence, the authors argue that these biomarkers will be listed in the NCCN "recommended" biomarker category in the near future (El-Deiry et al. 2019).

For example, some clinical trials recruit patients based on specific biomarker presence ("enrichment") and file for a companion diagnostics label, whereas other clinical trials use specific biomarkers for GO, NO-GO research decision-making.

In addition to prognostic and monitoring biomarkers, predictive biomarkers used in clinical trials are often based on the mechanisms targeted by targeted drugs and immunotherapies (Table 4).

For instance, HER2, EGFR, KRAS, BRCA1/2, and PD1/PD-L1 biomarkers are associated with therapeutic mechanisms in breast cancer, non-small cell lung cancer, and ovarian cancer. MRD and PSA are key biomarkers for monitoring heme and prostate cancers, respectively.

In recent years, pan-tumor biomarkers, including MSI and NTRK, both of which are associated with pan-tumor indication approvals, for pembrolizumab (MSI) and larotrectinib and entrectinib (NTRK), as well as TMB and FGFR have also been explored in clinical trials.

Pan-tumor biomarkers in future pan-tumor indications might guide personalized drug development independent of the more traditional organ-specific clinical trial approaches.

As such, biomarkers are becoming even more critical in cancer research, clinical development, and treatment management.

Despite the timely launch of therapies alongside their corresponding biomarker tests, many patients do not have access to these tests because of the inefficiencies within the clinical diagnostic testing system (Keeling et al. 2020).

These inefficiencies result in slow laboratory test adoption, lack of test reimbursement, deficient or inefficient physician biomarker training, and delays in the inclusion of diagnostic testing in associated clinical guidelines—ultimately hampering the use of these biomarker tests (Keeling et al. 2020).

15 Conclusions and Future Perspectives

In recent years, rapid improvements in various high-throughput molecular profiling technologies, computational biology, and bioinformatics, and more recently machine learning and artificial intelligence to analyze massive amounts of data sets generated from these efforts have improved our ability to go beyond single genetic mutations to other molecular alterations to construct causal network models, molecular signatures, and actionable targets that affect disease progression and tumor response to therapies and represent targets for new targeted therapeutics. These actionable targets also represent novel biomarkers to aid in better disease progress prognosis and predict patient's response to specific therapies.

However, better use of this technology will require creating supporting evidence through innovative clinical trial designs, integration of molecular profiling and clinical data in real-world databases, and improved analysis to identify novel target-drug pairings, eventually improving patient outcomes.

Precision medicine involves the collection and validation of population-wide data, including multi-omics molecular profiling, including epigenetic, genetic, genomic, proteomic, metabolomic changes to identify targetable/actionable molecular targets and patient's clinical data. This may also involve the use of artificial intelligence (AI) to both design a drug combination based on a patient's data and help with the selection of specific drug dosing regimens.

However, considerations for molecular profiling should include the type of tissues to utilize, timing of profiling in the disease course, comprehensiveness of the molecular panel, and the interpretation of molecular data, which remains a challenge. Because resistance to targeted therapies is also a concern, resampling and retesting of tumors, including the dynamic sampling of liquid biopsy specimens after clinical progression, may help make better treatment decisions.

Furthermore, deployment of both precision medicine and personalized medicine requires their successful integration, with genome-guided drug pairing (driven by population data) followed by AI-guided dynamic therapeutic intervention (driven by individual data). Consequently, all this requires significant resources ranging from large-scale data generation to analysis capabilities and coordination.

Several hundreds of drug labels have already used genetic or genomic biomarker information to pair them with targeted therapies. The FDA has already created resources to help the development and implementation of targeted therapeutics toward precision medicine.

The precision and personalized treatment strategies have resulted in improved patient outcomes in terms of response rate and progression-free survival in phase I clinical trials that selected patients using a specific biomarker as compared to those that did not (Krzyszczyk et al. 2018).

Nevertheless, there are many challenges to address, including the analysis and interpretation of large amounts of molecular profiling data sets to identify clinically relevant biomarkers for disease susceptibility and treatment efficacy. Likewise, the success of precision medicine depends on effective collaboration between all associated stakeholders, including scientists, clinicians, patients, biopharmaceutical companies, insurers, and regulatory agencies.

Acknowledgments AAS wrote the manuscript and agrees with the manuscript's results and conclusions.

Ethics Approval and Consent to ParticipateThis article does not contain any studies with human participants or animals performed by any of the authors. Informed consent was not required for the preparation of this review article as it used secondary sources only.

Competing Interests The author declares that he has no competing interests.

References

Acunzo M, et al. Selective targeting of point-mutated KRAS through artificial microRNAs. Proc Natl Acad Sci U S A. 2017;114:E4203–12. https://doi.org/10.1073/pnas.1620562114.

Alizadeh AA, et al. Distinct types of diffuse large B-cell lymphoma identified by gene expression profiling. Nature. 2000;403:503–11. https://doi.org/10.1038/35000501.

Barretina J, et al. The cancer cell line encyclopedia enables predictive modelling of anticancer drug sensitivity. Nature. 2012;483:603–7. https://doi.org/10.1038/nature11003.

Bashraheel SS, Domling A, Goda SK. Update on targeted cancer therapies, single or in combination, and their fine tuning for precision medicine. Biomed Pharmacother. 2020;125:110009. https://doi.org/10.1016/j.biopha.2020.110009.

Bednar EM, et al. A universal genetic testing initiative for patients with high-grade, non-mucinous epithelial ovarian cancer and the implications for cancer treatment. Gynecol Oncol. 2017;146:399–404. https://doi.org/10.1016/j.ygyno.2017.05.037.

Behan FM, et al. Prioritization of cancer therapeutic targets using CRISPR-Cas9 screens. Nature. 2019;568:511–6. https://doi.org/10.1038/s41586-019-1103-9.

Bhullar KS, et al. Kinase-targeted cancer therapies: progress, challenges and future directions. Mol Cancer. 2018;17:48. https://doi.org/10.1186/s12943-018-0804-2.

Boehm JS, et al. Cancer research needs a better map. Nature. 2021;589:514–6. https://doi.org/10.1038/d41586-021-00182-0.

Bray F, et al. Global cancer statistics 2018: GLOBOCAN estimates of incidence and mortality worldwide for 36 cancers in 185 countries. CA Cancer J Clin. 2018;68:394–424. https://doi.org/10.3322/caac.21492.

Canon J, et al. The clinical KRAS(G12C) inhibitor AMG 510 drives anti-tumour immunity. Nature. 2019;575:217–23. https://doi.org/10.1038/s41586-019-1694-1.
Cheng HH, Pritchard CC, Montgomery B, Lin DW, Nelson PS. Prostate cancer screening in a new Era of genetics. Clin Genitourin Cancer. 2017;15:625–8. https://doi.org/10.1016/j.clgc.2017.05.024.
Chia S, et al. Phenotype-driven precision oncology as a guide for clinical decisions one patient at a time. Nat Commun. 2017;8:435. https://doi.org/10.1038/s41467-017-00451-5.
Collins FS, Varmus H. A new initiative on precision medicine. N Engl J Med. 2015;372:793–5. https://doi.org/10.1056/NEJMp1500523.
Corsello SM, et al. Discovering the anti-cancer potential of non-oncology drugs by systematic viability profiling. Nat Cancer. 2020;1:235–48. https://doi.org/10.1038/s43018-019-0018-6.
Cox AD, Fesik SW, Kimmelman AC, Luo J, Der CJ. Drugging the undruggable RAS: mission possible? Nat Rev Drug Discov. 2014;13:828–51. https://doi.org/10.1038/nrd4389.
Das K, et al. Genomic predictors of chemotherapy efficacy in advanced or recurrent gastric cancer in the GC0301/TOP002 phase III clinical trial. Cancer Lett. 2018;412:208–15. https://doi.org/10.1016/j.canlet.2017.10.011.
Davies H, et al. Mutations of the BRAF gene in human cancer. Nature. 2002;417:949–54. https://doi.org/10.1038/nature00766.
Dempster JM, et al. Agreement between two large pan-cancer CRISPR-Cas9 gene dependency data sets. Nat Commun. 2019;10:5817. https://doi.org/10.1038/s41467-019-13805-y.
Dupont CA, Riegel K, Pompaiah M, Juhl H, Rajalingam K. Druggable genome and precision medicine in cancer: current challenges. FEBS J. 2021; https://doi.org/10.1111/febs.15788.
El-Deiry WS, et al. The current state of molecular testing in the treatment of patients with solid tumors, 2019. CA Cancer J Clin. 2019;69:305–43. https://doi.org/10.3322/caac.21560.
Eroglu Z, Ribas A. Combination therapy with BRAF and MEK inhibitors for melanoma: latest evidence and place in therapy. Ther Adv Med Oncol. 2016;8:48–56. https://doi.org/10.1177/1758834015616934.
Flaherty KT, et al. Combined BRAF and MEK inhibition in melanoma with BRAF V600 mutations. N Engl J Med. 2012;367:1694–703. https://doi.org/10.1056/NEJMoa1210093.
Garnett MJ, et al. Systematic identification of genomic markers of drug sensitivity in cancer cells. Nature. 2012;483:570–5. https://doi.org/10.1038/nature11005.
Grasso CS, et al. Genetic mechanisms of immune evasion in colorectal cancer. Cancer Discov. 2018;8:730–49. https://doi.org/10.1158/2159-8290.CD-17-1327.
Guinney J, et al. The consensus molecular subtypes of colorectal cancer. Nat Med. 2015;21:1350–6. https://doi.org/10.1038/nm.3967.
Hallin J, et al. The KRAS(G12C) inhibitor MRTX849 provides insight toward therapeutic susceptibility of KRAS-mutant cancers in mouse models and patients. Cancer Discov. 2020;10:54–71. https://doi.org/10.1158/2159-8290.CD-19-1167.
Hanahan D, Weinberg RA. Hallmarks of cancer: the next generation. Cell. 2011;144:646–74. https://doi.org/10.1016/j.cell.2011.02.013.
Hartmanshenn C, Scherholz M, Androulakis IP. Physiologically-based pharmacokinetic models: approaches for enabling personalized medicine. J Pharmacokinet Pharmacodyn. 2016;43:481–504. https://doi.org/10.1007/s10928-016-9492-y.
Hause RJ, Pritchard CC, Shendure J, Salipante SJ. Classification and characterization of microsatellite instability across 18 cancer types. Nat Med. 2016;22:1342–50. https://doi.org/10.1038/nm.4191.
Haverty PM, et al. Reproducible pharmacogenomic profiling of cancer cell line panels. Nature. 2016;533:333–7. https://doi.org/10.1038/nature17987.
Hawgood S, Hook-Barnard IG, O'Brien TC, Yamamoto KR. Precision medicine: Beyond the inflection point. Sci Transl Med. 2015;7.:300ps317 https://doi.org/10.1126/scitranslmed.aaa9970.
Hierro C, et al. Unveiling changes in the landscape of patient populations in cancer early drug development. Oncotarget. 2017;8:14158–72. https://doi.org/10.18632/oncotarget.13258.

Hierro C, Matos I, Martin-Liberal J, Ochoa de Olza M and Garralda E. Agnostic-histology approval of new drugs in oncology: are we already there? Clin Cancer Res 2019;25:3210. doi: https://doi.org/10.1158/1078-0432.CCR-18-3694.

Ho D, et al. Enabling technologies for personalized and precision medicine. Trends Biotechnol. 2020;2020(38):497–518. https://doi.org/10.1016/j.tibtech.2019.12.021.

Hong DS, et al. KRAS(G12C) Inhibition with Sotorasib in advanced solid tumors. N Engl J Med. 2020;383:1207–17. https://doi.org/10.1056/NEJMoa1917239.

Hopkins AL, Groom CR. The druggable genome. Nat Rev Drug Discov. 2002;1:727–30. https://doi.org/10.1038/nrd892.

Howitt BE, et al. Association of polymerase e-mutated and microsatellite-instable endometrial cancers with neoantigen load, number of tumor-infiltrating lymphocytes, and expression of PD-1 and PD-L1. JAMA Oncol. 2015;1:1319–23. https://doi.org/10.1001/jamaoncol.2015.2151.

Hunter DJ. Uncertainty in the Era of precision medicine. N Engl J Med. 2016;375:711–3. https://doi.org/10.1056/NEJMp1608282.

Iorio F, et al. A landscape of pharmacogenomic interactions in cancer. Cell. 2016;166:740–54. https://doi.org/10.1016/j.cell.2016.06.017.

Ischenko I, et al. KRAS drives immune evasion in a genetic model of pancreatic cancer. Nat Commun. 2021;12:1482. https://doi.org/10.1038/s41467-021-21736-w.

Johnson DB, et al. Enabling a genetically informed approach to cancer medicine: a retrospective evaluation of the impact of comprehensive tumor profiling using a targeted next-generation sequencing panel. Oncologist. 2014;19:616–22. https://doi.org/10.1634/theoncologist.2014-0011.

Kannaiyan R, Mahadevan D. A comprehensive review of protein kinase inhibitors for cancer therapy. Expert Rev Anticancer Ther. 2018;18:1249–70. https://doi.org/10.1080/14737140.2018.1527688.

Kato S, Subbiah V, Kurzrock R. Counterpoint: successes in the pursuit of precision medicine: biomarkers take credit. J Natl Compr Cancer Netw. 2017;15:863–6. https://doi.org/10.6004/jnccn.2017.0127.

Keeling P, Clark J, Finucane S. Challenges in the clinical implementation of precision medicine companion diagnostics. Expert Rev Mol Diagn. 2020;20:593–9. https://doi.org/10.1080/14737159.2020.1757436.

Kesisoglou F, Chung J, van Asperen J, Heimbach T. Physiologically based absorption modeling to impact biopharmaceutics and formulation strategies in drug development-industry case studies. J Pharm Sci. 2016;105:2723–34. https://doi.org/10.1016/j.xphs.2015.11.034.

Kinch MS. An analysis of FDA-approved drugs for oncology. Drug Discov Today. 2014;19:1831–5. https://doi.org/10.1016/j.drudis.2014.08.007.

Knox SS. From 'omics' to complex disease: a systems biology approach to gene-environment interactions in cancer. Cancer Cell Int. 2010;10:11. https://doi.org/10.1186/1475-2867-10-11.

Krzyszczyk P, et al. The growing role of precision and personalized medicine for cancer treatment. Technology (Singap World Sci). 2018;6:79–100. https://doi.org/10.1142/S2339547818300020.

Lacombe D, et al. The dream and reality of histology agnostic cancer clinical trials. Mol Oncol. 2014;8:1057–63. https://doi.org/10.1016/j.molonc.2014.06.002.

Le Tourneau C, et al. Molecularly targeted therapy based on tumour molecular profiling versus conventional therapy for advanced cancer (SHIVA): a multicentre, open-label, proof-of-concept, randomised, controlled phase 2 trial. Lancet Oncol. 2015;16:1324–34. https://doi.org/10.1016/S1470-2045(15)00188-6.

Ledermann J, et al. Olaparib maintenance therapy in patients with platinum-sensitive relapsed serous ovarian cancer: a preplanned retrospective analysis of outcomes by BRCA status in a randomised phase 2 trial. Lancet Oncol. 2014;15:852–61. https://doi.org/10.1016/S1470-2045(14)70228-1.

Liu Q, et al. Developing irreversible inhibitors of the protein kinase cysteinome. Chem Biol. 2013;20:146–59. https://doi.org/10.1016/j.chembiol.2012.12.006.

Luo J, Solimini NL, Elledge SJ. Principles of cancer therapy: oncogene and non-oncogene addiction. Cell. 2009;136:823–37. https://doi.org/10.1016/j.cell.2009.02.024.

Mashberg A, Boffetta P, Winkelman R and Garfinkel L. Tobacco smoking, alcohol drinking, and cancer of the oral cavity and oropharynx among U.S. veterans. Cancer 1993;72:1369–1375. doi: https://doi.org/10.1002/1097-0142(19930815)72:4<1369::aid-cncr2820720436>3.0.co;2-l.

Mateo J, et al. A framework to rank genomic alterations as targets for cancer precision medicine: the ESMO Scale for Clinical Actionability of molecular Targets (ESCAT). Ann Oncol. 2018;29:1895–902. https://doi.org/10.1093/annonc/mdy263.

Meghani SH, Gallagher RM. Disparity vs inequity: toward reconceptualization of pain treatment disparities. Pain Med. 2008;9:613–23. https://doi.org/10.1111/j.1526-4637.2007.00344.x.

Meric-Bernstam F, et al. Feasibility of large-scale genomic testing to facilitate enrollment onto genomically matched clinical trials. J Clin Oncol. 2015;33:2753–62. https://doi.org/10.1200/JCO.2014.60.4165.

Mohanty A, et al. CCND1 mutations increase protein stability and promote ibrutinib resistance in mantle cell lymphoma. Oncotarget. 2016;7:73558–72. https://doi.org/10.18632/oncotarget.12434.

Moore AR, Rosenberg SC, McCormick F, Malek S. RAS-targeted therapies: is the undruggable drugged? Nat Rev Drug Discov. 2020;19:533–52. https://doi.org/10.1038/s41573-020-0068-6.

Nicholson JK. Global systems biology, personalized medicine and molecular epidemiology. Mol Syst Biol. 2006;2:52. https://doi.org/10.1038/msb4100095.

Offin M, Liu D, Drilon A. Tumor-Agnostic drug development. Am Soc Clin Oncol Educ Book. 2018;38:184–7. https://doi.org/10.1200/EDBK_200831.

Panicker RC, Chattopadhaya S, Coyne AG, Srinivasan R. Allosteric small-molecule serine/threonine kinase inhibitors. Adv Exp Med Biol. 2019;1163:253–78. https://doi.org/10.1007/978-981-13-8719-7_11.

Pannone L, et al. Structural, functional, and clinical characterization of a Novel PTPN11 mutation cluster underlying Noonan syndrome. Hum Mutat. 2017;38:451–9. https://doi.org/10.1002/humu.23175.

Paul D. The systemic hallmarks of cancer. J Cancer Metastasis Treat. 2020;6:29. https://doi.org/10.20517/2394-4722.2020.63.

Perou CM, et al. Molecular portraits of human breast tumours. Nature. 2000;406:747–52. https://doi.org/10.1038/35021093.

Pottier C, et al. Tyrosine kinase inhibitors in cancer: breakthrough and challenges of targeted therapy. Cancers (Basel). 2020;12 https://doi.org/10.3390/cancers12030731.

Prasad V, Gale RP. What Precisely Is Precision Oncology—and Will It Work? The ASCO Post. 2017;

Ravoori, P. What Is precision medicine? Cancer research catalyst. American Association for Cancer Research; 2015.

Reeve E, Wiese MD, Mangoni AA. Alterations in drug disposition in older adults. Expert Opin Drug Metab Toxicol. 2015;11:491–508. https://doi.org/10.1517/17425255.2015.1004310.

Renfro LA, An MW, Mandrekar SJ. Precision oncology: a new era of cancer clinical trials. Cancer Lett. 2017;387:121–6. https://doi.org/10.1016/j.canlet.2016.03.015.

Rowland M, Peck C, Tucker G. Physiologically-based pharmacokinetics in drug development and regulatory science. Annu Rev Pharmacol Toxicol. 2011;51:45–73. https://doi.org/10.1146/annurev-pharmtox-010510-100540.

Schwaederle M, et al. On the road to precision cancer medicine: analysis of genomic biomarker actionability in 439 patients. Mol Cancer Ther. 2015;14:1488–94. https://doi.org/10.1158/1535-7163.Mct-14-1061.

Schwartzberg L, Kim ES, Liu D, Schrag D. Precision oncology: who, how, what, when, and when not? Am Soc Clin Oncol Educ Book. 2017;37:160–9. https://doi.org/10.1200/EDBK_174176.

Seyhan AA, Carini C. Are innovation and new technologies in precision medicine paving a new era in patients centric care? J Transl Med. 2019;17:114. https://doi.org/10.1186/s12967-019-1864-9.

Seyhan AA, et al. A genome-wide RNAi screen identifies novel targets of neratinib sensitivity leading to neratinib and paclitaxel combination drug treatments. Mol BioSyst. 2011;7:1974–89. https://doi.org/10.1039/c0mb00294a.

Seyhan AA, Varadarajan U, Choe S, Liu W, Ryan TE. A genome-wide RNAi screen identifies novel targets of neratinib resistance leading to identification of potential drug resistant genetic markers. Mol BioSyst. 2012;8:1553–70. https://doi.org/10.1039/c2mb05512k.

Sharma P, Hu-Lieskovan S, Wargo JA, Ribas A. Primary, adaptive, and acquired resistance to cancer immunotherapy. Cell. 2017;168:707–23. https://doi.org/10.1016/j.cell.2017.01.017.

Sholl LM, et al. Institutional implementation of clinical tumor profiling on an unselected cancer population. JCI Insight. 2016;1:e87062. https://doi.org/10.1172/jci.insight.87062.

Sicklick JK, et al. Molecular profiling of cancer patients enables personalized combination therapy: the I-PREDICT study. Nat Med. 2019;25:744–50. https://doi.org/10.1038/s41591-019-0407-5.

Simanshu DK, Nissley DV, McCormick F. RAS Proteins and their regulators in human disease. Cell. 2017;170:17–33. https://doi.org/10.1016/j.cell.2017.06.00.

Singh GK, Jemal A. Socioeconomic and racial/ethnic disparities in cancer mortality, incidence, and survival in the United States, 1950–2014: over six decades of changing patterns and widening inequalities. J Environ Public Health. 2017:2819372. https://doi.org/10.1155/2017/2819372.

Sjoblom T, et al. The consensus coding sequences of human breast and colorectal cancers. Science. 2006;314:268–74. https://doi.org/10.1126/science.1133427.

Smiech M, Leszczynski P, Kono H, Wardell C, Taniguchi H. Emerging BRAF mutations in cancer progression and their possible effects on transcriptional networks. Genes (Basel). 2020;11:1342. https://doi.org/10.3390/genes11111342.

Smith CEP, Prasad V. Targeted cancer therapies. Am Fam Physician. 2021;103:155–63.

Smyth E, et al. Pharmacogenetic analysis of the UK MRC (Medical Research Council) MAGIC trial: association of polymorphisms with toxicity and survival in patients treated with perioperative epirubicin, cisplatin, and 5-fluorouracil (ECF) chemotherapy. Clin Cancer Res. 2017;23:7543–9. https://doi.org/10.1158/1078-0432.CCR-16-3142.

Soda M, et al. Identification of the transforming EML4-ALK fusion gene in non-small-cell lung cancer. Nature. 2007;448:561–6. https://doi.org/10.1038/nature05945.

Stockley TL, et al. Molecular profiling of advanced solid tumors and patient outcomes with genotype-matched clinical trials: the Princess Margaret IMPACT/COMPACT trial. Genome Med. 2016;8:109. https://doi.org/10.1186/s13073-016-0364-2.

Stratton MR, Campbell PJ, Futreal PA. The cancer genome. Nature. 2009;458:719–24. https://doi.org/10.1038/nature07943.

Sunaga N, et al. Knockdown of oncogenic KRAS in non-small cell lung cancers suppresses tumor growth and sensitizes tumor cells to targeted therapy. Mol Cancer Ther. 2011;10:336–46. https://doi.org/10.1158/1535-7163.MCT-10-0750.

Teng KW, et al. Selective and noncovalent targeting of RAS mutants for inhibition and degradation. Nat Commun. 2021;12:2656. https://doi.org/10.1038/s41467-021-22969-5.

Tolcher AW, Peng W, Calvo E. Rational approaches for combination therapy strategies targeting the MAP kinase pathway in solid tumors. Mol Cancer Ther. 2018;17:3–16. https://doi.org/10.1158/1535-7163.MCT-17-0349.

Tsherniak A, et al. Defining a cancer dependency map. Cell. 2017;170:564–576.e16. https://doi.org/10.1016/j.cell.2017.06.010.

Vadas A, Bilodeau TJ, Oza C. Special report: the evolution of biomarker use in clinical trials for cancer treatments. J Precision Med. 2020;

van der Meel R, et al. Smart cancer nanomedicine. Nat Nanotechnol. 2019;14:1007–17. https://doi.org/10.1038/s41565-019-0567-y.

Vasan N, Baselga J, Hyman DM. A view on drug resistance in cancer. Nature. 2019;575:299–309. https://doi.org/10.1038/s41586-019-1730-1.

Venter JC, et al. The sequence of the human genome. Science. 2001;291:1304–51. https://doi.org/10.1126/science.1058040.

Verma M. Personalized medicine and cancer. J Pers Med. 2012;21-14 https://doi.org/10.3390/jpm2010001.

Weinshilboum RM, Wang L. Pharmacogenomics: precision medicine and drug response. Mayo Clin Proc. 2017;92:1711–22. https://doi.org/10.1016/j.mayocp.2017.09.001.

Weinstein IB, Joe AK. Mechanisms of disease: oncogene addiction--a rationale for molecular targeting in cancer therapy. Nat Clin Pract Oncol. 2006;3:448–57. https://doi.org/10.1038/ncponc0558.

Weinstein IB, Joe A. Oncogene addiction. Cancer Res. 2008;68:3077–80.; discussion 3080. https://doi.org/10.1158/0008-5472.CAN-07-3293.

What RP. Is precision medicine? Cancer research catalyst. American Association for Cancer Research; 2015.

Yan B, et al. Single-cell genomic profiling of acute myeloid leukemia for clinical use: a pilot study. Oncol Lett. 2017;13:1625–30. https://doi.org/10.3892/ol.2017.5669.

Zehir A, et al. Mutational landscape of metastatic cancer revealed from prospective clinical sequencing of 10,000 patients. Nat Med. 2017;23:703–13. https://doi.org/10.1038/nm.4333.

Metabolic Control Analysis for Drug Target Selection Against Human Diseases

Javier Belmont-Díaz, Citlali Vázquez, Rusely Encalada, Rafael Moreno-Sánchez, Paul A. M. Michels, and Emma Saavedra

Contents

Abstract

For identification of suitable therapeutic targets (enzymes/transporters) in intermediary metabolism of pathological and parasitic cells, the capacity of the target to govern the metabolic pathway flux should be considered. Metabolic Control

J. Belmont-Díaz · C. Vázquez · R. Encalada · E. Saavedra (✉)
Departamento de Bioquímica, Instituto Nacional de Cardiología Ignacio Chávez, Mexico City, Mexico
e-mail: emma_saavedra2002@yahoo.com

R. Moreno-Sánchez
Facultad de Estudios Superiores Iztacala, Universidad Nacional Autónoma de México: Tlalnepantla Estado de, Mexico, Mexico

Departamento de Bioquímica, Instituto Nacional de Cardiología Ignacio Chávez, Mexico City, Mexico

P. A. M. Michels (✉)
Centre for Immunity, Infection and Evolution (CIIE) and Centre for Translational and Chemical Biology (CTCB), School of Biological Sciences, The University of Edinburgh, Edinburgh, Scotland, UK
e-mail: paul.michels@ed.ac.uk

M. T. Scotti, C. L. Bellera (eds.), *Drug Target Selection and Validation*, Computer–Aided Drug Discovery and Design,
https://doi.org/10.1007/978-3-030-95895-4_8

Analysis (MCA) is a biochemical framework that enables one to quantitate the degree of control that the activity of a target *i* (*ai*) exerts on the pathway flux (*J*), defined as flux control coefficient (C^J_{ai}). Different experimental strategies are being used to determine the C^J_{ai} of individual pathway steps, and consequently, the distribution of control in the metabolic pathway. By applying MCA, the components with the highest control on flux can be identified, which are the targets with the highest therapeutic potential. In this chapter, we will review the MCA theoretical principles and experimental approaches to determine the C^J_{ai} in a range of metabolic pathways such as central carbon and antioxidant metabolism, with potential application to other pathways of diverse human diseases.

Keywords
Drug target · Metabolic control analysis · Flux control coefficient · Intermediary metabolism · Pathway modeling

1 Introduction

Targeting of metabolic pathways has emerged as an alternative approach with potential for drug discovery in order to find novel therapeutic strategies against a diversity of human diseases such as parasitic ones (Mukherjee et al. 2016; Müller and Hemphill 2016; De Rycker et al. 2018; Raj et al. 2020; Tyagi et al. 2019), immune disorders (Castegna et al. 2020), and cancer (Galluzzi et al. 2013; Moreno-Sánchez et al. 2010, 2014; Kaambre et al. 2013). Identification of suitable metabolic targets among the multitude of enzymes and transporters that constitute the cell metabolic networks, although a seemingly simple task, is a very challenging endeavor.

In general, the most divergent enzymes of intermediary metabolism between normal *versus* parasitic or pathological cells have been proposed as drug targets. For instance, pathway enzymes that are only present in parasites or that significantly diverge from those in the human cells. Specific examples are some enzymes of glycolysis (Müller et al. 2012; Saavedra et al. 2019a; Michels et al. 2021), the antioxidant defense in trypanosomatids (Saavedra et al. 2019b; Talevi et al. 2019), or nucleotide synthesis (Valente et al. 2019). For human diseases with the highest death tolls such as cancer, identification of drug targets becomes even more difficult due to the similarity of the enzymes in the normal *versus* pathological cells. Thus, differences in enzyme expression levels, isoform expression, and pathway regulation have been used to propose novel metabolic candidates for drug targeting (Marín-Hernández et al. 2009, 2014; Sukjoi et al. 2021).

Experimental strategies involving gene knockout or knockdown and more recently gene editing through CRISPR-CAS and RNAi libraries have been used to identify drug targets (reviewed in (De Rycker et al. 2018; Wyatt et al. 2011; Soares Medeiros et al. 2017; Kurata et al. 2018; Schuster et al. 2019)). In general, these strategies drastically lower the level of a protein, leading to the conclusion that the intervened gene is essential for cell survival or for a specific physiological function and therefore the gene product becomes validated as a suitable drug target. However, such large decreases in the content and activity of the targeted protein can be far from being achieved by pharmacological methods.

For these reasons, it seems necessary to include additional criteria to narrow the drug target options and identify the most convenient targets before embarking on a process of drug design or to screen a library of compounds. It is in this context that Metabolic Control Analysis (MCA), a biochemical framework to analyze the control of metabolism quantitatively, can be a valuable tool to refine drug target selection with the aim to improve existing therapeutic strategies.

2 Basic Principles of Metabolic Control Analysis

A criterion for drug target selection in the intermediary metabolism networks can be based on the capacity that the target (enzyme or transporter) has to influence the flux of the metabolic pathway of interest, and the MCA fundamentals (Saavedra et al. 2019b; Fell 1997; Nelson and Cox 2017; Moreno-Sánchez et al. 2008a) provide a quantitative way to assess it. MCA was developed independently by two groups, Kacser and Burns in Edinburgh, Scotland (Kacser and Burns 1973), and Heinrich and Rapoport in Berlin, Germany (Heinrich and Rapoport 1974). MCA is a theoretical framework that considers metabolic processes as a continuous flow in the steady state of matter and energy, in which each component of the metabolic pathway exerts some *control* over the pathway flux (Fell 1997; Nelson and Cox 2017).

Here it is necessary to distinguish the concepts of *control* and *regulation*, frequently found in biochemistry textbooks and the scientific literature as interchangeable, but that in MCA, similar to in engineering, have very different meanings. ***Regulation*** refers to the molecular mechanisms by which a cell maintains its homeostasis, (i.e., to maintain a physiological variable at constant level), to be able to change to a new metabolic status when it is exposed to a different environmental condition. Within these mechanisms are those of protein covalent modification, compartmentalization, protein synthesis and degradation and allosterism that can modify the activity of a preexisting enzyme; or transcriptional regulation to up or down regulate the expression of specific enzymes (Newsholme and Start 1973).

In contrast, ***Control*** is the capacity of an enzyme, transporter, or physiological process to affect the pathway flux under a defined metabolic steady state (Saavedra et al. 2019b; Fell 1997; Nelson and Cox 2017; Moreno-Sánchez et al. 2008a). To illustrate this, ATP phosphofructokinase (PFK-1) and pyruvate kinase (PyK) are highly *regulated* glycolytic enzymes able to drastically change the metabolic fluxes in different metabolic states to avoid futile cycles, e.g., during feed and starvation, which concur with changes of high and low blood glucose concentrations and thus with glycolysis and gluconeogenesis activation, respectively. Once a new stable metabolic steady state is reached, these enzymes do not necessarily exert significant *control* on the pathway flux any longer. As discussed below, glucose transport has significant control on stable steady state glycolysis fluxes of several cell models, whereas PFK-1 and PyK have low or negligible control (reviewed in (Saavedra et al. 2019a; Moreno-Sánchez et al. 2008a).

MCA is based on the following basic considerations:

1. The metabolic pathway must be at a stable steady state or quasi steady state. In order to achieve that condition, the system must be open and the initial substrates and final products have to be kept at constant levels.
2. At steady state, the rate (v) of each pathway component (enzyme, transporter) in a linear metabolic pathway is the same, and it is similar to the rate or production of the pathway's end product. Usually, the rate of the pathway enzymes under metabolic steady state is lower than their maximal activities ($Vmax$); hence, commonly there is a generalized excess of enzyme activity capacity in comparison to the pathway flux values. In a non-linear pathway, the individual step rates are not equal, particularly at and after the branching points, but they are constant.
3. Proper stoichiometric relationships of the pathway's substrates and end products have to be considered for the analysis of flux control distribution. This is, for example, relevant in branched pathways where unequal input and output of metabolic intermediates may occur.
4. Parameters are quantities that can be changed independently, and they typically remain constant during the evolution of the system toward its steady state, for example, kinetic constants, enzyme activity ($Vmax$).
5. Variables are quantities determined by the system, and they are time-dependent before reaching a given steady state. Variables are, for example, metabolic flux, metabolite concentration, enzyme or transport rates (v), or rates of any other physiological process (pathway branches).

To determine the degree of influence that a perturbation in a parameter (e.g., enzyme activity) has over a pathway variable at steady-state (e.g., pathway flux), the MCA theory relies on three types of coefficients (control, elasticity, and response) and two theorems that relate them (summation and connectivity). These constitute the fundamentals of MCA.

Control Coefficients

The control coefficient quantifies the impact that a small perturbation in the activity of an enzyme has on the pathway steady-state response, either on its flux (flux control coefficient) or on the concentration of a pathway intermediate metabolite (concentration control coefficient).

The flux control coefficient quantifies how much the pathway flux changes when the activity of an enzyme is changed; this coefficient is mathematically represented as:

$$C^{J}_{a_i} = \frac{\delta J}{\delta a_i}\left(\frac{a_{i0}}{J_0}\right) \tag{1}$$

where J is the pathway flux at steady state and a_i is the activity a of enzyme i. By multiplying by the scalar factor a_{io}/J_o, which corresponds to the ratio of the enzyme activity and pathway flux in the unperturbed (control) state, the $C^J{}_{ai}$ becomes dimensionless and independent of the units used.

On the other hand, the concentration control coefficient quantifies how much the concentration of a pathway intermediate metabolite changes when the activity of an enzyme is changed, and this coefficient is mathematically defined by:

$$C^{X}_{a_i} = \frac{\delta X}{\delta a_i}\left(\frac{a_{i0}}{X_0}\right) \tag{2}$$

where X and X_o are any pathway intermediate metabolite in the perturbed and unperturbed states, respectively.

From here, the chapter will focus on the experimental assessment and analysis of the flux control coefficients because it concerns drug target validation. For further information on concentration control coefficients and their theorems, which are relevant for biotechnological purposes, the reader is referred to the book by Fell (1997).

Summation Theorem

Kacser and Burns (1979) demonstrated that, regardless of the complexity of the kinetic mechanisms governing each enzymatic (and transport) component from a metabolic pathway, the sum of their control coefficients equals to unity. This relationship is named as the summation theorem and is expressed in Eq. (3) for pathway flux:

$$\sum_{i=1}^{n} C^{J}_{a_i} = 1 \tag{3}$$

This summation theorem reveals some important properties of $C^J{}_{ai}$ listed below:

- If one enzyme activity is changed by a substantial amount, its $C^J{}_{ai}$ is changed to a new value and the $C^J{}_{ai}$ of all other enzymes must have adjusted, so that the new sum, again, equals 1.
- Some of the individual $C^J{}_{ai}$ will be "negligibly small," however the sum of these "small" $C^J{}_{ai}$ may not be negligible.
- The larger the number of enzymes included in the pathway, the smaller the average $C^J{}_{ai}$ for all enzymes.

- Typical values of C^J_{ai} are between 0 and 1 ($0 < C^J_{ai} < 1$). However, the C^J_{ai} are negative for enzymes that branch off from the main pathway and withdraw metabolites, whereas the C^J_{ai} will be positive for branches that supply intermediates to the main pathway.

Flux Control Coefficient Estimation

In theory, C^J_{ai} can be determined by measuring the infinitesimal changes of the flux at steady state induced by an infinitesimal change in the activity of an enzyme. Thus, the C^J_{ai} can be estimated from the slope of the tangent line to the activity of the enzyme in the unperturbed steady state of interest (the basal value of activity or 100% activity). By analyzing the enzyme behavior shown in Fig. 1, it is clear that inhibition of the activity of an enzyme with a $C^J_{ai} = 0.5$ (panel A) induces larger decreases in flux than inhibition of another one with $C^J_{ai} = 0.1$ (panel B). Furthermore, 50% inhibition of the activity of the enzyme in panel A will lead almost to total control of the pathway flux (C^J_{ai} from 0.5 to 0.7), whereas the same inhibition of the enzyme in panel B will cause it to gain control, but still not meaningful (C^J_{ai} from 0.1 to 0.2). Hence, it becomes evident that the best therapeutic targets are enzymes or transporters that have a high C^J_{ai} (>0.5) under physiological conditions.

Figure 1 also shows that at enzyme activity inhibitions > 90% (orange lines) there is no difference on the pathway flux decrease among high and low flux

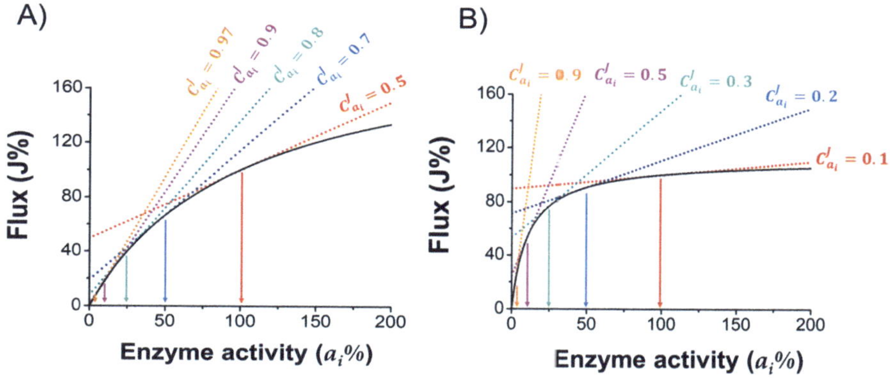

Fig. 1 Direct experimental determination of the flux control coefficient (C^J_{ai}). (**a**) The plot represents an enzyme that has a $C^J_{ai} = 0.5$ at physiological conditions (100% of enzyme activity and 100% of flux). As the activity of the enzyme is slightly inhibited, it gains significant control on the pathway flux. (**b**) The plot represents an enzyme with a $C^J_{ai} = 0.1$ at physiological conditions. The same inhibition percentages as used in panel A were used for this model and the respective C^J_{ai} is presented for each condition. For this low controlling enzyme to gain a control of 0.5 over the pathway flux, its inhibition must be greater than 90% (pink dot line). The plot was obtained by simulations using Eq. (9) considering $n = 1$, $Ji = 0$ and a $C^J_{ai} = 0.5$ or 0.1, and the software Microcal Origin v. 8.0. The straight dotted lines are the derivatives at the point indicated by the arrows

controlling enzymes, both reaching similar C^J_{ai} of 0.9. It should be realized that it is in that range of decreased expression and activity that drug targets are validated as essential by genetic methods.

In practice, experimental approaches using genetic engineering commonly induce large changes in both enzyme activity and pathway flux (Fell 1997) to visualize "phenotype changes." To circumvent this problem, a theoretical framework was created that allows accurate estimation of the control coefficients from the experimental data (Small and Kacser 1993a; Small and Kacser 1993b). In accordance with this theoretical framework, the C^J_{ai} in a system with large changes of enzyme activity in a linear metabolic pathway is defined as follows:

$$C^{J_0}_{a_{i0}} = \frac{\frac{\Delta J}{J_r}}{\frac{\Delta a_i}{a_{ir}}} \tag{4}$$

where ΔJ is the change of the flux from the original steady state (J_0) to the resulting one after the enzyme activity has changed by a factor r (J_r). Therefore, $\Delta J = J_r - J_0$, and Δa_i is the change of the enzyme activity from the original state (a_{i0}) to the new enzyme activity ($a_{ir} = r{*}a_{i0}$), hence ($\Delta a_i = a_{ir} - a_{i0}$) (Fell 1997; Small and Kacser 1993a). Rearranging Eq. (4), we have the following:

$$C^{J_0}_{a_{i0}} = \frac{\frac{(J_r - J_0)}{J_r}}{\frac{(ra_{i0} - a_{i0})}{ra_{i0}}} = \frac{1 - \left(\frac{J_0}{J_r}\right)}{\frac{(r-1)}{r}} \tag{5}$$

defining the amplification factor (f) as the ratio of the new flux (J_r) to the original one (J_0) ($f = J_r/J_0$) and rearranging Eq. (5), we obtain:

$$f = \frac{1}{1 - \frac{r-1}{r} C^J_{a_i}} \tag{6}$$

This last function shows that by increasing the activity of a single enzyme, the effect on the pathway flux will be low when its C^J_{ai} is below 0.5 (Fig. 2) (Fell 1997). For example, if a 10-times increase ($r = 10$) of enzyme activity is attained, and the enzyme has a $C^J_{ai} = 0.2$, then the flux only increases 25%. On the other hand, a decrease of enzyme activity (e.g., by a drug) always leads to a drop in the pathway flux which depends on the enzyme's C^J_{ai}. For example, a 10-times inhibition ($r = 0.1$) of an enzyme with $C^J_{ai} = 0.2$, induces a 65% decrease of the pathway flux, but lower inhibition levels do not relevantly affect the flux. The most interesting drug targets are enzymes with C^J_{ai} higher than 0.5, that require less inhibition degrees ($r = 0.5$) and lower drug doses to observe flux inhibition.

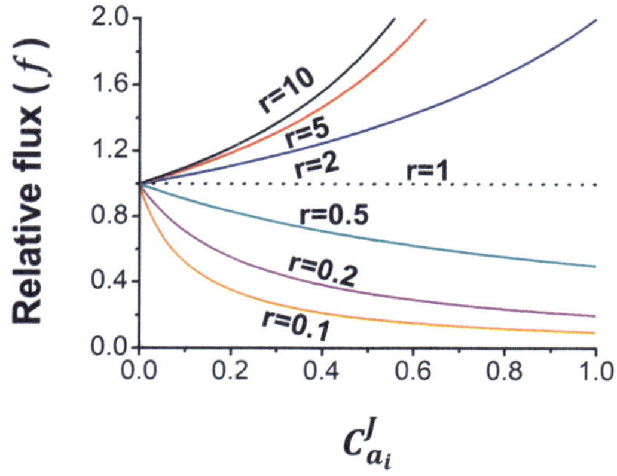

Fig. 2 Relative change of flux (f) for large changes in enzyme activity in a linear pathway
The plot was obtained by simulations with Eq. (6) using the software Microcal Origin v. 8.0. The degree of amplification, r, meaning overexpression or genetic down-regulation is shown at each curve

Relationship Between Flux and Enzyme Activity

There is no way to predict the behavior of the flux of a metabolic pathway as a function of a unique enzyme activity and therefore it is not possible to propose a general equation that describes this behavior. However, it has been empirically observed that the pathway flux shows a hyperbolic behavior as a function of the activity of the pathway enzymes, especially when the pathway is in vitro reconstituted (Fell 1997; Gellerich et al. 1990; González-Chávez et al. 2015, 2019). Therefore, the following equation has been proposed (Gellerich et al. 1990):

$$J = \frac{\propto a_i}{\beta + a_i} \tag{7}$$

where α and β are empiric constants. In a more general way, Eq. (7) can be rewritten as follows:

$$J = \frac{\propto a_i^n}{\beta + a_i^n} + J_i \tag{8}$$

where J_i represents an initial flux that persists even when the activity of the enzyme (a_i) is zero and "n" is an empirical constant that gives a sigmoidal or hyperbolic behavior (Gellerich et al. 1990). A more useful form of Eq. (8) is obtained when the constants α and β are expressed in terms of the control coefficient at the basal condition of 100% of flux and 100% of enzyme activity (Eq. 9) (Gellerich et al. 1990; Rodríguez-Enríquez et al. 2000):

$$J = \frac{n\left(100 - J_i\right)^2 a_i^n}{100^{n+1}\, C^{J_0}_{a_{i0}} + \left[n\left(100 - J_i\right) - 100\, C^{J_0}_{a_{i0}}\right] a_i^n} \tag{9}$$

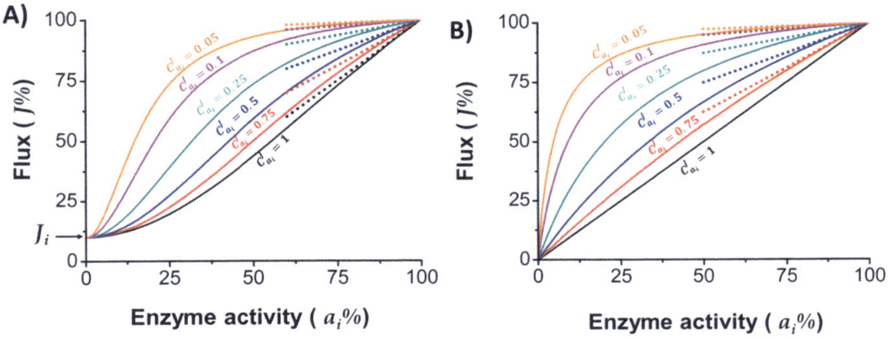

Fig. 3 Theoretical relationship between enzyme activity and flux. (**a**) The plot represents a case when the relationship between enzyme activity and flux is sigmoidal ($n > 1$ in Eq. 9) and a residual flux is obtained despite a null activity (Ji). For illustrative purpose, we use a model with $n = 2$ and $Ji = 10$. (**b**) The plot represents the most used fitting model where the relationship between activity and flux is hyperbolic ($n = 1$, $Ji = 0$). Both plots were obtained from simulations using Eq. (9) and the software Microcal Origin v. 8.0. The values of control coefficients (C^J_{ai}) are those of the slope of the tangent line at the basal condition of 100% of pathway flux and 100% of enzyme activity (marked with the dotted line)

The relationship between the flux and the enzyme activity expressed in Eq. (9) is graphically represented in Fig. 3. Again, almost linear decreases in flux are attained by inhibition of enzymes with C^J_{ai} greater than 0.5.

Experimental Determination of the Control Coefficient

In order to experimentally determine the C^J_{ai} it is necessary to modify the activity of a single enzyme at a time (without affecting the rest of the pathway enzymes), while determining the metabolic pathway flux. To achieve such condition, several experimental strategies have been successfully carried out: enzyme titrations with specific inhibitors, in vitro reconstitution of the pathway or pathway segments and genetic manipulation of enzyme expression in cells.

Enzyme Titration with Inhibitors

Titration of enzyme activity by using inhibitors to estimate control coefficients is restricted by the degree of specificity that inhibitors have. The availability of specific, potent, and permeable mitochondrial inhibitors has made it possible to use this approach to determine the C^J_{ai} of several complexes and enzymes of oxidative phosphorylation (OxPhos) in diverse organs. When the flux of ATP synthesis is titrated by adding increasing concentrations of each specific inhibitor, plots are generated in which the enzyme activity is progressively diminished at increasing inhibitor concentration (Moreno-Sánchez et al. 2008a) and the C^J_{ai} can be estimated from the initial slope (graphical determination).

However, the titration curves are in general sigmoidal, complicating the calculus of the initial slope by graphical determination, which can lead to overestimation of C^J_{ai} values (Gellerich et al. 1990). For this reason, equations for non-linear curve fitting have been developed depending on the type of inhibitor used (Moreno-Sánchez et al. 2008a). The values of C^J_{ai} estimated for OxPhos depend on experimental conditions (e.g., whole cells, isolated mitochondria), the tissue, and the equation used for each inhibitor. These considerations have been appropriately reviewed previously (Moreno-Sánchez et al. 2008a).

Two examples that use a variation of Eq. (9) are discussed below.

Antimycin A (inhibitor of cytochrome bc1 in complex III) and carboxyatractyloside (inhibitor of the adenine nucleotide translocator) are tightly-bound inhibitors that were used to estimate C^J_{ai} over total mitochondrial respiration in isolated rat liver mitochondria (Gellerich et al. 1990). The non-linear curve fitting of titrations with these inhibitors allowed determining the inhibitor affinity constants, $Kd = 7.8$ pM and $Kd = 0.86$ nM for antimycin A and carboxyatractyloside, respectively. The estimated C^J_{ai} was 0.2 for complex III and 0.34 for the adenine nucleotide translocator.

The same experimental approach and equation were used to determine the C^J_{ai} of mitochondrial components in OxPhos of hepatoma cells (Rodríguez-Enríquez et al. 2000). Specific mitochondrial inhibitors were used to estimate the C^J_{ai} of NADH dehydrogenase inhibited by rotenone ($C^J_{ai} = 0.3$), bc1 cytochrome complex inhibited by antimycin A ($C^J_{ai} = 0.23$), cytochrome c oxidase inhibited by cyanide ($C^J_{ai} = 0.04$), ATP synthase inhibited by oligomycin ($C^J_{ai} = 0.02$), ATP/ADP translocase inhibited by carboxyatractyloside ($C^J_{ai} = 0.3$), and pyruvate carrier inhibited by 3-hydroxycinnamate ($C^J_{ai} = 0.026$). Thus, control of OxPhos relied mainly on three reactions. Applying the summation theorem (Eq. 3) to this example, the processes studied control approximately 0.92 of OxPhos; therefore, all other reactions involved in OxPhos are responsible for the remaining 0.08; hence, the transport of substrates (pyruvate or phosphate carriers), dehydrogenases (succinate dehydrogenase), and ATP synthase exert negligible control over the OxPhos flux.

In Vitro Pathway Reconstitution

This approach is based on the in vitro reconstitution of segments of the metabolic pathway using purified enzymes. The in vitro pathway reconstitution must simulate as much and close as possible the in vivo conditions of the pathway. This should be carried out regarding (1) the ratio of enzyme activities present in the model under study, for which their actual *Vmax* values within the cell have to be determined at the physiological intracellular high K^+ medium, pH, and growth temperature of the biological model used; and (2) the physiological concentrations of substrates and coenzymes.

The titration curve of enzyme activity *versus* pathway flux (Fig. 1) has to be constructed by varying the activity of one enzyme at a time while keeping the rest of the system (partner enzymes, substrates, and cofactors) unaltered and in parallel determining changes in the pathway flux. A limitation of these ex vivo experiments is that only a quasi-steady state is attained because there is net substrate consumption and product accumulation. However, by using short times of the reaction system and adding saturating substrate concentrations, these problems can be circumvented.

This strategy was first applied for segments of glycolysis using mammalian enzymes due to the availability of commercial proteins or methods for their purification. Torres et al. (Torres et al. 1986, 1989) used pure enzymes added to cell extracts or in a cell free system, respectively, to determine the $C^J{}_{ai}$ for the first glycolytic segment, finding that hexokinase (HK) and PFK-1 have the highest $C^J{}_{ai}$. In these studies the summation theorem was demonstrated. On the other hand, Giersch (Giersch 1995) reconstituted the final pathway segment of glycolysis with commercial pure enzymes, where PyK was the controlling enzyme.

More recently, pathway reconstitution was performed for the pyrophosphate-dependent glycolysis of the intestinal parasitic protist *Entamoeba histolytica,* which does not have the typical ATP dependent PFK-1 but an inorganic pyrophosphate-dependent enzyme (PPi-PFK) and, instead of PyK, a PPi-dependent pyruvate phosphate dikinase (PPDK). Both PPi-dependent enzymes catalyze reversible reactions which lead to a different type of metabolic regulation of this pathway (Saavedra et al. 2019a; Pineda et al. 2015a). Pathway reconstitution was carried out using recombinant purified *E. histolytica* enzymes of the segments of HK to triosephosphate isomerase and from phosphoglycerate mutase (PGAM) to PPDK. It was found that HK and PGAM have the highest control of flux, with low control exerted by the PPi-dependent enzymes, indicating that despite their large structural divergence to the human enzymes, they are not suitable drug targets from the metabolic point of view (Moreno-Sánchez et al. 2008b). In addition, the dataset of the amoebal pathway reconstitutions were also used for kinetic pathway modeling, which helped to unveil previously unknown inhibition of pathway intermediates on the glycolytic enzymes (Moreno-Sánchez et al. 2008b) and also to improve other forms of pathway modeling by artificial neural networks (Lo-Thong et al. 2020).

Pathway reconstitution was also applied to the trypanothione-dependent peroxide detoxification system of *Trypanosoma cruzi,* the parasitic protist that is responsible for Chagas disease in Latin America and for which no adequate drugs are yet available. Trypanothione replaces glutathione as the main antioxidant system in the parasite and its entire peroxide detoxification machinery uses this metabolite as reductant (Fig. 4a). For several decades, this pathway has attracted attention for therapeutic intervention against the parasitic disease (Saavedra et al. 2019a; Talevi et al. 2019; Olin-Sandoval et al. 2010; Leroux and Krauth-Siegel 2016; Manta et al. 2018; Piñeyro et al. 2021). For pathway reconstitution, the recombinant enzymes trypanothione reductase (TryR), tryparedoxin (TXN), and tryparedoxin peroxidase (TXNPx) were used (González-Chávez et al. 2015, 2020) (Fig. 4a). In these studies, a low control by TryR (a popular target for therapeutic intervention against trypanosomatids) was determined, while TXN and TXNPx both have a $C^J{}_{ai}$ of 1; therefore,

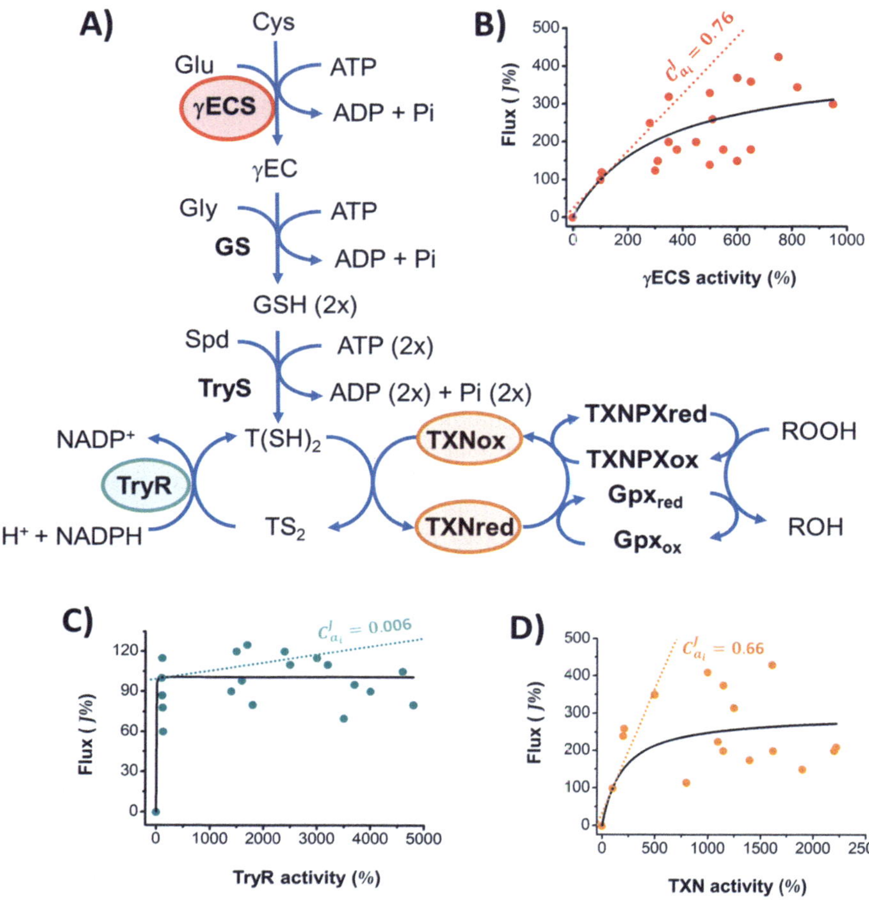

Fig. 4 Estimation of the control coefficient by enzyme overexpression. (**a**) *Trypanosoma cruzi* antioxidant pathway, the overexpressed proteins were γ-glutamylcysteine synthetase (γECS), trypanothione reductase (TryR), and tryparedoxin (TXN). For experimental determination of $C^J{}_{ai}$, overexpressing clones with different levels of activity (low, medium, and high) were used and their effect on pathway fluxes was evaluated for (**b**) γECS activity and its effect on $T(SH)_2$ synthesis; (**c**) TryR and (**d**) TXN activities and their effect on the hydroperoxide reduction flux. Dotted lines are the tangent to the 100% values (WT condition). Data from panels **b**, **c**, and **d** were taken from (González-Chávez et al. 2019) and fitted to Eq. (9) ($n = 1$). The estimated $C^J{}_{ai}$ with the non-linear curve fit (using Microcal Origin v. 8.0) were consistent with values previously reported (González-Chávez et al. 2019; Olin-Sandoval et al. 2012)

together producing a value of 2, in apparent conflict with the summation theorem. However, this difference is caused by the involvement of TXN and TXNPx in two processes, reduction and oxidation (González-Chávez et al. 2015), which contrasts with the situation for glycolytic enzymes in which each enzyme is involved in only one C-metabolite transformation.

Manipulation of Enzyme Expression in Cells

Manipulation of enzyme expression in cells by genetic engineering is a very useful tool for direct estimation of control coefficients (González-Chávez et al. 2020) because, in principle, just one enzyme is modified at the time. However, it is important to determine the actual level of enzyme activity (not just protein level resulting from altered gene expression). Furthermore, it is mandatory to determine the activity of the other enzymes that participate in the pathway to ascertain that they are not changed by undesirable or pleiotropic or regulation effects by the genetic modification of the single, targeted enzyme. Pleiotropic effects are the main limitations of the strategy because cells are open systems and can have responses to the genetic modification beyond the control of the experimenter; thus, a detailed biochemical characterization of each clone of cells expressing different levels of each pathway enzyme has to be performed for proper MCA application.

One of the first studies where this strategy was applied using traditional genetics was done for the arginine pathway in the fungus *Neurospora crassa* (Flint et al. 1981), where argininosuccinate synthetase was found to be the main controlling enzyme. For additional examples about other metabolic routes in different cell systems refer to Moreno-Sánchez et al. (2008a). Downregulation by RNAi of the activities of HK, PFK, PyK, PGAM, and enolase was performed in the glycolytic pathway of *Trypanosoma brucei*, the parasite responsible for sleeping sickness in sub-Saharan Africa. The study revealed that the first three enzymes have low flux control due to their overcapacity, as was also predicted by kinetic modeling (Albert et al. 2005). In fact, the glucose transporter accounted for 40% of the flux control.

Modulation of protein expression was recently applied in the trypanothione synthesis and trypanothione-dependent antioxidant system of *T. cruzi* (González-Chávez et al. 2019), where the expression of TryR and TXN of the peroxide detoxification system was modulated as well as that of two enzymes of the trypanothione synthesis pathway: γ-glutamylcysteine synthetase (γECS) and trypanothione synthetase (TryS) (Fig. 4a). It was demonstrated that γECS has a C^J_{ai} of 0.69 in the trypanothione synthesis flux (Fig. 4b), while TryS has at most a C^J_{ai} of 0.3. Furthermore, for the peroxide detoxification flux, TryR again showed a low C^J_{ai} of 0.15 (Fig. 4c), whereas TXN showed the highest C^J_{ai} of 0.73 (Fig. 4d). The latter two C^J_{ai} values were in agreement to those found by pathway reconstitution (see preceding section) and by kinetic modeling as described later (Sect. 4).

3 Elasticity Coefficients

The flux in a metabolic pathway is the result of the concerted reaction rates of all components that constitute the metabolic pathway. Since the partial rate (v) of each individual pathway component is the same in a linear metabolic pathway under steady-state conditions, it can be assumed that the rates of all the individual components are similar to the rate of production of the pathway's end product (pathway flux).

On the other hand, every enzyme within the pathway is connected to its upstream and downstream "neighbor" enzymes through its substrates and products. Furthermore, it can be connected with "distant" enzymes or cellular processes through metabolic effectors such as inhibitors and activators produced by enzymes within the same pathway or from different pathways.

Theoretically, a perturbation in the activity of any given enzyme in the pathway could affect the rate of reaction of the other pathway enzymes. The transmission and amplification of this change will depend on the intrinsic properties of each enzyme (e.g., kinetic parameters) and on the concentration of the molecules participating in the enzyme-catalyzed reaction. The elasticity coefficient quantifies this transmission response (Kacser 1983).

The elasticity coefficients are properties of the individual enzymes or transporters and quantify how much their rate is modulated by a variable (e.g., substrates, products, internal inhibitors/activators, pH, temperature) (Saavedra et al. 2019a; Fell 1997; Moreno-Sánchez et al. 2008a; Brand 1998). The mathematical expression of the elasticity coefficient is:

$$\varepsilon_{X_{i0}}^{v_{i0}} = \frac{\partial v_i}{\partial X_i}\left(\frac{X_{i0}}{v_{i0}}\right) \quad (10)$$

where v is the rate of a reaction (or transport) and X a variable that modifies the rate. Multiplication by X_{i0}/v_{i0}, which is the ligand concentration and rate in the unperturbed state, makes the coefficient dimensionless and thus independent of the units used.

The difference of the plots of enzyme rate *versus* substrate (or other ligand) concentration, typical of enzyme kinetic analyses using isolated enzymes, is that in the elasticity analysis, the concentrations of ligands (pathway intermediates) are determined by the functioning of the entire metabolic pathway. Moreover, in the in vitro kinetic analysis of enzymes usually (1) only one ligand at a time is varied whereas the other co-substrates are kept constant; and (2) experiments are performed in the absence of products or inhibitors ("initial rate conditions").

There are as many elasticity coefficients for each enzyme as there are substrates, products, and effectors interacting with it. For example, the rate of enzyme 1 (v_1) in Scheme 1, depends on three variables, the concentration of S_1, S_2, and S_3.

Scheme 1 Example of elasticity analysis for enzyme 1 in a short linear metabolic pathway

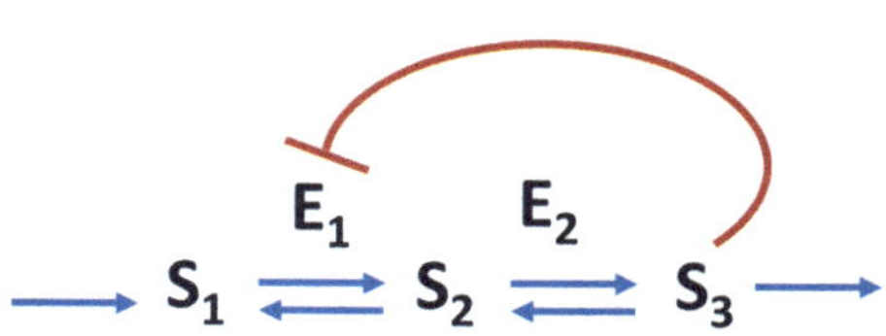

Hence, the elasticity coefficients of Scheme 1 for E_1 are:

$$\varepsilon_{S_1}^{vE_1} = \frac{\partial v_{E1}}{\partial S_1}\left(\frac{S_1}{v_{E1}}\right); \varepsilon_{S_2}^{vE_1} = \frac{\partial v_{E1}}{\partial S_2}\left(\frac{S_2}{v_{E1}}\right); \varepsilon_{S_3}^{vE_1} = \frac{\partial v_{E1}}{\partial S_3}\left(\frac{S_3}{v_{E1}}\right)$$

There are two ways to estimate the elasticity coefficient of a component from a metabolic pathway. (1) If the rate equation that describes the enzyme or transporter is known and it contains the variable to which the elasticity has to be calculated, the theoretical elasticity can be calculated from such equation; and (2) If the rate equation is not known, then the elasticity has to be experimentally estimated. These methods are described in the next section.

Estimation of the Elasticity Coefficient from the Rate Equation

The elasticities are intrinsically linked to the enzyme or transporter kinetics and can be calculated by solving the differential equation of the kinetic rate expression with respect to the variable or parameter of interest and subsequently interpolating the value of the steady-state concentration of the variable of interest (Fell 1997; Moreno-Sánchez et al. 2008a; Groen et al. 1986). For example, if the enzyme of interest follows reversible Michaelis–Menten kinetics, the expression of elasticity is:

$$\varepsilon_S^{v_{i0}} = \frac{\partial v_i}{\partial S}\left(\frac{S_0}{v_{i0}}\right) = \frac{1}{1-\frac{\Gamma}{Keq}} - \frac{\frac{[S]}{Ks}}{1+\frac{[S]}{Ks}+\frac{[P]}{Kp}} \quad (11)$$

where the term Γ (gamma uppercase) is the mass-action ratio, which is defined as the ratio between the concentration of products and the concentration of substrates at steady state ($\Gamma = [P]_{ss}/[S]_{ss}$). Keq is the equilibrium constant of the reaction (Keq = $[P]_{eq}/[S]_{eq}$). When analyzing Eq. (11), it is evident that as the value of Γ gets closer to the equilibrium conditions (Keq), then the elasticity value increases, and as the enzyme becomes more saturated with substrate ($[S]/Ks$ increases), the value of elasticity decreases. This behavior is illustrated in Fig. 5.

Estimation of the Elasticity Coefficient from the Substrate Saturation Curves

Direct in vitro measurement of elasticities is in theory difficult to accomplish because only one modulator has to be varied and the other molecules that affect the enzyme activity have to be kept constant and under steady-state conditions. If these conditions are met, then the elasticity coefficient can be estimated from the slope of the tangent line to the concentration of the variable at the steady state of interest, which is at the steady-state concentration of ligands in the cell (Fig. 6). The elasticity coefficients are positive ($\varepsilon^{v}{}_{X} > 0$) for those variables that increase the enzyme or

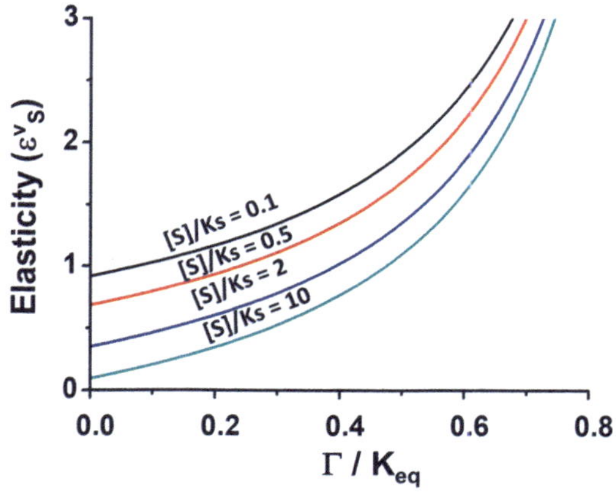

Fig. 5 Variation of elasticity with respect to the disequilibrium ratio (Γ/K_{eq})
The plot shows that the elasticity coefficient (ε^v_s) of an enzyme increases as the value of the mass-action ratio (Γ) approximates to the value of the equilibrium constant (K_{eq}). In this plot, it is also observed that when the saturation of the enzyme ($[S]/Ks$) increases, the elasticity value decreases. The graph was obtained by simulations using Eq. (11) and the software Microcal Origin v. 8.0 ($[P]/K_p = 0.1$ for all curves)

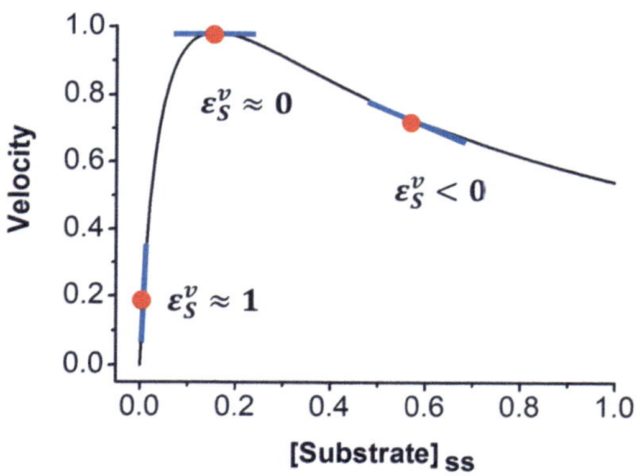

Fig. 6 Estimation of elasticity coefficients from the saturation curve
The plot represents an enzyme with substrate inhibition ($v = (Vmax\ [S])/(Ks(1 + [S]/Kis) + [S](1 + [S]/\alpha Kis))$). The elasticity coefficient of an enzyme toward a substrate (or any other ligand) is determined from the slope of the tangent line (or derivative) at the concentration of substrate at the control steady state. When the steady-state substrate concentration ($[S]_{ss}$) is below the Ks (*or* Km) value, the $\varepsilon^v_s \approx 1$, which means that the enzyme is highly responsive to substrate variation. In contrast, if $[S]_{ss}$ is near saturation, $\varepsilon^v_s \approx 0$, which means that the enzyme is saturated and cannot vary its rate. Finally, if [S]ss is in the zone of inhibition ($[S] \geq \alpha Kis$), the elasticity will have negative values ($\varepsilon^v_s < 0$).

transporter rate (substrate or activator), and they are negative ($\varepsilon^v_X < 0$) for the variables or parameters that decrease the enzyme or transporter rates (product or inhibitor).

Experimental Determination of the Elasticity Coefficient Within Cells

Elasticity analysis can be experimentally determined using live intact cells (Fig. 7). Elasticity coefficients of an enzyme or groups of enzymes around a metabolite are determined by monitoring the changes in pathway flux in response to changes in the concentration of an intermediate metabolite. The changes in the metabolite concentration are performed by manipulating the rates of the group of enzymes that supply it by feeding the pathway with the pathway's initial substrate (Fig. 7A). For

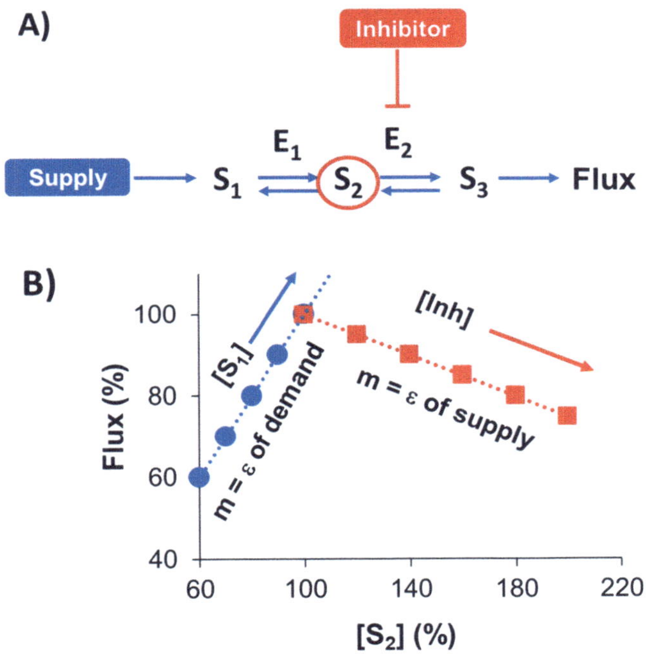

Fig. 7 Determination of the elasticity coefficient of one enzyme or a group of enzymes of a metabolic pathway. (**A**) Scheme that represents manipulation of the pathway intermediate S_2 to evaluate the elasticities of E1 (supply reactions) and E2 (demand reactions). Let us recall that, at steady state, the rate of reactions by these enzymes will be equal to the final flux in a linear metabolic pathway. In this example, S2 is a substrate of E2; thus, increases in S2 concentration attained by supplying E1 will increase the rate of E2. On the other side, S2 is a product of E1, thus increases in S2 concentration by E2 inhibition will promote rate inhibition of E1. (**B**) Plot of a theoretical behavior of the manipulation of a pathway intermediate S_2 by increasing the concentration of a supply metabolite (S1) or by the inhibition of an enzyme in the pathway after S_2. The elasticity coefficient of E2 (or S2 demand) is calculated from the slope of the tangent line closest to 100% resulting from the increase in $[S_1]$ concentration (blue dots). The elasticity coefficient of E1 (or S2 supply) is estimated from the slope of the tangent line to the curve closest to 100% resulting from the increase in [Inh] concentration (red dots)

example, this can be done by providing increasing concentrations of glucose for glycolysis to the cells. In another set of experiments, the intermediary metabolite concentration is manipulated by inhibition of a reaction downstream of the metabolite of interest, for example, inhibiting lactate dehydrogenase for glycolysis or E2 in Fig. 7A. From plots similar to those shown in Fig. 7B, the elasticity coefficients of the group of reactions that supply and demand a metabolite are determined. If similar experiments are carried out by monitoring different pathway intermediate metabolites, the elasticity coefficients of individual reactions can be determined and from these, the flux control coefficients calculated by the connectivity theorem as described next.

Relationship Between Elasticity Coefficient and Flux Control Coefficient

Intuitively, it can be deduced that the rate of an enzyme or a group of enzymes or reactions with elasticity coefficients approaching to zero, cannot be increased despite large variations in S (or P) concentration; in consequence, such enzyme(s) exert(s) a high flux control. In turn, an enzyme or a group of enzymes with a high elasticity can adjust its/their rate in response to the variation in S or P concentrations, and thus it does not constrain the metabolic flux, implying that it/they exert(s) a low flux control. The relationship between elasticity coefficient and the control of the pathway flux is related through the Connectivity theorem (Fell 1997; Moreno-Sánchez et al. 2008a; Saavedra et al. 2019a).

Connectivity Theorem

The change in a pathway flux is the result of a concerted response of each of the steps that constitute the pathway. In other words, the system response (the flux) arises from the local responses (elasticities) of the functional units (Kacser 1983). This relationship is reflected in the connectivity theorem, which relates the coefficients of elasticity and control as follows:

$$\sum_{i=1}^{n} C_{v_i}^{J} \varepsilon_X^{v_i} = 0 \tag{12}$$

In practice, two blocks of enzymes (supply and demand) connected by a common metabolite are used for the expression of the connectivity theorem. In Scheme 2, a branched pathway is presented along with the expression for its connectivity theorem with respect to metabolite S_2.

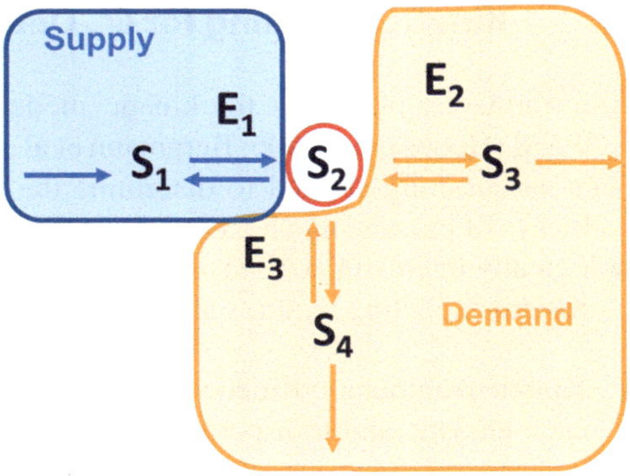

Scheme 2 Non-linear metabolic pathway to establish the connectivity theorem relationships for elasticity analysis

$$C^{J}_{vE1}\varepsilon^{v_{E1}}_{S2} + C^{J}_{vE2}\varepsilon^{vE2}_{S2} + C^{J}_{vE3}\varepsilon^{vE3}_{S2} = 0 \tag{13}$$

For unbranched pathways, the summation and connectivity theorems allow the direct calculation of the C^{J}_{ai} from the elasticities by solving a system of linear algebraic equations. For systems involving branches and cycles, additional relationships must be used to provide sufficient equations to solve the C^{J}_{ai} from the elasticity coefficients (Brand 1998; Groen et al. 1986; Westerhoff and Kell 1987).

One of the emblematic studies of elasticity analysis to determine the distribution of flux control was done for gluconeogenesis in rat liver cells where the connectivity and summation theorems were successfully used in order to propose a system of linear algebraic equations for the estimation of the control coefficients for the enzymes of this process (Groen et al. 1986). In that study, it was demonstrated that pyruvate carboxylase positively and PyK negatively controlled the gluconeogenesis flux in the presence and absence of glucagon, respectively.

Elasticity analysis of the mitochondrial OxPhos indicated that the control is shared between the proton leak and the respiratory chain, being higher for the former (Wanders et al. 1984; Hafner et al. 1990; Brown et al. 1990; Moreno-Sánchez et al. 1999). Also, experimental estimation of elasticity coefficients has been done for glycolysis in hepatoma cells where it was determined that the glucose transporter and HK are the most controlling steps (Marín-Hernández et al. 2006). In parasites, elasticity coefficients were used for the determination of the distribution of control of the PPi-dependent glycolysis in *E. histolytica*, where the group of reactions of glucose transport/HK/glycogen degradation had a C^{J}_{ai} of 0.72–0.86 followed by the bifunctional aldehyde–alcohol dehydrogenase (C^{J}_{ai} 0.18) (Pineda et al. 2015b). Furthermore, it was determined that near 90% inhibition of the low controlling enzyme pyruvate:ferredoxin oxidoreductase (C^{J}_{ai} 0.13), which is absent in humans, did not significantly decrease the pathway flux. Inhibition of the activity of the steps with the highest control with 2-deoxyglucose and disulfiram decreased the ATP content and cell viability by 60% and 50%, respectively, validating them as drug targets. Again, low control was attained for PPi-PFK ($C^{J}_{ai} = 0.2$).

4 Kinetic Modeling for C^J_{ai} Determination

Bioinformatics platforms for kinetic modeling of metabolic pathways, such as COPASI (Hoops et al. 2006; Bergmann et al. 2017) and SCAMP (Sauro 1993), have been successfully applied to determine the flux control distribution of metabolic pathways of parasite and host cells, healthy and pathological cells, and for biotechnologically interesting organisms.

Kinetic modeling requires at least four sets of information (Fig. 8):

1. Kinetic parameters: Kinetic models require thorough kinetic characterization of each enzyme and transporter from the pathway of interest in order to have the

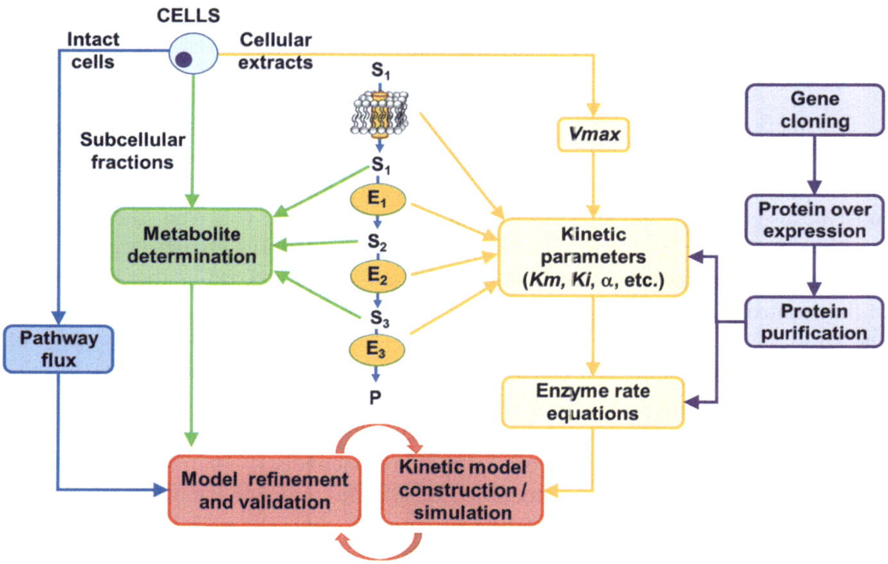

Fig. 8 Requirements for kinetic modeling

Kinetic modeling requires at least four sets of information, (1) kinetic parameters in rate equations (yellow boxes); (2) the content of active enzyme/transporter (*Vmax*); (3) concentration of pathway intermediate metabolites (green boxes); (4) pathway fluxes (blue box). The kinetic parameters and the rate equations are the main constituents for the construction of the kinetic model of the metabolic pathway using a software for metabolic modeling. The most important kinetic parameter for modeling is *Vmax*, which must be determined in cell extracts or permeabilized cells under the steady-state conditions of interest (initial concentration of substrates, temperature, pH). The other kinetic parameters such as *Km* and *Ki*, as well as the rate equations can be determined in cell extracts (yellow boxes) and/or on purified recombinant enzymes (purple boxes). Validation of the model predictions is carried out by comparison with the intracellular concentration of metabolites (green boxes) and the total flux of the pathway (blue box). It is important to note that there is constant feedback between in vitro experimentation and model simulations (circular red arrows) because inconsistencies in the model predictions could mean undescribed interactions that must be investigated to refine the kinetic model parameters (ligand affinity constants, *Vmax*) that define the rate equation where each effector (inhibitor or activator) must be included and considering the type of reaction mechanism specific for each pathway component (Segel 1975). Several ways to deal with the complexities to define or simplify the rate equations for kinetic modeling have been proposed (Tummler et al. 2014; Saa and Nielsen 2017). Many kinetic models are built after mining data of kinetic parameters reported in the literature (Hakenberg et al. 2004). Preferably, the kinetic data should all be experimentally determined in the same cell type and under experimental conditions similar to the physiological ones instead of at the optimal pH and temperature of each enzyme (Adamczyk et al. 2011; van Eunen et al. 2012). For irreversible reactions under physiological conditions, it is useful to use the Keq of the reaction in the rate equation.

2. *Vmax* in the cell: The actual activity of each enzyme (rate determined under *Vmax* conditions) in the cells or tissue is one of the most important parameters in kinetic models. A close relationship has been found between the *Vmax* or *Vmax/Km* value in the cell and the degree of control. Enzymes with low *Vmax* or *Vmax/Km* very frequently have high C^J_{ai}, although an enzyme with high activity in the cell but potent feedback inhibition may also have high control on the flux; however, this cannot be a priori inferred from pathway inspection.
3. Concentration of pathway intermediate metabolites. These variables of the system should be best determined in the same cell type and under a specific steady state.
4. Pathway fluxes. The main flux and also the fluxes of the pathway's branches should be determined under the same conditions.

Items one and two are the parameters that serve to build the models in the pathway simulator of choice, whereas items three and four are the system variables of reference to validate the accuracy of the simulations obtained by the kinetic model built. Usually, several iterative rounds of modeling and experimentation are required to obtain a kinetic model that simulates to close proximity the fluxes and pathway intermediates.

Kinetic modeling of glycolysis has been used to search for therapeutic targets in *T. brucei* (Bakker et al. 1997, 2000), *E. histolytica* (Moreno-Sánchez et al. 2008b; Lo-Thong et al. 2020; Saavedra et al. 2007), and cancer cells (Marín-Hernández et al. 2011, 2020); for the latter also involving an analysis under hypoxia conditions (Marín-Hernández et al. 2014). These studies have shown that, in general, the first part of the pathway constituted by glucose transporter, HK, hexose-6-phosphate isomerase, and glycogen metabolism exerts the greatest control (≈50%). Furthermore, inhibition of low controlling steps that produce metabolites that inhibit high controlling steps was shown to be a way to arrest glycolysis in cancer cells (Marín-Hernández et al. 2016).

The pathways involved in handling oxidative stress have also been studied using kinetic modeling in *T. cruzi* and tumor cells. In the case of *T. cruzi*, two models have been developed, one for the synthesis of trypanothione (González-Chávez et al. 2019; Olin-Sandoval et al. 2012) and one for the peroxide detoxification (González-Chávez et al. 2019). The main potential therapeutic targets found in these studies were γECS and TryS for trypanothione synthesis, and TXN1 for the peroxide detoxification. In the study by González-Chávez et al. (González-Chávez et al. 2019), the C^J_{ai} obtained by kinetic modeling and genetic manipulation of the enzyme activity (see Sect. 2) were remarkably similar, mutually validating both methodologies for C^J_{ai} determination.

In the case of tumor cells, the NADPH supply for oxidative stress handling has been modeled and revealed that peroxide detoxification was mainly controlled by glutathione peroxidase-1 and the cytosolic NADPH supply was mainly controlled by glucose-6-phosphate dehydrogenase (Moreno-Sánchez et al. 2017).

Kinetic modeling is useful not only to find potential drug targets but also to reveal some biochemical aspects that remain unknown or that are unexpected, but which emerge when the interaction among all pathway components is analyzed. For example, the intracellular steady-state concentration of reactive oxygen species (ROS) and even more the ROS concentration in different cellular compartments are very difficult to assess with the currently available techniques. However, it is possible to have a solid approximation by using kinetic models, as was illustrated in the studies of oxidative stress handling in cancer cells (Moreno-Sánchez et al. 2017, 2018). Kinetic modeling of OxPhos has also been reported, where rather similar C^J_{ai} *values* among complexes I, III, IV, and ATP synthase were found (Heiske et al. 2017).

5 Conclusion

Metabolic control analysis is a very powerful tool that provides insight into how metabolic pathways work. Understanding the control of metabolic pathways can be used for various purposes, for example, basic science, improvement of biotechnological processes, and for drug target selection. From the several examples reviewed in this chapter, one can recognize the usefulness and rationality of the approach to identify the best candidates for therapeutic intervention in the metabolism of parasitic and pathological cells. In addition, MCA provides an additional strategy in the discovery of new drugs or for repurposing existing drugs by making target selection more efficient. Thus, MCA contributes to reaching the goal of developing therapeutics against diseases afflicting human beings, accelerating the translation of basic science knowledge toward its immediate use and application in humans (translational medicine and science).

References

Adamczyk M, van Eunen K, Bakker BM, Westerhoff HV. Enzyme kinetics for systems biology when, why and how. Methods Enzymol. 2011;500:233–57.

Albert MA, Haanstra JR, Hannaert V, Van Roy J, Opperdoes FR, Bakker BM, Michels PA. Experimental and in silico analyses of glycolytic flux control in bloodstream form *Trypanosoma brucei*. J Biol Chem. 2005;280(31):28306–15.

Bakker BM, Michels PA, Opperdoes FR, Westerhoff HV. Glycolysis in bloodstream form *Trypanosoma brucei* can be understood in terms of the kinetics of the glycolytic enzymes. J Biol Chem. 1997;272(6):3207–15.

Bakker BM, Westerhoff HV, Opperdoes FR, Michels PA. Metabolic control analysis of glycolysis in trypanosomes as an approach to improve selectivity and effectiveness of drugs. Mol Biochem Parasitol. 2000;106(1):1–10.
Bergmann FT, Hoops S, Klahn B, Kummer U, Mendes P, Pahle J, Sahle S. COPASI and its applications in biotechnology. J Biotechnol. 2017;261:215–20.
Brand MD. Top-down elasticity analysis and its application to energy metabolism in isolated mitochondria and intact cells. Mol Cell Biochem. 1998;184(1–2):13–20.
Brown GC, Lakin-Thomas PL, Brand MD. Control of respiration and oxidative phosphorylation in isolated rat liver cells. Eur J Biochem. 1990;192(2):355–62.
Castegna A, Gissi R, Menga A, Montopoli M, Favia M, Viola A, Canton M. Pharmacological targets of metabolism in disease: opportunities from macrophages. Pharmacol Ther. 2020;210:107521.
De Rycker M, Baragaña B, Duce SL, Gilbert IH. Challenges and recent progress in drug discovery for tropical diseases. Nature. 2018;559(7715):498–506.
Fell D. Understanding the control of metabolism. London: Portland Press; 1997. ISBN: 978 1 85578 047 7. ISSN: 1353 6516
Flint HJ, Tateson RW, Barthelmess IB, Porteous DJ, Donachie WD, Kacser H. Control of the flux in the arginine pathway of *Neurospora crassa*. Modulations of enzyme activity and concentration. Biochem J. 1981;200(2):231–46.
Galluzzi L, Kepp O, Vander Heiden MG, Kroemer G. Metabolic targets for cancer therapy. Nat Rev Drug Discov. 2013;12(11):829–46.
Gellerich FN, Kunz WS, Bohnensack R. Estimation of flux control coefficients from inhibitor titrations by non-linear regression. FEBS Lett. 1990;274(1–2):167–70.
Giersch C. Determining elasticities from multiple measurements of flux rates and metabolite concentrations. Application of the multiple modulation method to a reconstituted pathway. Eur J Biochem. 1995;227(1–2):194–201.
González-Chávez Z, Olin-Sandoval V, Rodíguez-Zavala JS, Moreno-Sánchez R, Saavedra E. Metabolic control analysis of the *Trypanosoma cruzi* peroxide detoxification pathway identifies tryparedoxin as a suitable drug target. Biochim Biophys Acta. 2015;1850(2):263–73.
González-Chávez Z, Vázquez C, Mejia-Tlachi M, Márquez-Dueñas C, Manning-Cela R, Encalada R, Rodríguez-Enríquez S, Michels PAM, Moreno-Sánchez R, Saavedra E. Gamma-glutamylcysteine synthetase and tryparedoxin 1 exert high control on the antioxidant system in *Trypanosoma cruzi* contributing to drug resistance and infectivity. Redox Biol. 2019;26:101231.
González-Chávez Z, Vázquez C, Moreno-Sánchez R, Saavedra E. Metabolic control analysis for drug target prioritization in trypanosomatids. Methods Mol Biol. 2020;2116:689–718.
Groen AK, van Roermund CW, Vervoorn RC, Tager JM. Control of gluconeogenesis in rat liver cells. Flux control coefficients of the enzymes in the gluconeogenic pathway in the absence and presence of glucagon. Biochem J. 1986;237(2):379–89.
Hafner RP, Brown GC, Brand MD. Analysis of the control of respiration rate, phosphorylation rate, proton leak rate and protonmotive force in isolated mitochondria using the 'top-down' approach of metabolic control theory. Eur J Biochem. 1990;188(2):313–9.
Hakenberg J, Schmeier S, Kowald A, Klipp E, Leser U. Finding kinetic parameters using text mining. OMICS. 2004;8(2):131–52.
Heinrich R, Rapoport TA. A linear steady-state treatment of enzymatic chains. General properties, control and effector strength. Eur J Biochem. 1974;42(1):89–95.
Heiske M, Letellier T, Klipp E. Comprehensive mathematical model of oxidative phosphorylation valid for physiological and pathological conditions. FEBS J. 2017;284(17):2802–2828.
Hoops S, Sahle S, Gauges R, Lee C, Pahle J, Simus N, Singhal M, Xu L, Mendes P, Kummer U. COPASI--a COmplex PAthway SImulator. Bioinformatics. 2006;22(24):3067–74.
Kaambre T, Chekulayev V, Shevchuk I, Tepp K, Timohhina N, Varikmaa M, Bagur R, Klepinin A, Anmann T, Koit A, Kaldma A, Guzun R, Valvere V, Saks V. Metabolic control analysis of respiration in human cancer tissue. Front Physiol. 2013;4:151.
Kacser H. The control of enzyme systems in vivo: elasticity analysis of the steady state. Biochem Soc Trans. 1983;11(1):35–40.
Kacser H, Burns JA. The control of flux. Symp Soc Exp Biol. 1973;27:65–104.

Kacser H, Burns JA. Molecular democracy: who shares the controls? Biochem Soc Trans. 1979;7(5):1149–60.

Kurata M, Yamamoto K, Moriarity BS, Kitagawa M, Largaespada DA. CRISPR/Cas9 library screening for drug target discovery. J Hum Genet. 2018;63(2):179–86.

Leroux AE, Krauth-Siegel RL. Thiol redox biology of trypanosomatids and potential targets for chemotherapy. Mol Biochem Parasitol. 2016;206(1–2):67–74.

Lo-Thong O, Charton P, Cadet XF, Grondin-Perez B, Saavedra E, Damour C, Cadet F. Identification of flux checkpoints in a metabolic pathway through white-box, grey-box and black-box modeling approaches. Sci Rep. 2020;10(1):13446.

Manta B, Bonilla M, Fiestas L, Sturlese M, Salinas G, Bellanda M, Comini MA. Polyamine-based thiols in trypanosomatids: evolution, protein structural adaptations, and biological functions. Antioxid Redox Signal. 2018;28(6):463–86.

Marín-Hernández A, Rodríguez-Enríquez S, Vital-González PA, Flores-Rodríguez FL, Macías-Silva M, Sosa-Garrocho M, Moreno-Sánchez R. Determining and understanding the control of glycolysis in fast-growth tumor cells. Flux control by an over-expressed but strongly product-inhibited hexokinase. FEBS J. 2006;273(9):1975–88.

Marín-Hernández A, Gallardo-Pérez JC, Ralph SJ, Rodríguez-Enríquez S, Moreno-Sánchez R. HIF-1alpha modulates energy metabolism in cancer cells by inducing over-expression of specific glycolytic isoforms. Mini Rev Med Chem. 2009;9(9):1084–101.

Marín-Hernández A, Gallardo-Pérez JC, Rodríguez-Enríquez S, Encalada R, Moreno-Sánchez R, Saavedra E. Modeling cancer glycolysis. Biochim Biophys Acta. 2011;1807(6):755–67.

Marín-Hernández A, López-Ramírez SY, Del Mazo-Monsalvo I, Gallardo-Pérez JC, Rodríguez-Enríquez S, Moreno-Sánchez R, Saavedra E. Modeling cancer glycolysis under hypoglycemia, and the role played by the differential expression of glycolytic isoforms. FEBS J. 2014;281(15):3325–45.

Marín-Hernández Á, Rodríguez-Zavala JS, Del Mazo-Monsalvo I, Rodríguez-Enríquez S, Moreno-Sánchez R, Saavedra E. Inhibition of non-flux-controlling enzymes deters cancer glycolysis by accumulation of regulatory metabolites of controlling steps. Front Physiol. 2016;7:412.

Marín-Hernández Á, Gallardo-Pérez JC, Reyes-García MA, Sosa-Garrocho M, Macías-Silva M, Rodríguez-Enríquez S, Moreno-Sánchez R, Saavedra E. Kinetic modeling of glucose central metabolism in hepatocytes and hepatoma cells. Biochim Biophys Acta Gen Subj. 2020;1864(11):129687.

Michels PAM, Villafraz O, Pineda E, Alencar MB, Cáceres AJ, Silber AM, Bringaud F. Carbohydrate metabolism in trypanosomatids: new insights revealing novel complexity, diversity and species-unique features. Exp Parasitol. 2021;224:108102.

Moreno-Sánchez R, Bravo C, Westerhoff HV. Determining and understanding the control of flux. An illustration in submitochondrial particles of how to validate schemes of metabolic control. Eur J Biochem. 1999;264(2):427–33.

Moreno-Sánchez R, Saavedra E, Rodríguez-Enríquez S, Olín-Sandoval V. Metabolic control analysis: a tool for designing strategies to manipulate metabolic pathways. J Biomed Biotechnol. 2008a;2008:597913.

Moreno-Sánchez R, Encalada R, Marín-Hernández A, Saavedra E. Experimental validation of metabolic pathway modeling. FEBS J. 2008b;275(13):3454–69.

Moreno-Sánchez R, Saavedra E, Rodríguez-Enríquez S, Gallardo-Pérez JC, Quezada H, Westerhoff HV. Metabolic control analysis indicates a change of strategy in the treatment of cancer. Mitochondrion. 2010;10(6):626–39.

Moreno-Sánchez R, Marín-Hernández A, Saavedra E, Pardo JP, Ralph SJ, Rodríguez-Enríquez S. Who controls the ATP supply in cancer cells? Biochemistry lessons to understand cancer energy metabolism. Int J Biochem Cell Biol. 2014;50:10–23.

Moreno-Sánchez R, Gallardo-Pérez JC, Rodríguez-Enríquez S, Saavedra E, Marín-Hernández Á. Control of the NADPH supply for oxidative stress handling in cancer cells. Free Radic Biol Med. 2017;112:149–61.

Moreno-Sánchez R, Marín-Hernández Á, Gallardo-Pérez JC, Vázquez C, Rodríguez-Enríquez S, Saavedra E. Control of the NADPH supply and GSH recycling for oxidative stress management in hepatoma and liver mitochondria. Biochim Biophys Acta Bioenerg. 2018;1859(10):1138–50.

Mukherjee S, Mukherjee N, Gayen P, Roy P, Babu SP. Metabolic inhibitors as antiparasitic drugs: pharmacological, biochemical and molecular perspectives. Curr Drug Metab. 2016;17(10):937–70.

Müller J, Hemphill A. Drug target identification in protozoan parasites. Expert Opin Drug Discov. 2016;11(8):815–24.

Müller M, Mentel M, van Hellemond JJ, Henze K, Woehle C, Gould SB, Yu RY, van der Giezen M, Tielens AG, Martin WF. Biochemistry and evolution of anaerobic energy metabolism in eukaryotes. Microbiol Mol Biol Rev. 2012;76(2):444–95.

Nelson DL, Cox MM. Lehninger principles of biochemistry. 7th ed. New York: W.H. Freeman; 2017. Chapter 15.2

Newsholme EA, Start C. Regulation of metabolism. London, UK: John Wiley & Sons; 1973.

Olin-Sandoval V, Moreno-Sánchez R, Saavedra E. Targeting trypanothione metabolism in trypanosomatid human parasites. Curr Drug Targets. 2010;11(12):1614–30.

Olin-Sandoval V, González-Chávez Z, Berzunza-Cruz M, Martínez I, Jasso-Chávez R, Becker I, Espinoza B, Moreno-Sánchez R, Saavedra E. Drug target validation of the trypanothione pathway enzymes through metabolic modelling. FEBS J. 2012;279(10):1811–33.

Pineda E, Encalada R, Vázquez C, González Z, Moreno-Sánchez R, Saavedra E. Glucose metabolism and its controlling mechanisms in *Entamoeba histolytica*. In: Nozaki T, Bhattacharya A, editors. Amebiasis. Tokyo: Springer; 2015a.

Pineda E, Encalada R, Vázquez C, Néquiz M, Olivos-García A, Moreno-Sánchez R, Saavedra E. In vivo identification of the steps that control energy metabolism and survival of *Entamoeba histolytica*. FEBS. 2015b;282(2):318–31.

Piñeyro MD, Arias D, Parodi-Talice A, Guerrero S, Robello C. Trypanothione metabolism as drug target for trypanosomatids. Curr Pharm Des. 2021;27(15):1834–46.

Raj S, Sasidharan S, Balaji SN, Saudagar P. An overview of biochemically characterized drug targets in metabolic pathways of Leishmania parasite. Parasitol Res. 2020;119(7):2025–37.

Rodríguez-Enríquez S, Torres-Márquez ME, Moreno-Sánchez R. Substrate oxidation and ATP supply in AS-30D hepatoma cells. Arch Biochem Biophys. 2000;375(1):21–30.

Saa PA, Nielsen LK. Formulation, construction and analysis of kinetic models of metabolism: a review of modelling frameworks. Biotechnol Adv. 2017;35(8):981–1003.

Saavedra E, Marín-Hernández A, Encalada R, Olivos A, Mendoza-Hernández G, Moreno-Sánchez R. Kinetic modeling can describe in vivo glycolysis in *Entamoeba histolytica*. FEBS J. 2007;274(18):4922–40.

Saavedra E, González-Chávez Z, Moreno-Sánchez R, Michels PAM. Drug target selection for *Trypanosoma cruzi* metabolism by metabolic control analysis and kinetic modeling. Curr Med Chem. 2019a;26(36):6652–71.

Saavedra E, Encalada R, Vázquez C, Olivos-García A, Michels PAM, Moreno-Sánchez R. Control and regulation of the pyrophosphate-dependent glucose metabolism in *Entamoeba histolytica*. Mol Biochem Parasitol. 2019b;229:75–87.

Sauro HM. SCAMP: a general-purpose simulator and metabolic control analysis program. Comput Appl Biosci. 1993;9(4):441–50.

Schuster A, Erasimus H, Fritah S, Nazarov PV, van Dyck E, Niclou SP, Golebiewska A. RNAi/CRISPR screens: from a pool to a valid hit. Trends Biotechnol. 2019;37(1):38–55.

Segel IH. Enzyme kinetics. New York, USA: Wiley; 1975.

Small JR, Kacser H. Responses of metabolic systems to large changes in enzyme activities and effectors. 1. The linear treatment of unbranched chains. Eur J Biochem. 1993a;213(1):613–24.

Small JR, Kacser H. Responses of metabolic systems to large changes in enzyme activities and effectors. 2. The linear treatment of branched pathways and metabolite concentrations. Assessment of the general non-linear case. Eur J Biochem. 1993b;213(1):625–40.

Soares Medeiros LC, South L, Peng D, Bustamante JM, Wang W, Bunkofske M, Perumal N, Sanchez-Valdez F, Tarleton RL. Rapid, selection-free, high-efficiency genome editing in protozoan parasites using CRISPR-Cas9 ribonucleoproteins. MBio. 2017;8(6):e01788–17.

Sukjoi W, Ngamkham J, Attwood PV, Jitrapakdee S. Targeting cancer metabolism and current anticancer drugs. Adv Exp Med Biol. 2021;1286:15–48.

Talevi A, Carrillo C, Comini M. The thiol-polyamine metabolism of *Trypanosoma cruzi*: molecular targets and drug repurposing strategies. Curr Med Chem. 2019;26(36):6614–35.

Torres NV, Mateo F, Meléndez-Hevia E, Kacser H. Kinetics of metabolic pathways. A system in vitro to study the control of flux. Biochem J. 1986;234(1):169–74.

Torres NV, Souto R, Meléndez-Hevia E. Study of the flux and transition time control coefficient profiles in a metabolic system in vitro and the effect of an external stimulator. Biochem J. 1989;260(3):763–9.

Tummler K, Lubitz T, Schelker M, Klipp E. New types of experimental data shape the use of enzyme kinetics for dynamic network modeling. FEBS J. 2014;281(2):549–71.

Tyagi R, Elfawal MA, Wildman SA, Helander J, Bulman CA, Sakanari J, Rosa BA, Brindley PJ, Janetka JW, Aroian RV, Mitreva M. Identification of small molecule enzyme inhibitors as broad-spectrum anthelmintics. Sci Rep. 2019;9(1):9085.

Valente M, Vidal AE, González-Pacanowska D. Targeting kinetoplastid and apicomplexan thymidylate biosynthesis as an antiprotozoal strategy. Curr Med Chem. 2019;26(22):4262–79.

van Eunen K, Kiewiet JA, Westerhoff HV, Bakker BM. Testing biochemistry revisited: how in vivo metabolism can be understood from in vitro enzyme kinetics. PLoS Comput Biol. 2012;8(4):e1002483.

Wanders RJ, Groen AK, Van Roermund CW, Tager JM. Factors determining the relative contribution of the adenine-nucleotide translocator and the ADP-regenerating system to the control of oxidative phosphorylation in isolated rat-liver mitochondria. Eur J Biochem. 1984;142(2):417–24.

Westerhoff HV, Kell DB. Matrix method for determining steps most rate-limiting to metabolic fluxes in biotechnological processes. Biotechnol Bioeng. 1987;30(1):101–7.

Wyatt PG, Gilbert IH, Read KD, Fairlamb AH. Target validation: linking target and chemical properties to desired product profile. Curr Top Med Chem. 2011;11:1275–83.

Progress on Open Chemoinformatic Tools for Drug Discovery

José L. Medina-Franco, Rodrigo Gutiérrez-Nieto,
and Homero Gómez-Velasco

Contents

Abstract

Informatics plays a fundamental role in many chemistry applications giving rise to the consolidation of well-established disciplines such as bioinformatics and chemoinformatics. It has also led to the maturation of subdisciplines such as food informatics, epi-informatics, and more recently, to the so-called natural products informatics. The extensive practice of informatics across different disciplines and subdisciplines has been boosted by the large and increasing availability of open and well-documented resources. A number of them have been implemented as web-applications that further encourage the use by the scientific community. In this chapter, we review the recent progress on the development of public chemoinformatic resources for different tasks, with special focus/emphasis on drug discovery applications. Due to the current COVID-19 pandemic, we emphasize resources that have been developed and released over the past few months to support drug discovery efforts worldwide.

J. L. Medina-Franco (✉) · R. Gutiérrez-Nieto · H. Gómez-Velasco
DIFACQUIM Research Group, Department of Pharmacy, School of Chemistry, Universidad Nacional Autónoma de México, Mexico City, Mexico
e-mail: medinajl@unam.mx

M. T. Scotti, C. L. Bellera (eds.), *Drug Target Selection and Validation*, Computer–Aided Drug Discovery and Design,
https://doi.org/10.1007/978-3-030-95895-4_9

Keywords
COVID-19 · Chemical space · Chemoinformatics · Drug discovery · Education · Natural products informatics · Open science · SARS-CoV-2 · Structure-activity relationships · Webserver

Abbreviations

AI	Artificial intelligence
COCONUT	Collection of Open Natural Products
CoVs	Coronaviruses
ETP	Epigenetic Target Profiler
LANaPD	Latin America Natural Product Database
NP	Natural products
SAR	Structure-activity relationships
SARS-CoV-2	Severe Acute Respiratory Syndrome Coronavirus 2
SMILES	Simplified Molecular Input Line Entry System

1 Introduction

Chemoinformatics has played a crucial role in drug discovery, organic synthesis (planning and analysis), analytical chemistry, and more recently in food chemistry and natural product (NP) chemistry, among many other areas. The impact of chemoinformatics in chemistry in general, and drug discovery in particular, has been discussed extensively in the literature (López-López et al. 2021; Gasteiger 2020; Martinez-Mayorga et al. 2020; Medina-Franco and Saldívar-González 2020). The conceptual and methodological progress in chemoinformatics have been advanced in the past decade in parallel with an increasing number of open-source programs (including public web servers) well documented and available as stand-alone software or free customizable code. Several available public repositories facilitate the storage and maintenance of free code. Furthermore, specialized journals in chemoinformatics and other computational areas are now requiring the proper documentation and availability of the software to facilitate data and method sharing and reproducibility (Merz et al. 2020).

There are extensive reviews of open chemoinformatics resources for numerous applications. For instance, Singh et al. recently reviewed online web servers to perform virtual screening of small molecules and docking (Singh et al. 2020). Wu et al. reviewed databases and software typically used to predict ADME/Tox (Absorption, Distribution, Metabolism, Excretion, and Toxicity) (Wu et al. 2020). The authors of this chapter surveyed open resources to analyze structure–activity relationships (SAR) of compound data sets. That review included web servers and workflows that can be implemented with public software (Gonzalez-Medina et al. 2017).

The availability of free computational resources is useful for any disease but it is particularly valuable to address complex, rare, or neglected diseases (Ekins et al. 2011). Also, in the emergence of major health outbreaks such as the COVID-19 disease affecting the worldwide population. Such diseases demand the use and implementation of resources that assist every step of the drug discovery process (Martinez-Mayorga et al. 2020).

In this chapter, we focus on recent developments of free and open-source chemoinformatics resources to assist the main activities in drug discovery, emphasizing the tools to support COVID-19 drug discovery. We also cover free tools to quantify NP-likeness that is an emerging need in natural product-based drug development. The chapter is organized into three main sections. First, we present an overview of free compound databases for virtual screening, emphasizing NP databases and compound collections for COVID-19 drug discovery. The next section discusses tools for virtual screening and target fishing. The third section survey advances in chemoinformatic resources for NP-based drug discovery, emphasizing free tools to measure NP-likeness. The last section presents summary conclusions and future directions in the field. For readers interested in additional reading of the topics covered in this chapter, we suggest references cited in each section.

2 Compound Databases

Compound databases are at the core of chemoinformatics and represent a systematic and organized manner of collecting the chemical space. The chemical space is huge, and novel chemical databases are assembled and under continued development. This has been exemplified by the large number of on-demand and virtual libraries commented recently (Walters 2019; Saldívar-González et al. 2020). Reviews of recent chemical libraries for drug discovery had been published elsewhere (Gong et al. 2017; Wang et al. 2019). A particular set of chemical libraries are compound collections focused on therapeutic targets or therapeutic indications (Wassermann et al. 2014; Harris et al. 2011). For instance, the authors have recently reviewed compound libraries focused on epigenetic targets (Sessions et al. 2020). Herein, we elaborate on recent developments in this area, emphasizing the focused libraries dedicated to COVID-19 drug discovery.

Databases to Support COVID-19 Research and Drug Discovery

Coronaviruses (CoVs) are a family of enveloped RNA viruses and have a characteristic crown-shaped appearance caused by the surface glycoproteins that decorate the virus (Mousavizadeh and Ghasemi 2021). A novel CoV currently named Severe Acute Respiratory Syndrome Coronavirus 2 (SARS-CoV-2) was associated with a cluster of respiratory tract infections called COVID-19 in December 2019. It has rapidly spread across the world, causing the World Health Organization to recognize it as a pandemic in March 2020 (Cucinotta and Vanelli 2020).

Extensive structural, microbiology, proteomics, screening of chemical libraries, and many other experimental investigations of SARS-CoV-2 have provided substantial and highly valuable inputs for multiple computational approaches, including bio- and cheminformatics tools and means of molecular modeling, data mining, and, especially, drug discovery and development. As of March 2021, more than 110,000 peer-reviewed papers have been published on COVID-19, according to PubMed, and more than 12,500 preprints have been deposited into MedRxiv or BioRxiv.

As part of the worldwide effort, chemical and structural databases have been rapidly released in the public domain. Table 1 summarizes representative free compound databases aimed to support COVID-19 drug discovery.

Overall, the databases summarized in Table 1 provide information to explore and analyze some parts of different types of coronavirus, compounds, and approved drugs with activity on them. Table 1 indicates that among the databases focused on COVID-19 protein sequence and viral replication research, NCBI SARS-CoV-2 is the largest with 54,572 sequences. Within databases of compounds with activity on SARS-CoV-2, CAS COVID-19 antiviral candidate compound dataset is the largest with nearly 50,000 molecules. Depending on what the user is looking for, most databases have browsing and filtering capabilities. All the databases supply a table with the search results, but the information differs across them. It may include identifiers or codes for proteins, the name of the virus, the host, the name of the drug or compound, and if it is approved for clinical use or not, etc. In most of the databases, the user has the option to download the results table. Usually, it is available in CSV, PDF, or Excel format, depending on the database. Some databases include tools for a better analysis of the results, such as 3D viewers, a tool to compare chemical structures, a sequence logo generator, phylogenetic trees builder, etc. In DBCOVP and NCBI SARS-CoV-2 databases, the users can submit a new protein sequence for analysis. They only have to fill in a submission form and enter the protein sequence in FASTA format.

Compound Databases of General Interest

Table 2 compiles databases not necessarily focused on COVID-19 research but with general application. Although some of the databases listed in this table have been around for several years, some such as PubChem and ZINC have significant updates.

A notable recent development is the new release of the database ZINC, one of the largest assemblies of compound databases used in virtual screening (using either structure-based or ligand-based approaches) (Table 2). The new version is ZINC20, which collects compounds for virtual screening, ligand discovery, pharmacophore screens, benchmarking, and force field development. A distinctive feature of the database is that it now incorporates billions of new molecules and online approaches to search the information. Of note, besides the significant amount of new molecules compared to the previous version (ZINC15) is the inclusion of make-on-demand

Table 1 Examples of free databases aimed to support COVID-19 drug discovery

Database	Contents	URL	Ref.
CAS COVID-19 antiviral candidate compounds dataset	Nearly 50,000 chemical substances, including antiviral drugs and structurally similar compounds.	https://www.cas.org/covid-19-antiviral-compounds-dataset	https://datascience.nih.gov/covid-19-open-access-resources
SARS-CoV-2 Related Structures	About 941 carefully validated SARS-CoV-2 protein structures, including structural models that have been re-processed or re-defined. It also contains 64 additional structures of other coronaviruses.	https://covid19.bioreproducibility.org	https://datascience.nih.gov/covid-19-open-access-resources
CoronaVirus Explorer	Twenty-seven viral proteins, 332 host proteins, and 332 experimentally validated virus-host interactions for both SARS-CoV-1 and SARS-CoV-2. It is a resource to understand molecular mechanisms of pathogenicity and prioritize compounds.	https://exbio.wzw.tum.de/covex	Sahoo et al. (2021), Sadegh et al. (2020)
Corona Drug Interactions database CORDITE	Collects and aggregates data for SARS-CoV-2 from PubMed, MedRxiv, BioRxiv, ChemRxiv. Furthermore, it collects and provides up-to-date information on computational predictions, in vitro, in vivo study data as well as clinical trials.	https://cordite.mathematik.uni-marburg.de/#/	Sahoo et al. (2021)
Database of Coronavirus Virulent Glycoproteins DBCOVP	It provides information on structural glycoproteins from coronavirus genomes of betacoronavirus genera, including SARS-1, MERS-CoV, and SARS-CoV-2 strains. Conserved promiscuous T-cell and B-cell were predicted for each protein.	http://covp.immt.res.in/Default.aspx	Sahoo et al. (2021)
Coronavirus 3D Structure Database CoV3D	Up-to-date reference for SARS-CoV-2, SARS-CoV, and MERS-CoV structures, focused on antibody recognition, host receptor binding, drug targeting, and diversity. It contains a complete annotated set of structures of coronavirus proteins, updated weekly from the Protein Data Bank.	https://cov3d.ibbr.umd.edu/about	Sahoo et al. (2021)
NCBI SARS-CoV-2	More than 244,000 SRA runs, 66,044 nucleotide records, 4762 clinical trials, 54,572 viral genomes and protein sequences. It includes links to analyze protein and nucleotide sequences.	https://www.ncbi.nlm.nih.gov/sars-cov-2/	

Table 2 Examples of free compound databases of general application

Database	Contents	URL	Ref.
PubChem	It contains more than 109 million compounds, more than 270 million substances, and more than 1.2 million BioAssays. It mostly includes small molecules but also larger ones such as nucleotides, carbohydrates, lipids, peptides, and macromolecules. PubChem is integrated with other literature and biomedical databases.	https://pubchem.ncbi.nlm.nih.gov	Cheng et al. (2014)
Natural products database of the Bahia semi-arid region NatProDB	It has a broad chemical diversity, suited for the discovery of new chemical entities. It contains naturally occurring diverse compounds, which have not yet been explored in the process of discovery of bioactive molecules through virtual screening.	http://natprodb.uefs.br	da Paixão and Pita (2019)
COCONUT	An aggregated dataset of natural products in different open sources. The user can search, analyze, and download chemical structures.	https://coconut.naturalproducts.net/	Sorokina and Steinbeck (2020)
ChEMBL	A manually curated database of bioactive molecules. It contains more than 1.9 distinct compounds and more than 16 million activities.	https://www.ebi.ac.uk/chembl/	Mendez et al. (2019)
ZINC	It is a compound database that connects biological activities by gene products, drugs, and NP with commercial availability. The current version has more than 1.4 billion compounds, from which 1.3 billion are purchasable.	http://zinc15.docking.org	Sterling and Irwin (2015), Irwin et al. (2020)
Protein Data Bank RCSB PDB	Database that provides access to 3D structure data for more than 170,000 large biological molecules, such as proteins or nucleic acids.	https://www.rcsb.org	Berman et al. (2000)
Nuclear Receptor Activity, NURA	Collects curated information on small molecules that modulate nuclear receptors for both pharmacological and toxicological applications. It contains bioactivity annotations for more than 15,000 molecules and 11 selected nuclear receptors.	https://zenodo.org/record/3999420#.YCc-Di2xBQI	Valsecchi et al. (2020)

molecules. The newest version of ZINC is freely available at http://zinc20.docking.org/, and the details are published in (Irwin et al. 2020).

Recently, Bobrowski et al. developed an online free platform to expedite the search of commercially available molecules as annotated in ZINC15. The web site is named ZINC Express, and it is freely available at https://zincexpress.mml.unc.edu/ . The details of ZINC Express have also been published (Bobrowski et al. 2021).

Another major development in compound databases falls within NP research. Recently the Collection of Open Natural Products (COCONUT) was assembled, curated, and publicly released (Table 2). COCONUT is a compendium of 50 open-access databases collecting more than 400,000 compounds (Sorokina and Steinbeck 2020). The developers of COCONUT implemented the collection in a freely accessible website (Sorokina and Steinbeck 2020). A related database that includes information of the structure–organism relationships has also been recently released: the naturaL prOducTs occUrrences databaSe (LOTUS) contains more than 500,000 curated and referenced structure–organism pairs. All the programs to develop this web resource, as well as the data of LOTUS, are freely available at https://lotus.naturalproducts.net (Rutz et al. 2021).

Chávez-Hernández et al. have published a comprehensive fragment library with approximately 206,000 fragments derived from a drug-like subset of COCONUT. In that work, the NP's fragment library was compared to therein generated fragment libraries of ChEMBL as representative of biologically relevant compounds and a vast on-demand database of synthetic molecules. Chávez-Hernández et al. made public the fragment library (Chávez-Hernández et al. 2020). In this area, a multidisciplinary and multi-national effort to assemble NP collections from different Latin American countries is underway under the collaborative project named LANaPD (Latin American Natural Product Database) (Medina-Franco 2020).

3 Virtual Screening

Singh et al. have recently collected and discussed online web servers to perform virtual screening of small molecules and docking (Singh et al. 2020) The authors reported 68 web tools classified divided into target-fishing, ligand-based, and structure-based virtual screening. The authors also included 15 databases, approximately. The databases provide different information, such as approved drugs (i.e., DrugBank), patented molecules (SureCheMBL), or the capability to buy compounds with different vendors (ZINC). The tools designed for ligand-based virtual screening include pharmacophore generators (PharmaGist), similarity search (ChemMine Tools), computation of molecular descriptors and fingerprints (ChemDes); machine learning models encompass SAR builders (ChemSAR), QSAR (DPubChem), and deep learning model construction (DeepScreening); the target fishing can be performed using fingerprints (Anglerfish), using 2D similarity (ChemProt-3), etc. Recent example tools for virtual screening are summarized in Table 3.

The tools shown in Table 3 enable users to perform virtual screening. Each tool can be used in different cases. PyRx identifies and analyzes compound properties, such as logP, number of different atom types, molecular weight, etc. Virtual Flow includes various docking programs, in particular, the members of the "AutoDock family": Autodock Vina, QuickVina 2, Smina, QuickVina-W, etc. The docking can be performed in different scenarios per ligand. The user is also able to see the results in real-time. Pharmit identifies all pharmacophore features present in the ligand if a

Table 3 Examples of free resources for virtual screening

Tool name	Brief description	Ref.
PyRx	This software can be used for computational drug designing by screening libraries of compounds against potential drug targets. PyRx helps users in every step of the process, from data preparation to analysis of results. The newest PyRx version is available for free, or you can purchase it. Older versions are available for free with no bug fixes or support.	Raj (2021)
VirtualFlow for Virtual Screening.	It is a versatile, parallel workflow platform for carrying out virtual screening-related tasks on Linux-based computer clusters of any type. It can be used for hit identification by screening ligand libraries of any size, hit optimization by screening custom hit-based analog libraries, and thorough and extensive dockings of one or more molecules.	Gorgulla et al. (2020)
Pharmit	An online environment for virtual screening compound databases using pharmacophores, molecular shape, and energy minimization. Users can import, create, and edit virtual screenings queries in an interactive browser-based interface. This service enables users to search for small-molecules based on their structural and chemical similarity to another small molecule. Search results can be analyzed using energy minimization.	Sunseri and Koes (2016)
ParaDockS	This is a flexible and easily extensible program for performing protein–ligand docking. This tool supplies an interaction-based classifier, which can be used in a post-docking filter step. The code is available to download, reuse, modify, and redistribute any source code file.	Scharfe et al. (2013), Pippel et al. (2011)
AutoDock Vina	The user can perform molecular docking. It searches for a docking pose using both a gradient calculation and Monte Carlo steps, based on a scoring function used for the searching and the ranking.	Masters et al. (2020)
rDock	This is a fast and versatile docking program of small molecules against proteins and nucleic acids. It provides several pseudo-energy scoring functions that are added to the total scoring function under optimization.	Ruiz-Carmona et al. (2014)
Open Drug Discovery Toolkit (ODDT)	This is a modular and comprehensive toolkit for use in chemoinformatics, molecular modeling, etc. It implements many state-of-the-art methods, such as machine learning scoring functions, and wraps other external software to ease the process of developing CADD pipelines.	Wójcikowski et al. (2015)
NCBI BLAST	This program finds regions of similarity between biological sequences. The program compares nucleotide or protein sequences to sequence databases and calculates the statistical significance of matches. BLAST can be used to infer functional and evolutionary relationships between sequences and help identify members of gene families.	Johnson et al. (2008)
DockThor-VS	It is a virtual screening platform with curated structures of potential therapeutic targets from SARS-CoV-2. DockThor-VS facilitate repurposing virtual screening experiments providing curated libraries of currently available drugs in the market. The user can perform blind docking as well as re-docking and large-scale docking experiments exploring multiple binding modes.	Guedes et al. (2021)

ligand structure file is provided. Autodock Vina searches for a docking pose using both a gradient calculation and Monte Carlo steps, based on a scoring function used for the searching and the ranking. ParaDockS is used to perform protein–ligand docking. rDock is a docking program to screen small molecules against proteins and nucleic acids. NCBI BLAST is a sequence similarity search program that can be used via a web interface or as a stand-alone tool to compare a user's query to a sequence database. Finally, DockThor-VS is used for protein–ligand docking; the user can perform large-scale docking experiments exploring multiple binding modes, perform blind docking on the proteins to find cavities, and validate docking protocols with redocking experiments. Furthermore, other tools help the virtual screening process, such as Gypsum DL, which converts 1D or 2D small-molecule representations into 3D models, or AMdock, that prepares the input structure files, and optimally defines the search space, offering several alternatives and degrees of user supervision.

Recently, the authors developed a free web server for target fishing focused on epigenetic targets of therapeutic interest, i.e., for epigenetic target fishing. The web server is called Epigenetic Target Profiler (ETP), and it is freely available at http://www.epigenetictargetprofiler.com/. This is an easy-to-use and open web application to predict small molecules' bioactivity profile over a panel of 55 epigenetic targets. ETP implements the best performing model for epigenetic target prediction, as identified from a systematic comparison of machine learning models built on molecular fingerprints of different designs described in detail in (Sánchez-Cruz and Medina-Franco 2021a, b). This web server is part of the D-TOOLS (DIFACQUIM Tools for Chemoinformatics), a collection of free resources to support drug discovery research from an academic setting (Naveja et al. 2018). ETP is illustrated in Fig. 1. ETP has two main features: a home or entry page (panel A) and a results page (panel B). The user can draw a query molecule on the home page using the built-in JavaScript-based JSME molecular editor and from there generate its SMILES (Simplified Molecular Input Line Entry System) string by pressing the "Get SMILES" button. Else, the user can generate the SMILES from an external source and paste it in the space provided (Fig. 1a). After entering the query SMILES, the target prediction can be performed by clicking on the "Predict Targets" button, and the user will be directed to the results page (Fig. 1b). After the predictions have been made on the results page, the user will see two images at the top and a table below (Fig. 1b). The image on the left side shows the query compound's chemical structure as interpreted from the SMILES submitted by the user. In contrast, the image on the right side depicts the standardized compound's chemical structure as processed by the server. The results table displays the known and predicted targets for the query compound structure, including five columns with additional information. The first three columns contain the targets and external links to ChEMBL and GeneCards (Name, ChEMBL ID, Gene). The last two columns contain information about the predictions, status, and quartile, respectively. The "status column" indicates if the association is known or predicted, while the "quartile column" indicates the distance quartile (Q1–Q4) to which the query compound belongs for each of the predictions as an estimation of its reliability. The complete list of results that

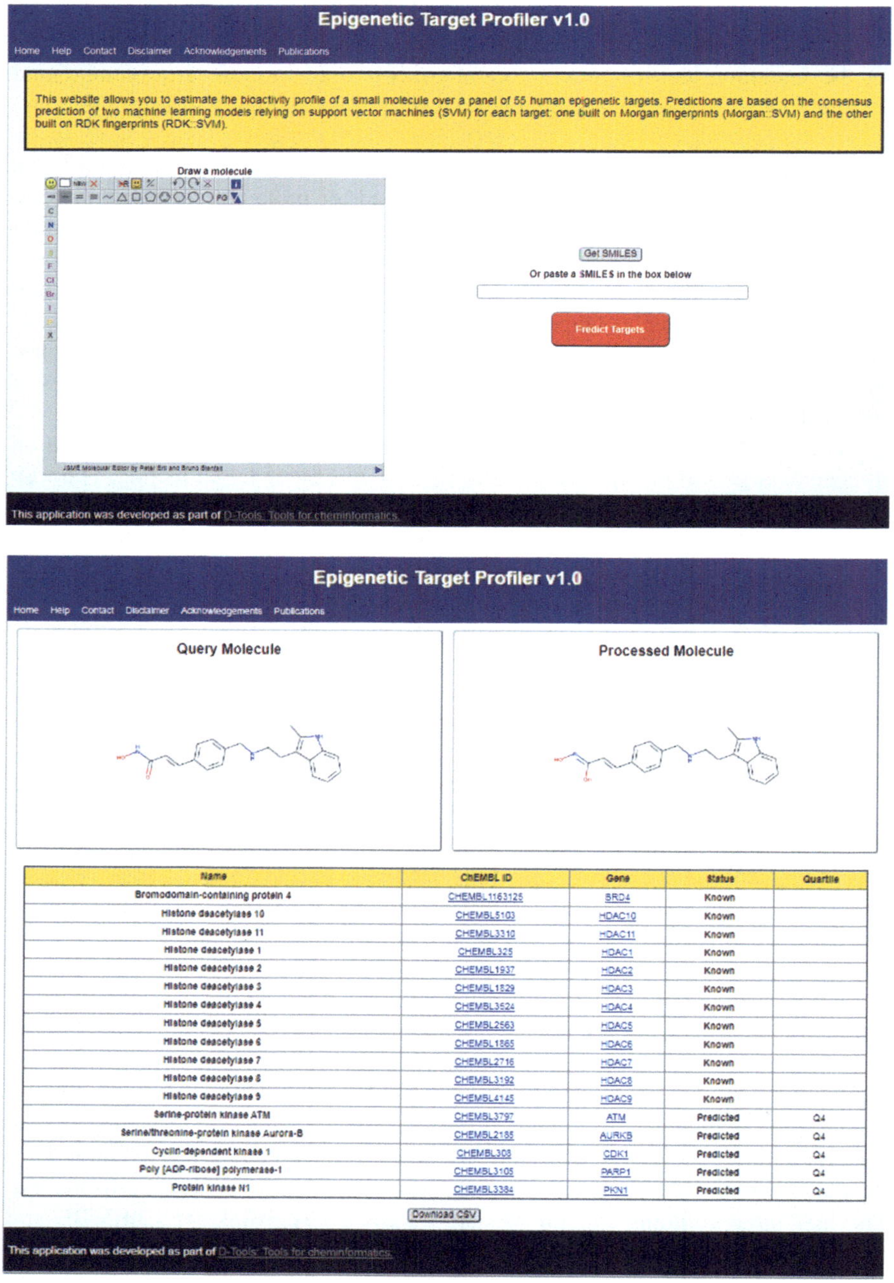

Fig. 1 The graphical user interface of Epigenetic Target Profiler. (**a**) Home page: the user can draw a query molecule using the built-in JavaScript-based JSME molecular editor or enter the SMILES string directly for the query compound. (**b**) The results page includes the known and predicted targets for the query compound structure (input structure on the left and standardized compound on the right). The complete list of predictions for all 55 targets can be downloaded by clicking on the "Download CSV" button

contains the predictions from the individual models for all 55 targets can be downloaded by clicking on the "Download CSV" button below the table.

In a separate contribution, the project SwissDrugDesign, led by the Molecular Modeling Group of the SIB Swiss Institute of Bioinformatics, provides a collection of freely available online tools for computer-aided drug design (Daina and Zoete 2019). Some of these tools, such as SwissSimilarity and SwissTargetPrediction, were designed to perform ligand-based virtual screening. SwissSimilarity focuses on the screening of small to very large libraries of drugs, small bioactive molecules, and commercially available compounds. The latter focuses on predicting the most probable protein targets of small bioactive molecules. SwissDrugDesign provides other tools that can be used for different activities related to drug discovery: SwissADME can find applications in predictive models for physicochemical properties, pharmacokinetics drug-likeness, and the so-called "medicinal chemistry friendliness"; SwissDock is a ligand–protein docking web service and interface to predict molecular interactions; SwissParam is a service that provides topologies and parameters of small organic molecules. SwissBioisostere is a database that has collected over 4.5 million molecular sub-structural replacements extracted from the literature, along with information on how frequently such replacements were applied in the past and the impact on the biological activity.

Applications to COVID-19 Drug Discovery

Table 4 summarizes exemplary advancements of virtual screening efforts to identify candidate compounds for the potential treatment of COVID-19. Thus far, most virtual screenings have been performed against well-validated molecular targets of SARS-CoV-2, whose three-dimensional structures became readily available. Best practices for virtual screening campaigns, particularly against SARS-CoV-2 molecular targets, have been discussed recently and extensively in the literature. As summarized in the table, most of the virtual screenings still need experimental validation. However, some reports incorporate experimental characterization of the virtual screening hits.

4 Natural Products

For many years, NP or secondary metabolites (Abegaz and Kinfe 2019) have acted as a source of therapeutic agents. NP are considered as valuable starting points for the development of drugs and other bioactive synthetic products (Sorokina and Steinbeck 2019) Many drugs in the current pharmacopeias are NP, and many others have NP origin (Newman and Cragg 2020; Atanasov et al. 2021).

Many free chemoinformatic resources are being used to support NP-based drug discovery and other aspects of NP research, such as biodiversity and metabolomics. This is giving rise to the emerging subdiscipline "natural products informatics" (López-López et al. 2021). Such resources have recently been discussed in several

Table 4 Examples of recent successful virtual screenings vs. SARS-CoV-2

Study	Computational tool	Main outcome	Ref.
Molecular docking, molecular dynamics, and in vitro studies reveal angiotensin ll receptor blockers' potential to inhibit the COVID-19 main protease.	Molecular Operating Environment (MOE) was used for docking and Desmond for molecular dynamic simulations.	Nine angiotensin receptor blocker drugs were docking against M^{pro}. The compounds with the best docking score were fimastaran, candesartan, and olmesartan. Olmestaran showed the best CC_{50}, IC_{50} values, and selectivity index against SARS-CoV-2.	Alnajjar et al. (2020)
Interaction of small molecules with the SARS-CoV-2 papain-like protease: In silico studies and in vitro validation of protease activity inhibition using an enzymatic inhibition assay.	AutoDock Vina was used for blind docking with the SARS-CoV-2 papain-like protease (PL^{pro}). Glide docking was used to screen the library of 300 compounds. PrankWeb was used to identify potential ligand-binding sites within PxL^{pro}.	Three hundred compounds from OliveNet™ and synthetic and dietary compounds with reported antiviral activities were investigated. Docking led to select compounds that were evaluated in vitro *as* PL^{pro} inhibitors using an enzymatic inhibition assay. Hypericin was active. Rutin and cyanidin-3-O-glucoside showed micromolar activity in a concentration-dependent manner.	Pitsillou et al. (2021)
Identification of small-molecule inhibitors of the Deubiquitinating activity of the SARS-CoV-2	Schrodinger Suite was used for docking. Autodock Vina was used to perform blind docking. HDOCK was used for ab initio free protein–protein docking. PrankWeb was used to identify potential binding pockets.	Docking was performed to examine the potential inhibition of four dietary compounds and two naphthalene-based inhibitors against the papain-like protease, PL^{pro}. The deubiquitinating activity was measured using an enzymatic activity assay. The dietary compounds showed stronger affinities and interatomic contacts with surrounding residues than the naphthalene-based inhibitors. Enzymatic assay indicated that three compounds had a concentration-dependent inhibition of PL^{pro} deubiquitinase activity. Hypericin was the most potent compound.	Pitsillou et al. (2020)

(continued)

Table 4 (continued)

Study	Computational tool	Main outcome	Ref.
A novel virtual screening procedure identifies Pralatrexate as an inhibitor of SARS-CoV-2 RdRp, and it reduces viral replication in vitro.	DFCNN was used for predicting protein–drug binding probability. Autodock Vina was used for docking.	One thousand nine hundred six approved drugs for clinical use were screened. Four compounds, azithromycin, pralatrexate, amoxicillin, and sofosbuvir, were validated experimentally. Pralatrexate and azithromycin inhibited the replication of SARS-CoV-2. Pralatrexate was more potent than remdesivir.	Zhang et al. (2020)
Discovery of novel inhibitors against M^{pro} of SARS-CoV-2 via virtual screening and biochemical evaluation.	Discovery Studio was used to prepare protein. LibDock and GOLD were used to do the docking.		Guo et al. (2021)

publications (Medina-Franco and Saldívar-González 2020; Chen and Kirchmair 2020). It is notable in the review of NP databases in the public domain that, at the time of writing this manuscript, it has collected nearly 400,000 natural products (Sorokina et al. 2021). Such efforts include regional and international efforts exemplified by the "Latin American Natural Product Database" (LANaPD) project (Medina-Franco 2020).

The broad and increasing availability of the chemical structures of NP in the public domain, along with the advances in the development of open resources to conduct virtual screening (discussed in Sect. 2), has encouraged the application of such techniques to perform virtual screening of NP databases to identify potential hit compounds for different targets. A clear example has been the numerous virtual screening studies to suggest NP as potential active compounds for different molecular targets of SARS CoV-2.

Natural Product-Likeness

The concept of "natural product-likeness" (NP-likeness) connotes a molecule's similarity to the structure space covered by NP and is a useful criterion in screening compound libraries (Ntie-Kang et al. 2019). NP-likeness is a measure of similarity to the NP molecules (Ertl et al. 2008). NP-likeness can be used to identify NP in compound libraries, guide the de novo generation of NP mimetics, optimize synthetic bioactive compounds, or design new drugs (Sorokina and Steinbeck 2019; Chen et al. 2019; Jayaseelan et al. 2012).

Computational methods were recently created to identify NP with high accuracy. These are open-source web applications based on open data to compute NP likeness scores for chemical libraries. One of these applications is NaPLeS—Natural Product Likeness Score calculator. NP-likeness scorer has been trained with NP from various public databases and randomly selected synthetic molecules (SM) from the ZINC database (Sorokina and Steinbeck 2019). Atom signatures of compounds from NP and SM datasets are indexed separately to look for the frequency of molecule fragments in question, employing Bayesian statistics (Ertl et al. 2008).

NaPLeS is freely available at https://naples.naturalproducts.net. Submission of molecules to compute their NP-likeness scores is possible in three ways: uploading a file (SDF, MOL, or SMI format, for a maximum of 1000 molecules), pasting a SMILES string, or drawing a molecule in the chemical editor (Sorokina and Steinbeck 2019; Ertl et al. 2008; Jayaseelan et al. 2012). The stereochemistry and sugar are removed from the user-submitted molecules. The NP-likeness scores results are reported in a table, where the minimal NP-likeness score is −3.63, and the maximum is 5.15. The higher the score is, the higher the probability is that the molecule is an NP. The result table can be exported in comma-delimited (.CSV) and Excel formats (Sorokina and Steinbeck 2019).

NP-scout is another free web application that is also trained on public data and allows NP identification. This tool computes a probability of a molecule to be an NP based on two-dimensional molecular descriptors, MACCS keys, and Morgan2 fingerprints (Sorokina and Steinbeck 2019; Chen et al. 2019; Stork et al. 2020). NP-scout web service allows pasting a SMILES string, uploading a file with a list of SMILES, or drawing the molecule. Salt components and stereochemical information are removed from the submitted molecules. The results can be downloaded in a comma-delimited (.CSV) file format. Prediction results are not calculated for molecules with a molecular weight below 150 Da or above 1500 Da. The results page displays a table that summarizes the NP class probability and similarity maps. In this similarity map, green highlights mark atoms contributing to the classification of a molecule as NP. In contrast, orange highlights mark atoms contributing to the classification of a molecule as SM. NP-scout is available at https://nerdd.zbh.uni-hamburg.de/npscout.

To illustrate the application of the two open web servers to profile the NP-likeness to compound databases, here we used both computational methods for the identification of NP of eight compound databases: BIOFACQUIM (NP from Mexico), $NuBBE_{DB}$ (NP from Brazil), Cyanobacteria metabolites, Fungi metabolites, Marine NP, purified natural product screening compounds (MEGx), Approved drugs and semisynthetic compounds (NATx). These structure files, already curated, were

Table 5 Compound databases to illustrate the application of web servers to quantify natural product likeness

Database	Size	Processed by NaPLeS	Processed by NP-scout
BIOFACQUIM	530	530	516
Cyanobacteria	473	464	456
Fungi metabolites	206	204	188
Marine NP	6253	6251	6195
MEGx	4103	4102	4046
NATx	26,318	26,295	26,318
$NuBBE_{DB}$	2214	2210	2134
Approved drugs	1806	1806	1691

taken from previous comparisons and chemoinformatic analysis (Sánchez-Cruz et al. 2071; Saldívar-González et al. 2019). The compound databases used in this study, and the molecules processed by both servers, NaPLeS and NP-scout, are summarized in Table 5.

Figure 2 shows box plots of the distribution of the NP-likeness scores computed with NaPLeS. In this figure, the boxes enclose the data points with values within the first and third quartile; the line that divides the box denotes the distributions' median, the "X" represents the mean. The lines above and below indicate the upper and

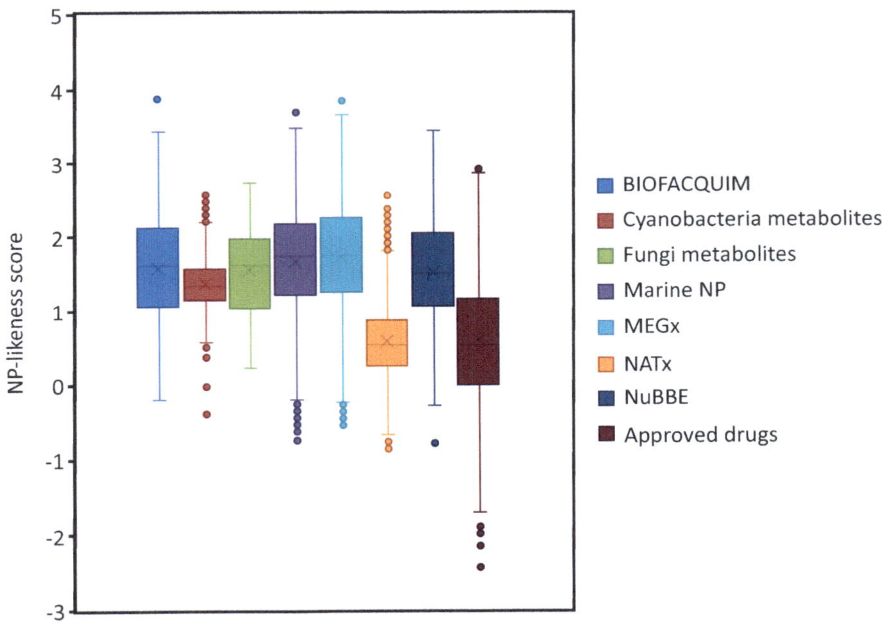

	BIOFACQUIM	Cyanobacteria	Fungi	Marine	MEGx	NATx	NuBBE	Approved
Min	-0.19	-0.39	0.24	-0.77	-0.57	-0.85	-0.79	-2.44
1Q	1.06	1.15	1.04	1.22	1.25	0.25	1.07	-0.01
Median	1.6	1.33	1.61	1.74	1.74	0.54	1.49	0.52
Mean	1.57	1.37	1.55	1.65	1.72	0.58	1.51	0.59
3Q	2.11	1.57	1.96	2.17	2.25	0.88	2.04	1.16
Max	3.87	2.59	2.74	3.69	3.87	2.56	3.43	2.97
Std Dev	0.64	0.42	0.55	0.67	0.66	0.45	0.65	0.90

Fig. 2 Box plots for the NP-likeness score distributions for BIOFACQUIM, Cyanobacteria, Fungi, Marine NP, MEGx, NATx, $NuBBE_{DB}$, and Approved, were determined using the NaPLeS method. The boxes enclose data points with values within the first and third quartile, the "X" symbol corresponds to the mean, and the points above or below of the boxes indicate outliers. Summary statistics are included below the plot

lower adjacent values. The table below indicates the minimum, first and third quartile, median, mean, standard deviation, and maximum values for each database

According to Fig. 2, except for Approved and NATx datasets, all databases have the first quartile of NP-likeness score at one or above one. Notably, MEGx and Marine NP have a very similar mean NP-likeness score, 1.7 and 1.6, respectively, and the same median, 1.7. Moreover, BIOFACQUIM, Fungi, and NuBBE have a very similar mean NP-likeness score, around a value of 1.5. BIOFACQUIM and Fungi have the same median, 1.6, whereas NuBBE has a median of 1.5. The mean and median NP-likeness score for Cyanobacteria is the same, 1.3. Approved and NATx are the only databases that show the first quartile, median, and mean NP-likeness score below one, and both databases have the same mean and median NP-likeness score, 0.6 and 0.5, respectively. As expected, the six NP databases (BIOFACQUIM, Cyanobacteria, Fungi, Marine NP, MEGx, and $NUBBE_{DB}$) showed the highest NP-likeness score, whereas for Approved and NATx datasets, where the compounds are mostly SM, NP-likeness score was lower.

Figure 3 summarizes the calculated NP class probabilities distributions of all eight databases. The figure also shows two similarity maps of the highest (Pinitol, on the left-hand side of the figure) and lowest (Aristoloxazine A) compounds of BIOFACQUIM, obtained by NP-scout. More than 93% of BIOFACQUIM, Fungi,

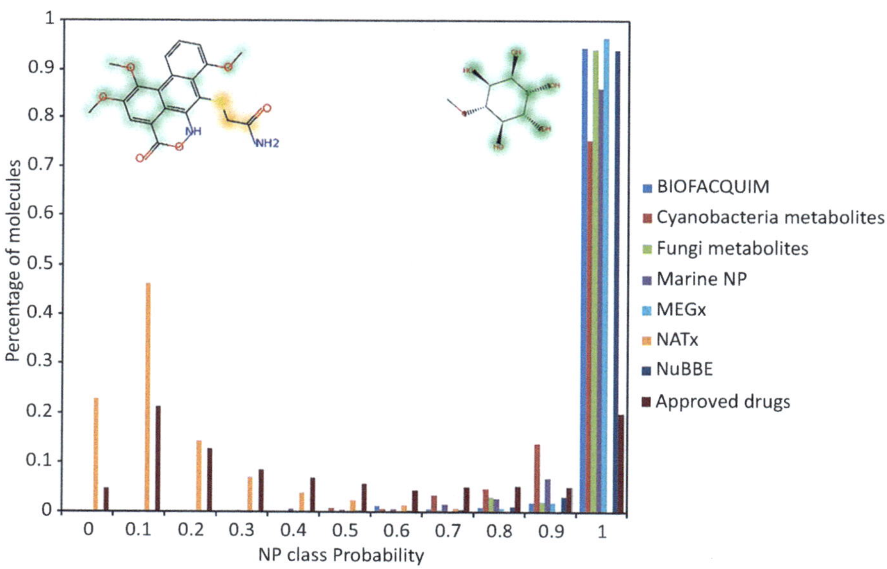

Fig. 3 Distribution of NP class probabilities for BIOFACQUIM, Cyanobacteria, Fungi, Marine, MEGx, NATx, $NuBBE_{DB}$, and Approved determined using the NP-scout method. As an example of the similarity map results, in the plot, two compounds from the BIOFACQUIM library are shown, Aristoloxazine A (left) and Pinitol (right), corresponding to the lowest (0.12) and highest (1.0) score, respectively. Green highlights mark atoms contributing to the classification of a molecule as NP, whereas orange highlights mark atoms contributing to the classification of a molecule as SM

MEGx, and NuBBE databases, have an NP class probability of 1.0 (the maximum score), whereas 75% and 86% of the compounds of Cyanobacteria and Marine, respectively, were assigned an NP class probability of 1.0. In comparison, NATx and Approved have a lower score: 54% of the compounds of the Approved database have an NP class probability between 0.0 and 0.4, and only 20% have a probability of 1.0. NATx shows the lowest score, ~95% of their compounds have an NP class probability between 0.0 and 0.4. More than 94% of all compounds of each NP database have an NP class probability between 0.8 and 1.0. This means that all six NP databases could be identified by NP-scout.

Taking the results together, NaPLeS and NP-scout showed similar conclusions. Both servers were able to differentiate between NP and SM. All NP databases showed a higher score. The opposite is observed for SM databases (Approved and NATx). That is, they showed a lower score.

These computational methods allow the identification of NP based on its physicochemical properties, so virtual libraries with many compounds can be processed easily and with high accuracy. Both tools can be used in a complementary manner to yield a robust result.

5 On the Rational Use of Free Resources

The benefits of cheminformatics and several other disciplines that form part of computer-aided drug design (CADD) during different drug discovery processes are abundantly clear. Its potential has been increased by the renaissance of artificial intelligence (AI) through machine and deep learning. The increased importance and contributions of CADD across different states of drug development, plus the availability of open-source or free software and applications for academia, has boosted the widespread use of such techniques. Other factors, such as the lock-down recently imposed by the COVID-19 pandemic, and the shifting from some early career and more established groups toward the use of computational tools, have also contributed to the recent exacerbated use of computational methods. Indeed, the limited or lack of funding to equip experimental research laboratories are pushing early career and established scientists initially trained and highly experienced in an experimental-based discipline (e.g., chemical synthesis, biochemical or biophysical assays) to shift to in silico analysis. This shift (and not drift) is perfectly fine, but it is convenient to emphasize that CADD is not "an easy" field. The work is not purely technical; few-hour workshops are not enough to fully understand the computational models' full scope and complexity. Newcomers to the field (students, early career, and scientists either shifting or expanding their practice to computational approaches) should be aware that, similar to the experiments, one needs to be critical of the computational results and include positive and negative controls as virtually in any experiment. Also, keep in mind that computational approaches *aid* drug discovery. Experimental validation of the results will always be required (no matter the quantity and accuracy of the computational approaches used). The same as with any experiment, it is crucial to watch out for the "data purity" that is entered into the computer.

The sharp increase in the application of in silico methods has not been accompanied by proper or formal training. Consequently, many newcomers to the CADD methodologies start using "easy-to-use software" with no realization that proper use of such techniques is no different from what is often seen or believed to be the "difficult disciplines" such as experimental-driven ones (e.g., chemical synthesis or biological testing). The dangerous over and misuse of computational techniques may have serious collateral effects such as the set-up and increase of false expectations of CADD, and misconceptions and propagation of terminologies.

General Recommendations for the Proper Use of Open Source and Other Resources

Listed below are general and specific considerations that, in the author's opinion, can help the students, newcomers to CADD, and multidisciplinary research teams using (or planning to use) computational approaches to aid drug discovery projects. The list is by no means exhaustive, and it is also intended to stimulate an open discussion with the scientific community.

- Study the literature as in-depth as possible to become a real expert. Similar to many experimental techniques, computational methods also require proper training to be familiar with the scope, limitations, and approximations of CADD methods.
- The research projects should be problem-oriented and not technique-oriented. Projects (thesis dissertations, for example) should not be a compilation of techniques applied for the same data set but should be directed to answer a research question.
- Before deciding what computational tools (and approaches) to use, conduct an exhaustive and proper literature search. Then, based on the experimental information available, decide on the technique and approach to be used. The method selection has been extensively discussed in the literature and books. It is not within the scope of this manuscript (Jorgensen 2004; Saldívar-González et al. 2017). The excitement and rush to run a calculation and rapidly obtain data from a computer often lead to an incorrect technique.
- Do not choose a computational technique just because it is "popular": use it for the right reasons. Misuse or interpretation of highly used approaches or empirical rules has the risk of leading to error propagation. In drug discovery, arguably one of the frequently misused guides is the Lipinski rule of five (Lipinski et al. 1997). It was initially proposed to estimate bioavailability. It has many exceptions and cautions (Lipinski 2004; Shultz 2019), but it is wrongly used as a guide if a compound can become a drug or even if it is going to be active!
- Similar to the previous point, a proper literature search is mandatory to become familiar with the research done on a particular field. For example, before running

virtual screening with highly explored targets (e.g., SARS-CoV-2), a proper literature search will tell the practitioner what virtual screening techniques, molecular database, etc., have been used and studied to avoid re-doing similar or the same work. In this regard, in particular, the students and newcomers to a new technique should be aware of the study to be done for training and practice purposes (what is valid) and what is novel.

- Avoid setting the goals of a research project based on a technique (e.g., the "tool-oriented approach"). This will help to force using a technique that is not suited to solve a research question with the experimental information available. In other words, learning a technique or using a specific web server and then wanting to apply such technique to *any* problem just to use the technique. A typical pitfall in this area is molecular docking and molecular dynamics: useful techniques but cannot always be used in CADD if there is not enough data available. For instance, if the molecular target is known. Pitfalls in QSAR and virtual screening, which are common techniques used in CADD, have been discussed extensively elsewhere (Scior et al. 2009, 2012).
- Students should keep in mind that learning and applying different techniques across disciplines is desirable; they should be used in harmony to answer research questions.
- Get supervision or advice from experts: do not be afraid or hesitate to ask. It is more common that computational experts approach experimentalists to collaborate than otherwise.

6 Conclusions and Future Directions

Over the past few years, there has been a substantial increase in the development and applications of open code and free software (either stand-alone or web servers) to perform a wide range of chemoinformatic studies. Epigenetic Target Profiler is an example of a recently developed open resource for epigenetic target fishing. With the occurrence of the recent COVID-19 pandemic, many groups have used the free software (frequently in combination with commercial software) to propose potential drug candidates. In a few instances, experimental validation of the computational hits has been reported. Also, in recent years there has been a substantial increase in NP databases in the public domain. Thus far, one of the largest compilations of NP databases has approximately 400,000 compounds. LANaPD is an international collaborative project underway aimed to assemble NP databases from different Latin American countries. There are open resources to quantify the NP-likeness of compound libraries. As illustrated in this chapter, these tools enable profiling NP-likeness of compound databases in the public domain, leading to similar conclusions. This analysis concluded that databases such as MEGx and Marine have the largest number of "NP-like compounds." In contrast, Cyanobacteria metabolites have the lowest NP-likeness scores as compared to other natural product databases. The analysis also revealed that a collection of semisynthetic NP has a strong bias to non-NP-like molecules. We also want to emphasize the importance and relevance of open resources' rational use following well and well-established practices. This is particularly important for correctly employing easy-to-use web servers. The user should be aware of the approximations and potential limitations that are not always apparent. This is for the correct interpretation and use of the data generated that can ultimately be transformed into useful knowledge. To this end, it is also important to keep promoting and increasing students' formal training and practitioners of cheminformatics tools (López-López et al. 2021).

We encourage students, newcomers, practitioners, or CADD users with expertise in non-computational areas to use the tools and informatics resources for the right reasons, not just because they are easily accessible or easy-to-use. Similarly, we highly encourage them not to select approaches and computational methods just because they are "popular" but because they have been thoroughly validated and properly documented, preferably in peer-reviewed journals. Include positive and negative controls in the computational analysis as if you were experimenting. If using a web server, come to know what the server is doing: what are the applications' limitations and learn to interpret results correctly, be critical of the results. For instance, in organic synthesis, it is common to become familiar with the reaction mechanism to properly characterize the reaction's products. Something similar should be expected in a server or program in general: be critical and become familiar with what the program is doing so you can properly interpret the results and detect "impurities" in your results.

Acknowledgments We thank the support of DGAPA, UNAM, *Programa de Apoyo a Proyectos de Investigación e Innovación Tecnológica* (PAPIIT), grant IN201321.

References

Abegaz BM, Kinfe HH. Secondary metabolites, their structural diversity, bioactivity, and ecological functions: an overview. Phys Sci Rev. 2019;4:20180100.

Alnajjar R, Mostafa A, Kandeil A, Al-Karmalawy AA. Molecular docking, molecular dynamics, and in vitro studies reveal the potential of angiotensin II receptor blockers to inhibit the Covid-19 main protease. Heliyon. 2020;6:e05641.

Atanasov AG, Zotchev SB, Dirsch VM, Orhan IE, Banach M, Rollinger JM, Barreca D, Weckwerth W, Bauer R, Bayer EA, Majeed M, Bishayee A, Bochkov V, Bonn GK, Braidy N, Bucar F, Cifuentes A, D'Onofrio G, Bodkin M, Diederich M, Dinkova-Kostova AT, Efferth T, El Bairi K, Arkells N, Fan T-P, Fiebich BL, Freissmuth M, Georgiev MI, Gibbons S, Godfrey KM, Gruber CW, Heer J, Huber LA, Ibanez E, Kijjoa A, Kiss AK, Lu A, Macias FA, Miller MJS, Mocan A, Müller R, Nicoletti F, Perry G, Pittalà V, Rastrelli L, Ristow M, Russo GL, Silva AS, Schuster D, Sheridan H, Skalicka-Woźniak K, Skaltsounis L, Sobarzo-Sánchez E, Bredt DS, Stuppner H, Sureda A, Tzvetkov NT, Vacca RA, Aggarwal BB, Battino M, Giampieri F, Wink M, Wolfender J-L, Xiao J, Yeung AWK, Lizard G, Popp MA, Heinrich M, Berindan-Neagoe I, Stadler M, Daglia M, Verpoorte R, Supuran CT. The International Natural Product Sciences T natural products in drug discovery: advances and opportunities. Nat Rev Drug Discov. 2021;20:200–16.

Berman HM, Westbrook J, Feng Z, Gilliland G, Bhat TN, Weissig H, Shindyalov IN, Bourne PE. The protein data bank. Nucleicl Acids Res. 2000;28:235–42.

Bobrowski TM, Korn DR, Muratov EN, Tropsha A. ZINC Express: a virtual assistant for purchasing compounds annotated in the zinc database. J Chem Inf Model. 2021;61:1033–6.

Chávez-Hernández AL, Sánchez-Cruz N, Medina-Franco JL. A fragment library of natural products and its comparative chemoinformatic characterization. Mol Inf. 2020;39:2000050.

Chen Y, Kirchmair J. Cheminformatics in natural product-based drug discovery. Mol Inf. 2020;39:2000171.

Chen Y, Stork C, Hirte S, Kirchmair J. Np-Scout: machine learning approach for the quantification and visualization of the natural product-likeness of small molecules. Biomol Ther. 2019;9:43.

Cheng T, Pan Y, Hao M, Wang Y, Bryant SH. Pubchem applications in drug discovery: a bibliometric analysis. Drug Discov Today. 2014;19:1751–6.

Cucinotta D, Vanelli M. Who declares Covid-19 a pandemic. Acta Bio Medica Atenei Parmensis. 2020;91:157–60.

da Paixão VG, Pita S. In silico identification and evaluation of new trypanosoma cruzi trypanothione reductase (Tctr) inhibitors obtained from natural products database of the Bahia Semi-Arid Region (Natprodb). Comput Biol Chem. 2019;79:36–47.

Daina A, Zoete V. Application of the Swissdrugdesign online resources in virtual screening. Int J Mol Sci. 2019;20

Ekins S, Williams AJ, Krasowski MD, Freundlich JS. In silico repositioning of approved drugs for rare and neglected diseases. Drug Discov Today. 2011;16:298–310.

Ertl P, Roggo S, Schuffenhauer A. Natural product-likeness score and its application for prioritization of compound libraries. J Chem Inf Model. 2008;48:68–74.

Gasteiger J. Chemistry in times of artificial intelligence. ChemPhysChem. 2020;21:2233–42.

Gong Z, Hu G, Li Q, Liu Z, Wang F, Zhang X, Xiong J, Li P, Xu Y, Ma R, Chen S. Li J Compound libraries: recent advances and their applications in drug discovery. Curr Drug Discov Technol. 2017;14:216–28.

Gonzalez-Medina M, Naveja JJ, Sanchez-Cruz N, Medina-Franco JL. Open chemoinformatic resources to explore the structure, properties and chemical space of molecules. RSC Adv. 2017;7:54153–63.

Gorgulla C, Boeszoermenyi A, Wang ZF, Fischer PD, Coote PW, Padmanabha Das KM, Malets YS, Radchenko DS, Moroz YS, Scott DA, Fackeldey K, Hoffmann M, Iavniuk I, Wagner G, Arthanari H. An open-source drug discovery platform enables ultra-large virtual screens. Nature. 2020;580:663–8.

Guedes IA, Costa LSC, Dos Santos KB, Karl ALM, Rocha GK, Teixeira IM, Galheigo MM, Medeiros V, Krempser E, Custódio FL, Barbosa HJC, Nicolás MF, Dardenne LE. Drug design and repurposing with Dockthor-Vs web server focusing on Sars-Cov-2 therapeutic targets and their non-synonym variants. Sci Rep. 2021;11:5543.

Guo S, Xie H, Lei Y, Liu B, Zhang L, Xu Y. Zuo Z Discovery of Novel inhibitors against main protease (Mpro) of Sars-Cov-2 via virtual screening and biochemical evaluation. Bioorg Chem. 2021;110:104767.

Harris CJ, Hill RD, Sheppard DW, Slater MJ, Stouten PF. The design and application of target-focused compound libraries. Comb Chem High Throughput Screen. 2011;14:521–31.

Irwin JJ, Tang KG, Young J, Dandarchuluun C, Wong BR, Khurelbaatar M, Moroz YS, Mayfield J, Sayle RA. Zinc20-a Free ultralarge-scale chemical database for ligand discovery. J Chem Inf Model. 2020;60:6065–73.

Jayaseelan KV, Moreno P, Truszkowski A, Ertl P, Steinbeck C. Natural product-likeness score revisited: an open-source, open-data implementation. BMC Bioinformatics. 2012;13:106.

Johnson M, Zaretskaya I, Raytselis Y, Merezhuk Y, McGinnis S, Madden TL. NCBI blast: a better web interface. Nucleic Acids Res. 2008;36:W5–9.

Jorgensen WL. The Many roles of computation in drug discovery. Science. 2004;303:1813–8.

Lipinski CA. Lead- and druglike compounds: the rule-of-five revolution. Drug Discov Today Technol. 2004;1:337–41.

Lipinski CA, Lombardo F, Dominy BW, Feeney PJ. Experimental and computational approaches to estimate solubility and permeability in drug discovery and development settings. Adv Drug Deliv Rev. 1997;23:3–25.

López-López E, Bajorath J, Medina-Franco JL. Informatics for chemistry, biology, and biomedical sciences. J Chem Inf Model. 2021;61:26–35.
Martinez-Mayorga K, Madariaga-Mazon A, Medina-Franco JL, Maggiora G. The impact of chemoinformatics on drug discovery in the pharmaceutical industry. Exp Opin Drug Discov. 2020;15:293–306.
Masters L, Eagon S, Heying M. Evaluation of consensus scoring methods for Autodock Vina, Smina and Idock. J Mol Graph Model. 2020;96:107532.
Medina-Franco JL. Towards a unified Latin American natural products database: Lanapd. Future Sci OA. 2020;6:FSO468.
Medina-Franco JL, Saldívar-González FI. Cheminformatics to characterize pharmacologically active natural products. Biomol Ther. 2020;10:1566.
Mendez D, Gaulton A, Bento AP, Chambers J, De Veij M, Félix E, Magariños María P, Mosquera Juan F, Mutowo P, Nowotka M, Gordillo-Marañón M, Hunter F, Junco L, Mugumbate G, Rodriguez-Lopez M, Atkinson F, Bosc N, Radoux Chris J, Segura-Cabrera A, Hersey A, Leach AR. Chembl: towards direct deposition of bioassay data. Nucl Acids Res. 2019;47:D930–40.
Merz KM, Amaro R, Cournia Z, Rarey M, Soares T, Tropsha A, Wahab HA, Wang R. Editorial: method and data sharing and reproducibility of scientific results. J Chem Inf Mod. 2020;60:5868–9.
Mousavizadeh L, Ghasemi S. Genotype and phenotype of Covid-19: their roles in pathogenesis. J Microbiol Immunol Infect. 2021;54:159–163. https://doi.org/10.1016/j.jmii.2020.03.022.
Naveja JJ, Oviedo-Osornio CI, Trujillo-Minero NN, Medina-Franco JL. Chemoinformatics: a perspective from an academic setting in Latin America. Mol Divers. 2018;22:247–58.
Newman DJ, Cragg GM. Natural products as sources of new drugs over the nearly four decades from 01/1981 to 09/2019. J Nat Prod. 2020;83:770–803.
Ntie-Kang F, Nyongbela KD, Ayimele GA, Shekfeh S. "Drug-Likeness" properties of natural compounds. Phys Sci Rev. 2019;4:20180169.
Pippel M, Meier R, Sippl W. Paradocks—a framework for molecular docking. J Cheminf. 2011;3:P35.
Pitsillou E, Liang J, Ververis K, Lim KW, Hung A, Karagiannis TC. Identification of small molecule inhibitors of the deubiquitinating activity of the Sars-Cov-2 Papain-like protease: in silico molecular docking studies and in vitro enzymatic activity assay. Front Chem. 2020;8:623971.
Pitsillou E, Liang J, Ververis K, Hung A, Karagiannis TC. Interaction of small molecules with the Sars-Cov-2 Papain-like protease: in silico studies and in vitro validation of protease activity inhibition using an enzymatic inhibition assay. J Mol Graph Model. 2021;104:107851.
Raj R. Analysis of non-structural proteins, Nsps of Sars-Cov-2 as targets for computational drug designing. Biochem Biophys Rep. 2021;25:100847.
Ruiz-Carmona S, Alvarez-Garcia D, Foloppe N, Garmendia-Doval AB, Juhos S, Schmidtke P, Barril X, Hubbard RE, Morley SD. Rdock: a fast, versatile and open source program for docking ligands to proteins and nucleic acids. PLoS Comput Biol. 2014;10:e1003571.
Rutz A, Sorokina M, Galgonek J, Mietchen D, Willighagen E, Graham J, Stephan R, Page R, Vondrášek J, Steinbeck C, Pauli GF, Wolfender J-L, Bisson J, Allard P-M. Open natural products research: curation and dissemination of biological occurrences of chemical structures through Wikidata. bioRxiv. 2021; 2021.02.28.433265
Sadegh S, Matschinske J, Blumenthal DB, Galindez G, Kacprowski T, List M, Nasirigerdeh R, Oubounyt M, Pichlmair A, Rose TD, Salgado-Albarrán M, Späth J, Stukalov A, Wenke NK, Yuan K, Pauling JK, Baumbach J. Exploring the SARS-Cov-2 virus-host-drug interactome for drug repurposing. Nat Commun. 2020;11:3518.
Sahoo S, Mahapatra SR, Parida BK, Rath S, Dehury B, Raina V, Mohakud NK, Misra N, Suar M. Dbcovp: a database of coronavirus virulent glycoproteins. Comput Biol Med. 2021;129:104131.
Saldívar-González F, Prieto-Martínez FD, Medina-Franco JL, Descubrimiento Y. Desarrollo De Fármacos: Un Enfoque Computacional. Educ Quim. 2017;28:51–8.

Saldívar-González FI, Valli M, Andricopulo AD, da Silva BV, Medina-Franco JL. Chemical space and diversity of the Nubbe database: a chemoinformatic characterization. J Chem Inf Model. 2019;59:74–85.

Saldívar-González FI, Huerta-García CS, Medina-Franco JL. Chemoinformatics-based enumeration of chemical libraries: a tutorial. J Chem Inf. 2020;12:64.

Sánchez-Cruz N, Medina-Franco JL. Epigenetic target profiler: a web server to predict epigenetic targets of small molecules. J Chem Inf Model. 2021a;61:1550–1554. https://doi.org/10.1021/acs.jcim.1c00045.

Sánchez-Cruz N, Medina-Franco JL. Epigenetic target fishing with accurate machine learning models. J Med Chem. 2021b;64:8208–8220. https://doi.org/10.1021/acs.jmedchem.1c00020.

Sánchez-Cruz N, Pilón-Jiménez B, Medina-Franco J. Functional group and diversity analysis of Biofacquim: a mexican natural product database [Version 2; Peer Review: 3 Approved]. F1000Research. 2020;8:2071.

Scharfe M, Pippel M, Sippl W. Paradocks—an open-source framework for molecular docking: implementation of target-class-specific scoring methods. J Chem Inf. 2013;5:P11.

Scior T, Medina-Franco JL, Do Q-T, Martínez-Mayorga K, Yunes Rojas JA, Bernard P. How to recognize and workaround pitfalls in Qsar studies: a critical review. Curr Med Chem. 2009;16:4297–313.

Scior T, Bender A, Tresadern G, Medina-Franco JL, Martínez-Mayorga K, Langer T, Cuanalo-Contreras K, Agrafiotis DK. Recognizing pitfalls in virtual screening: a critical review. J Chem Inf Model. 2012;52:867–81.

Sessions Z, Sánchez-Cruz N, Prieto-Martínez FD, Alves VM, Santos HP Jr, Muratov E, Tropsha A, Medina-Franco JL. Recent progress on cheminformatics approaches to epigenetic drug discovery. Drug Discov Today. 2020;25:2268–76.

Shultz MD. Two decades under the influence of the rule of five and the changing properties of approved oral drugs. J Med Chem. 2019;62:1701–14.

Singh N, Chaput L, Villoutreix BO. Virtual screening web servers: designing chemical probes and drug candidates in the cyberspace. Brief Bioinform. 2020;22:1790–818.

Sorokina M, Steinbeck C. Naples: a natural products likeness scorer-web application and database. J Cheminform. 2019;11:55.

Sorokina M, Steinbeck C. Review on natural products databases: where to find data in 2020. J Chem Inf. 2020;12:20.

Sorokina M, Merseburger P, Rajan K, Yirik MA, Steinbeck C. Coconut online: collection of open natural products database. J Cheminformatics. 2021;13:2.

Sterling T, Irwin JJ. Zinc 15—ligand discovery for everyone. J Chem Inf Model. 2015;55:2324–37.

Stork C, Embruch G, Šícho M, de Bruyn KC, Chen Y, Svozil D, Kirchmair J. Nerdd: a web portal providing access to in silico tools for drug discovery. Bioinformatics. 2020;36:1291–2.

Sunseri J, Koes DR. Pharmit: interactive exploration of chemical space. Nucleic Acids Res. 2016;44:W442–8.

Valsecchi C, Grisoni F, Motta S, Bonati L, Ballabio D. Nura: a curated dataset of nuclear receptor modulators. Toxicol Appl Pharmacol. 2020;407:115244.

Walters WP. Virtual chemical libraries. J Med Chem. 2019;62:1116–24.

Wang J, Ge Y, Xie XQ. Development and testing of druglike screening libraries. J Chem Inf Model. 2019;59:53–65.

Wassermann AM, Camargo LM, Auld DS. Composition and applications of focus libraries to phenotypic assays. Front Pharmacol. 2014;5:164.

Wójcikowski M, Zielenkiewicz P, Siedlecki P. Open drug discovery toolkit (Oddt): a new open-source player in the drug discovery field. J Chem Inf. 2015;7:26.

Wu F, Zhou Y, Li L, Shen X, Chen G, Wang X, Liang X, Tan M, Huang Z. Computational approaches in preclinical studies on drug discovery and development. Front Chem. 2020;8:726.

Zhang H, Yang Y, Li J, Wang M, Saravanan KM, Wei J, Tze-Yang Ng J, Tofazzal Hossain M, Liu M, Zhang H, Ren X, Pan Y, Peng Y, Shi Y, Wan X, Liu Y, Wei Y. A Novel virtual screening procedure identifies pralatrexate as inhibitor of Sars-Cov-2 RDRP and it reduces viral replication in vitro. PLoS Comput Biol. 2020;16:e1008489.

Index

M. T. Scotti, C. L. Bellera (eds.), *Drug Target Selection and Validation*, Computer–Aided Drug Discovery and Design,
https://doi.org/10.1007/978-3-030-95895-4

The manufacturer's authorised representative in the EU is Springer Nature Customer Service Centre GmbH, Europaplatz 3, 69115 Heidelberg, Germany. If you have any concerns regarding our products, please contact ProductSafety@springernature.com

Printed and bound by CPI Group (UK) Ltd, Croydon, CR0 4YY
22/09/2026
02220961-0007